GROUND WATER

Hydrogeology • Ground Water Survey and Pumping Tests • Rural Water Supply and Irrigation Systems

(Second Edition)

H.M. Raghunath
Water Resources Specialist

JOHN WILEY & SONS
New York   Chichester   Brisbane   Toronto   Singapore

First published in 1982
Second Edition, 1987
WILEY EASTERN LIMITED
4835/24 Ansari Road, Daryaganj
New Delhi 110 002, India

Distributors:

Australia and New Zealand:
Jacaranda-Wiley Ltd., Jacaranda Press,
JOHN WILEY & SONS, INC.
GPO Box 859, Brisbane, Queensland 4001, Australia

Canada:
JOHN WILEY & SONS CANADA LIMITED
22 Worcester Road, Rexdale, Ontario, Canada

Europe and Africa:
JOHN WILEY & SONS LIMITED
Baffins Lane, Chichester, West Sussex, England

South East Asia:
JOHN WILEY & SONS, INC.
05–05 Block B, Union Industrial Building
37 Jalan Pemimpin, Singapore 2057

Africa and South Asia:
WILEY EASTERN LIMITED
4835/24 Ansari Road, Daryaganj
New Delhi 110 002, India

North and South America and rest of the world:
JOHN WILEY & SONS, INC.
605 Third Avenue, New York, N.Y. 10158, USA

Copyright © 1982, 1987, WILEY EASTERN LIMITED
New Delhi, India

Library of Congress Cataloging in Publication Data

Raghunath, H.M., 1938–
    Ground Water.

    1. Hydrogeology. 2. Water supply. I. Title.
GB1003.2.R33   1987   553.7'9   86-24526

ISBN 0-470-20698-5 John Wiley & Sons, Inc.
ISBN 0-85226-298-1 Wiley Eastern Limited

Printed in India at Prabhat Press, Meerut.

Preface to the Second Edition

The book has been revised as follows:
  i. Special cases of ground water flow by the method of images, one- and two-dimensional flows, numerical analysis, Glover's method for seawater intrusion, etc. have been included.
  ii. Completely written in SI Units.
  iii. More problems on ground water flow analysis and design, and case histories have been included.

With the above, it is expected to be a Standard Text Book and a valuable reference for men in profession.

May 1987   H.M. RAGHUNATH

Preface to the First Edition

Water is a precious and most commonly used resource. Surface water resources, being exploited from time to time, may become short of supply or may not be easily available at site. Ground water commonly occurs and is widely distributed. There has been an increase in ground water development and utilisation, particularly in developing countries, for agriculture, industry and rural water supply schemes. In this present book an attempt is made to introduce to the reader all the aspects of ground water—its assessment, development, utilisation and management, in a lucid style; the information on different topics which is usually found scattered in various technical books, journals and seminar volumes has been pooled with an emphasis on basic principles and without enlarging on the derivation of the formulae. The practical application of the different formulae for the field conditions, data collection and processing, test procedures and principles of design are elaborated and actual field problems are worked out to illustrate the theory and the design procedure. Objective and intelligence questions and problems for assignment (with answers for some) are given at the end of each chapter. The metric system SI Units has been used and the unit conversion factors are given in the Appendix F.

Graduate students, research scholars and professionals associated with Ground Water Development and Management will find the book useful.

It is difficult to acknowledge all original sources in a book of this type, but where appropriate they are given. A list of references is given at the end for details and derivations and for supplementary reading.

Constructive criticisms are always welcome, will be appreciated and will be incorporated in the next edition.

H.M. RAGHUNATH

Contents

Preface to the Second Edition   iii
Preface to the First Edition   v

1. Introduction   1
2. Hydrometeorology   22
3. Hydrogeology and Aerial Photography   58
4. Aquifer Properties and Ground Water Flow   78
5. Well Hydraulics   135
6. Ground Water Analog Models   255
7. Sea Water Intrusion   288
8. Ground Water Geophysics   304
9. Geochemical Survey and Water Quality   344
10. Water Well Design   370
11. Water Well Drilling   403
12. Water Well Construction   423
13. Yield Test and Selection of Pumpsets   438
14. Ground Water Pollution and Legislation   457
15. Ground Water Recharge   463
16. Ground Water Basin Management and Conjunctive Use   474
17. Irrigation Systems   493

Appendix A. Crop Details   523
Appendix B. Hydraulic Details   532
Appendix C. Equipment Dealers   536
Appendix D. Tubewell Estimates   539
Appendix E. Geological Time Scale   544
Appendix F. Unit Conversion Factors   546
Selected References   551
Bibliography   557
Index   561

1

Introduction

Ground water is a precious and the most widely distributed resource of the earth and unlike any other mineral resource, it gets its annual replenishment from the meteoric precipitation. The world's total water resources are estimated at 1.37×10^8 million ha-m. Of these global water resources about 97.2% is salt water, mainly in oceans, and only 2.8% is available as fresh water at any time on the planet earth. Out of this 2.8%, about 2.2% is available as surface water and 0.6% as ground water. Even out of this 2.2% of surface water, 2.15% is fresh water in glaciers and icecaps and only of the order of 0.01% (1.36×10^4 M ha-m) is available in lakes and reservoirs, and 0.0001% in streams; the remaining being in other forms—0.001% as water vapour in atmosphere, and 0.002% as soil moisture in the top 0.6 m. Out of 0.6% of stored ground water, only about 0.3% (41.1×10^4 M ha-m) can be economically extracted with the present drilling technology, the remaining being unavailable as it is situated below a depth of 800 m.

Thus ground water is the largest source of fresh water on the planet excluding the polar icecaps and glaciers. The amount of ground water within 800 m from the ground surface is over 30 times the amount in all fresh water lakes and reservoirs, and about 3000 times the amount in stream channels, at any one time.

At present nearly one fifth of all the water used in the world is obtained from ground water resources. Agriculture is the greatest user of water accounting for 80% of all consumption. It takes, roughly speaking, 1000 tons of water to grow one ton of grain and 2000 tons to grow one ton of rice. Animal husbandry and fisheries all require abundant water. Some 15% of world's crop land is irrigated. The present irrigated area in India is 60 million hectares (M ha) of which about 40% is from ground water.

The average annual rainfall (a.a.r.) of India is around 114 cm. Based on this a.a.r., Dr. K. L. Rao has estimated that the total annual rainfall over the entire country is of the order of 370 M ha-m and one third of this is lost in evaporation. Of the remaining 247 M ha-m of water, 167 M ha-m goes as runoff and the rest of the 80 M ha-m goes as subsoil water. Out of this 80 M ha-m of subsoil water that seeps down annually into the soil, about 43 M ha-m gets absorbed in the top layer, thereby contributing to the soil moisture; the balance of 37 M ha-m is the contribution to ground water from rainfall.

The average annual ground water recharge from rainfall and seepage from canals and irrigation systems is of the order of 67 M ha-m of which 40% i.e. 27 M ha-m, is extractable economically. The present utilisation of ground water is roughly half of this (13 M ha-m), and about 14 M ha-m is available for further exploitation and utilisation.

Ground Water Development in India

The excavations at Mohanjodaro have revealed brick-lined dug wells existing as early as 3000 B.C. during the Indus Valley Civilisation. The writings of Vishnu Kautilya (in the reign of Chandragupta Maurya—300 B.C.) indicate that ground water was being used for irrigation purposes at that time.

Sinking of wells and a variety of water devices were well known from Vedic times.

The first Irrigation Commission in 1903 affirmed the importance of irrigation wells. The Well Sinking Department of the Government of Nizam at Hyderabad made interesting studies on ground water in the Deccan Basaltic Terrains.

In 1934 a project for construction of 1500 community tubewells in the Ganga basin was initiated in U.P. The success of this project led to the constitution of a Sub-Soil Water Section in the Government of India in 1944 which was converted later into the Central Ground Water Organisation which functioned till 1949. During this period a Central Drilling School at Roorkee was established which trained more than 100 officers of the Central and State Government.

The Exploratory Tube-wells Organisation (ETO) was set up during 1954 under Indo-US Technical Cooperation Operational Agreement No. 12, under the Ministry of Agriculture and concomitantly the Ground Water Exploration Section was set up in the Geological Survey of India.

In October 1970, the Ministry of Agriculture upgraded the Exploratory Tube-wells Organisation into the Central Ground Water Board (CGWB) merging it with the Ground Water Regional Directorates and District Offices of the Geological Survey of India to effectively shoulder the ground water investigation programmes; it started functioning from August 1972. As an apex body at the national level, the Board is concerned with all matters relating to exploration, assessment, development, management and regulation of the country's ground water resources.

Large scale ground water investigation programmes have been taken up

Introduction

since 1967 in Rajasthan, Gujarat, and Tamil Nadu with the assistance of the UNDP, the Canadian assisted project in AP, the Upper Betwa River Basin Project in MP and UP with UK assistance, Narmada Valley Project in MP, Vedavathi and Tungabhadra River Basin in Karnataka under UK assistance and many such projects.

During the middle and late sixties, the Government of India urged all State Governments to set up a State Level Ground Water Organisation to

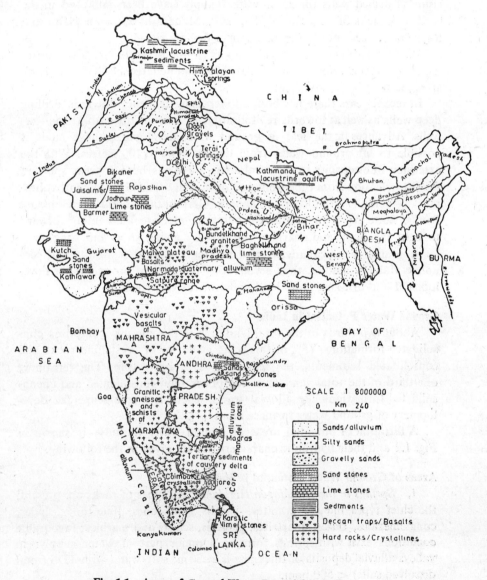

Fig. 1.1 Areas of Ground Water Potential in India.

deal with problems of ground water surveys and development and utilisation for minor irrigation and eventually they have been set up as State Directorates of Ground Water or as a Ground Water Cell (in Karnataka). The Central Ground Water Board is contemplating special measures for ensuring coordination of work among the various States and between the Centre and States so that overlapping or duplication is avoided.

Since 1970, major programmes with the assistance of UNICEF for provision of drilled wells for rural water supply have been launched in the hard rock areas of AP, Karnataka, MP, Maharashtra, Tamil Nadu and Rajasthan. These utilise the air hammer drilling rigs.

The updated hydrogeological map of India of scale 1:5,000,000 (released by GSI in 1969 to scale 1:2,000,000 and updated by CGWB in 1976) gives many hydrological details.

In recent years there has been an increasing tendency towards drilling deep wells as well as towards revitalisation of existing open (dug, shallow) wells. Advances in the field of ground water development have made it possible to lift ground water from depths of 60 to 100 metres. With the extension of electricity in the rural areas, there has been a great spurt in the lift irrigation from tube wells and open wells. The Government, voluntary agencies, Agricultural Refinance Corporation, Land Development Banks, State Agro Industries Corporation, etc. are all coming forward to help the poor and marginal farmers by giving short and long term loans, grants, technical advice, and making technical feasibility and economic viability studies, thus accelerating the pace of ground water development and bringing more land under intensive irrigation.

Ground Water Potential in India

About two-thirds of the total land area in the country comprises consolidated formations, 75% of this being made up of crystalline rocks and consolidated sediments, the remaining 25% being trap. The remaining one-third of the total land area comprises semi-consolidated and unconsolidated formations like alluvial tracts. There is ample scope for development of ground water in these areas.

A flirtation of potential areas of ground water in India is shown in Fig. 1.1 and their hydraulic characteristics are given in the following.

Areas of Ground Water Potential in India

1. *Springs in the Himalayan Highlands*: All types of rocks are present; the chief types include granites, basalts, sandstones, limestones, shales, conglomerates, slates, quartzites, gneisses, schists and marbles; favourable conditions exist with springs forming a major part of water supply; in valleys alluvial deposits of thickness is ≈ 30 m; soft water with TDS (total dissolved salts) < 500 ppm.

2. *Fresh water sediments of Kashmir valley*: The Kashmir valley which

Introduction

was a vast lake during the Pleistocene times, shows a large scale development of fresh water sediments of lacustrine, fluvial and glacial origin ≈ 600 m thick.

3. *Indo-Gangetic alluvium (vast reservoir of fresh sweet water)*: Coarse sands, gravel and boulders of variable thickness—3 to 60 m. TDS < 400 ppm, water commonly hard; shallow and deep aquifers interconnected (leaky); aquifer soil: D_{10} (effective size) $= 0.075$-0.38 mm, $D_{50} = 0.17$-0.30 mm, C_u (uniformity coefficient) $= 1.1$-3.3, $T = 1.7 \times 10^5$-5.0×10^6 lpd/m (litres per day per metre), $S = 2 \times 10^{-2}$ to 4×10^{-6}, leakance $K'/b' \approx 0.3$ lpd/m³, accretion to g.w.t. (ground water table) $\approx 21\%$ of a.a.r., tubewell yield ≈ 50-100 m³/hr. (Here T is the transmissibility coefficient; K', the permeability and b', the thickness of the semi-pervious layer, and S, the storage coefficient.) The exploitation of ground water is usually done by using spiral augers, hand boring (H.B.) sets, cable tool and rotary rigs.

The alluvial material of Punjab constitutes an extensive heterogeneous and anisotropic unconfined aquifer with lateral permeability ranging from 26 to 156 m³/day/m². There are about 120,000 tubewells in this area and the extraction of ground water should be limited to the annual recharge to avoid undue depletion of the aquifer.

4. *Coastal alluvium—Malabar and Coromandel coastal areas*: Depth 15-150 m, yield ≈ 12-50 m³/hr; low TDS; water in tertiary aquifer associated with lignite or carbonaceous clays is sulphuretted (H_2S) and contains iron > 1 ppm. Extensive saline patches occur in Ramnad, Tirunelveli, Ongole, Nellore and Krishna districts.

In Ramanathapuram and Tirunelveli, the ground water in the unconfined aquifers is generally of poor quality with Cl > 1000 ppm and at some places even > 3000 ppm.

In the west coast areas of Kerala and Karnataka, the substratum is mostly lateritic and a good yield of ground water may be expected.

Saline water influx in response to tides is noticed at places in Goa, up to distances of 25-40 km inland. In the upper reaches, the tidal streams show cyclic fluctuation in salinity, the salinity flows corresponding to low tides when waters are utilisable.

5. *Cretaceous sandstones of Kathiawar and Kutch areas*: Moderately potential aquifers; depth 100-300 m, yield ≈ 10-120 m³/hr; water commonly brackish with TDS 2000-5000 ppm.

In Gujarat sandstone aquifers are of depth 60-200 m; yield ≈ 10-50 m³/hr, TDS 1000-2500 ppm.

6. *Mesozoic sandstones of the Lathi region in Rajasthan (Jaisalmer, Barmer, Bikaner)*: Moderately potential aquifers, depths 100-150 m, yield ≈ 45-150 m³/hr; water is generally brackish to saline, TDS 1000-5000 ppm, Cl 1000-5000 ppm, EC (electrical conductivity) $> 3000\mu$ mhos/cm, Na% > 80, SAR 25-55, waters C_4-S_4 or C_4-S_3 types. (For explanation of these notations see Fig. 9.7.)

7. *Cavernous limestones of Vindhyan system in Borunda and Ransingaon areas in Jodhpur district*: Potential aquifers, yield 40-70 m^3/hr. Potable water, TDS < 2000 ppm; fractured up to 150 m.

In arid zones of Rajasthan and Gujarat excess concentration of fluoride (5 to 20 ppm) has been observed, resulting in mottling of teeth.

Recharge of some arid zones in Haryana and Rajasthan can be done by diverting the flood waters of Yamuna river through Saraswathi and Ghaggar rivers by making suitable connections.

8. *Doon valley gravels*: Boulders, pebbles, gravel, sand and clay possibly of fanglomeratic and collovial origin. Major portion of the valley is hilly, sloping ground; only the central part ($\approx$388 km^2 which is approximately one-fifth of the total area of 2090 km^2) can be developed. As a rough estimate of the potential of Doon gravels;

$$\begin{aligned} \text{Ground water storage} &= \text{Annual fluctuation of g.w.t.} \times \text{Involved area of aquifer} \times \text{Specific yield } (\approx 15\%) \\ &\approx 6.4 \text{ m} \times (388 \text{ km}^2 \times 1000^2) \times 0.15 \\ &\approx 37{,}248 \times 10^4 \text{ m}^3 \\ &\approx 37{,}248 \text{ ha-m} \end{aligned}$$

which can support more than 200 tubewells yielding $\approx$150 m^3/hr each; a.a.r.=216 cm, therefore accretion to g.w.t. $\approx$44%. Thickness of fill 150-200 m; TDS 100 to 500 ppm, Cl > 30 ppm, pH 7.8; water—bicarbonate to sulphate type. Lower areas yield $\approx$30-50 m^3/hr.

In the Terai zone ground water is available under artesian conditions and at shallow depths of 3-50 m. Sands and gravels confined in the silts and clays make good aquifers under confined conditions. Wells are usually bored with augers and/or hand boring sets.

9. *Quaternary alluvium of Narmada, Purna, Tapti, Chambal and Mahanadi rivers*: Thickness 75-150 m (lenses of sand and gravel); tubewell yield 20-150 m^3/hr; good quality water with TDS 100-500 ppm.

10. *Vesicular basalts in the Deccan trap formations of Maharashtra and Madhya Pradesh*: Form good aquifers; ground water occurs under both confined and unconfined conditions in the Satpura range and Malwa plateau; tubewell yields in Indore, Bhopal, Raisen, Vidisha and Sagar districts $\approx$10-40 m^3/hr.

In central Maharashtra the tubewells drilled in weathered basalts yield $\approx$2-10 m^3/hr while in exceptional cases the yields are $\approx$25 m^3/hr, mostly within depths of 50-100 m. Borewells, due to their low yield, are mainly a source of drinking water supply and only in exceptional cases can they be used for irrigation; TDS < 1000 ppm.

11. *Carbonate rocks with solution cavities in Madhya Pradesh*: In the Vindhyan, Cuddapah and Bijawar region, the carbonate rocks with interconnected solution cavities and caverns form good aquifers. The lime-

Introduction

stones of Raipur, Charmuria, Kajrahat (Sidhi district), Karstic areas of Chhatisgarh basin and Baghelkhand region of MP yield $\approx$ 10-60 m³/hr.

12. *Dharwarian and Bundelkhand granite region of Madhya Pradesh*: Igneous and metamorphic rocks; the movement of water is mainly through joints and openings. Tubewells in Tikamgarh, Chattarpur, Balaghat and Gwalior area yield $\approx$ 10-30 m³/hr, mostly under water table conditions.

The water quality in all regions of MP is generally good, except in the water logged areas of the Chambal valley (due to seepage from canals).

13. *Tertiary sandstones and quaternary sand to pebble beds in the Godavari-Krishna interstream area*: Form potential aquifers with artesian conditions. Near Muppavaram the piezometric surface is 19 m above the land surface and falls towards the coast at a gradient of 4-25 m/km. Aquifer thickness 3-184 m, $T = 80\text{-}6485$ m³/day/m, $K = 1\text{-}80$ m³/day/m². Tubewell yield $\approx$ 20-120 m³/hr for a drawdown of 6 m. For a similar drawdown, Rajahmundry sandstones yield $\approx$42 m³/hr for a screened thickness of 20-140 m. Similar yields are in the Tirupathi, Gollapalli and Chitalpudi sandstones.

As a rough estimate of the ground water flow (Q), with a hydraulic gradient (i) of 1/304 near Tadikalapudi, the length of the aquifer (w) of 90 km between Viravalli and Guddigudem and an average transmissibility (T) of 945 m³/day/m, it follows from Darcy's law that

$$Q = Tiw = 945 \times \frac{1}{304} \times (90 \times 10^3) = 280 \times 10^3 \text{ m}^3/\text{day}$$

With two-thirds of this yield (allowing for short duration data) about 375 to 400 tubewells can be constructed with an average yield per tubewell of 500 m³/day.

The quality of ground water in the sandstones is fresh while that in alluvium is highly saline in the vicinity of Kolleru lake, along the coast and at depths; TDS 1800-15,000 ppm, Cl 600-8000 ppm, making the ground water unsuitable for any purpose.

14. *Alluvium in Palar and Kortallaiyar-Araniyar rivers in Tamil Nadu*: Form potential aquifers; water of good quality within 50 m; Cl $<$ 250 ppm, EC 750-2,000 μmhos/cm.

15. *Tertiary sediments of Cauvery delta*: The tertiary sediments in Tanjore and South Arcot districts form extensive aquifers up to 200 m depth. Water is of good quality; Cl $<$ 150 ppm, EC $<$ 1,500 μmhos/cm. Many of the tubewells have free flow, some exceeding 2 m³/hr.

In the Cauvery delta rocks ranging in age from Precambrian Crystallines to Quaternary Sediments are encountered. Multiple aquifer systems are quite prevalent in a sufficiently thick sedimentary basin. The deep-seated aquifers are generally under confined conditions and there is hydraulic interconnection (vertical leakage) between aquifers. Recharge facilities are more for the top aquifers than the bottom aquifers. A fair yield of 76,500 m³/day may be expected in the Cauvery delta of Tanjore as per UNDP investigations.

$$T = 1.2 \times 10^5 - 8.2 \times 10^5 \text{ lpd/m}$$
$$S = 3.3 \times 10^{-4} - 6.8 \times 10^{-5}$$
$$EC = 800 - 1100 \text{ }\mu\text{mhos/cm}$$
$$Q = 30 - 60 \text{ m}^3/\text{hr}$$

Although artesian wells are quite prevalent, large scale development will lower the piezometric head and free flow condition would cease. For example, a decade ago there were many flowing wells in and around Neyveli. But now the piezometric head has been lowered and many flowing wells have become sub-artesian wells.

In Coimbatore and the central districts of Tamil Nadu, the substratum consists of rock at a moderate depth and the yield is very poor.

Near Pondichery, deep wells even very close to the coast yield potable water.

Ground water potential in Madras environs (UNDP)

Well Field	G.W. Potential, m^3/day
Minjur	33750
Duranallur—Panjetti	40500
Tamarapakkam—Villanur	49500

These well fields presently supply 45000 m^3/day to industrial units at Manali (Madras).

16. *Granitic gneisses and schists of Karnataka*: The principal rock types of Karnataka are igneous and metamorphic granites, gneisses and schists of Precambrian age and basalt of the Deccan trap of Eocene-Upper Cretaceous age in the extreme northern part of the state. The yield is very low and the borewells drilled up to depths of 30-75 m yield 5.40 m^3/hr.

The yield of wells in the crystalline rocks depends on the presence of weathered pockets, joints and fractures, of which there may be no indication at the surface. The yield of a well may be strikingly different from that of another well a few metres away. Surface geophysical resistivity survey may however indicate the depth and extent of concealed weathered pockets, which may ensure against risk of failure.

17. *Upper Gondwana sandstones and the alluvial tract of Orissa*: Form

Introduction

potential aquifers. The a.a.r. ≈ 142 cm and about 20% of this, i.e, ≈ 28 cm may be assumed as the recharge; also $\approx 50\%$ of the recharge potential can be utilised by the installation of filter point tubewells with a spacing of ≈ 330-580 m and the remaining 50% utilised by dug wells and deep tubewells. The yield of filter point tubewell of size 7.5-10 cm penetrating aquifers of 7-12 m thick (located within 50 m b.g.l.) is in the range of 20-50 m^3/hr which provide irrigation for 4-6 ha of land. The draw-down usually does not exceed 7 m and a low head centrifugal pump (2-4 kW) coupled directly to the filterpoint tubewell. The capital cost of installation of a filter point tubewell varies from Rs. 8000-12,000 depending on the depth and size of the filter point (see Appendix A.3) and require a simple hand boring set with a tripod for drilling the bore. At present there are about 800-1,000 filter point tubewells in Orissa state, most of them being located in the Balasore district and under the World Bank loan assistance programme there is a proposal to take up installation of 15,000 filter point tubewells.

In the alluvial tract where the granular aquifer material occurs within 8-10 m below ground level and also in the semi-consolidated sedimentary sandstones, weathered within 5 m b.g.l. open wells fitted with 2-4 kW centrifugal pump can be installed for irrigation purpose, with a minimum spacing of 150-200 m in alluvial tracts. Such wells can irrigate 2-2.5 ha of land in Kharif and 1-1.25 ha in Rabi season.

The capital cost of such wells varies from Rs. 10,000-15,000 with benefit cost ratio ranging from 1.5 to 3.1. Small wells of 1.5-2 m diameter fitted with indigenous type of water-lifting devices can irrigate about 0.4 ha of land in Kharif and 0.2 ha in the Rabi season. The capital cost of such small wells varies from Rs. 3,500-5,000 and the benefit cost ratio is very low due to the involvement of high cost of man power; yet it is a prized possession for a small farmer and has a significant influence on his economic condition.

It is estimated that 65% of the ground water potential of the state can be developed by installation of open wells. At present there are about 2 lakh open wells meant for irrigation in Orissa, and there is a World Bank loan assistance for the installation of 2 lakh open wells for the irrigation during the next five years.

18. *The Quaternary sediments in the deltaic tract around Digha, district Midnapur, West Bengal*: These are of depth 140 m and yield fresh water.

19. *The multilayered Lacustrine aquifer in Nepal in the centre of the Kathmandu valley basin*: Extends to 350 km^2 out of the total area of a roughly circular basin of 607 km^2. It has a depth of sediment of >450 m, deposits becoming coarser towards the north where most of the catchment is mountainous. The sediment consists of silt, coarse sand and cobbles with electrical resistivity ranging from 40-120 ohm-m. The a.a.r. = 174 cm, mostly during summer monsoons. Wells with rather low yields could be

constructed in the northern potential areas, the recharge being only 3,600 m³/day. Mud flush rotary drilling rigs are found suitable. $T = 92\text{-}301$ m²/day, $K = 2\text{-}12$ m/day, 48 hour specific capacity $\approx 113\text{-}137$ m³/day/m, $S = 2.3 \times 10^{-4}\text{-}3.7 \times 10^{-3}$, these rather high values indicate a high degree of elasticity of the confined aquifer. For unconfined aquifer $S_y \approx 0.1$.

20. *Karstic limestones in the north coastal belt of Sri Lanka*: Nine-tenths of the area of the island are underlain by the crystalline rocks such as gneisses, schists, quartzites and crystalline limestones of the Precambrian basement complex; low yields are obtained from the locally developed fissures and fractures. The soil overburden is 2-15 m and large diameter dug wells are generally suitable.

The remaining one-tenth of the island, in the north and north-western coastal belt, consists of deep sedimentary formations, where the Miocene limestone formations provide the major Karst aquifers under artesian and phreatic conditions; depth 90 m (average); piezometric level 15-33 m b.g.l. (free flowing wells at low ground levels). Tubewell yield 50-150 m³/hr, specific capacity 36-72 m³/hr/m, Cl 100-770 ppm, $T \approx 7900$ m³/day/m.

The a.a.r. of the island is 220 cm and the ground water potential ≈ 24.6 Mm³.

21. *Thermal and mineral springs*: They are found in many parts of India—Bombay, Punjab, Bihar, Assam, in the foothills of Himalayas and Kashmir.

Neyveli Artesian Aquifer (T.N.)—A Case History

A highly potential confined aquifer occurs in the South Arcot District of Tamil Nadu (South India) covering large areas of Cuddalore, Chidambaram and Vriddhachalam Taluks, and is referred to generally as 'Neyveli Aquifer' Fig. 1.2. After extensive investigations for a number of years by different agencies like Neyveli Lignite Corporation (NLC), Public Works Department of Tamil Nadu, UNDP, GSI, etc. it has been estimated that the recharge rate is of the order of 90 to 150 M m³/yr with a safe average of 120 M m³/yr. The recharge is due to river flow percolations and rainfall. The recharge area has been assessed to be 350 to 360 km². The average rainfall in the area for 42 years prior to 1964 was about 1100 mm, and the recharge is about 15 to 20% of rainfall.

It is called a 'coastal aquifer' as it lies adjacent to the sea and there are indications that it lies underneath the sea also between cuddalore and Alapakkam. The aquifer outcrops in a NE-SW direction, the length along the normal being about 30 km and the average width at right angles to normal being 12 km, Fig. 1.3. The river Manimuktar near Vriddhachalam is also flowing right through the recharge basin and contributes to the recharge of the aquifer. River Gadilam in the north also embraces the aquifer area.

The aquifer which is exposed in the recharge area gradually dips in E-SE

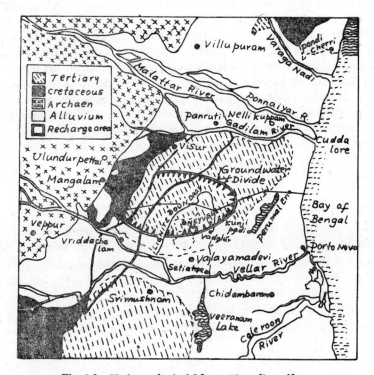

Fig. 1.2 Hydrogeological Map—Neyveli aquifer.

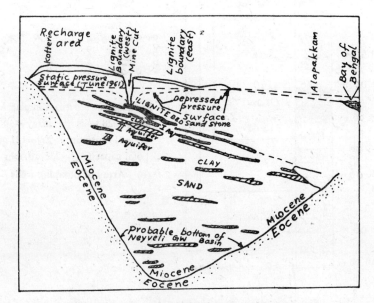

Fig. 1.3 Neyveli artesian aquifer—Section.

direction. The top of this aquifer occurs at about 73.1 m bgl in the first mine area and about 91.46 m bgl in the second mine area, and dips steeply in S-E direction in the Sethiatope area to 167.7 m or more at the roadside park. The aquifer swl was more or less steady at 30 m above msl except for a small slope towards the sea, the hydraulic gradient being due to frictional loss during movement of ground water. So in places where the ground surface was lower than + 30 m, when a bore hole was drilled to the top of the aquifer the water would start flowing from the pipe (flowing wells). A survey before the commencement of pumping had indicated that agriculture was drawing about 180000 m³/day from the aquifer. The average hydraulic characteristics of the aquifer are $S = 2.5 \times 10^{-4}$, $K = 40$ m/day; for $b = 30$ m, $T = Kb = 1200$ m²/day.

Lignite Deposits: An idealised vertical section (bore log) is shown in Fig. 1.4. The lignite deposit occurring at Neyvelisits on the top of the first aquifer separated by a thin layer of clay 1.5 to 3 m in thickness. While excavating the open cast mine it was calculated that the critical depth was 42.68 m below which it was dangerous to proceed unless the aquifer pressure was depressed to below the bottom of lignite. Thus commenced the now well known 'Ground Water Control Operations' of NLC on July 6, 1961. In the beginning stages, about 28 pumps were installed in a rectangular

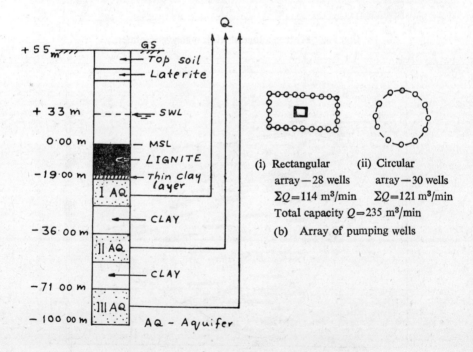

Fig. 1.4 Idealised vertical section—Neyveli aquifer.

Introduction

array to give a total discharge of about 114 m³/min. As the mine was deepened and lignite removed, more water had to be pumped to keep the mine floor from bursting. During the peak period of pumping (1962-64), the pumping rate was nearly 235 m³/min by installing a circular array of 30 more wells. For fear of salt water intrusion, ultimately in 1969, the pump wells were installed in the deeper portions of the mine in the lignite bench itself and the pumping was reduced to about 137 m³/min. This scheme not only saved unnecessary pumpage of fresh water but also saved power consumption of 3300 kW which amounted to roughly 80000 units per day.

At the deepest portion, lignite bottom occurs at about -100 m. Since the aquifer is lying beneath or very close to the sea, there is danger of sea water contamination.

The NLC now pumps about 180000 to 205000 m³/day of water and the free flowing agricultural wells account for another 160000-180000 m³/day giving a total 'abstraction from aquifer of about 140 M m³/yr. Thus there is a precarious balance compared to annual recharge, and steady state conditions.'

Neyveli Water Requirements: The first power station, fertilizer, industrial units, township, etc. need about 220000 m³/day which is now being obtained by mines pumping as well as township water supply bore holes. The second power station of 1470 MW would need about 260000 m³/day, the second super fertilizer factory about 100000 m³/day, and the additional population about 5000 m³/day. The total water supply requirement is about 360000 m²/day apart from agricultural abstraction which is about 180000 m³/day. Thus the future water requirement will be of the order of 540000 to 600000 m³/day or 200 M m³/yr. Since this is more than the estimated annual recharge of 120 M m³/yr, the additional requirement can be obtained from Veeranam Lake which is nearby where the surplus water of Coleroon river is stored.

Method of Exploitation of Ground Water

Any programme of ground water exploitation should have the following equipment for well sinking (boring or drilling) or revitalisation.

(i) Tractor/compressor for blasting or extension drilling.
(ii) Compressors (VT-4, VT-5, etc.) for drilling rigs and development of wells.
(iii) Bencher units for extension drilling.
(iv) Cobra units for drilling blast holes.
(v) Auger rigs and hand boring sets for boring shallow wells, say, in terai region, in coastal aquifers of Cauvery data, in the alluvial tract of Orissa, or boring cavity wells as in Delhi IARI Pusa area.

(vi) Cable tool or percussion rigs as may be suitable in the areas of Indo-Gangetic alluvium, sediments of Jammu and Kashmir valley, unconsolidated formations in Bengal, Gujarat and Madhya Pradesh, and soft and boulder formations.

(vii) Rotary rigs (straight rotary) in semi-consolidated formations and reverse rotary (reverse circulation) for large diameter and deep holes in soft consolidated formations.

(viii) Air rotary is specially suitable for limestones and air foam is used to remove cuttings.

(ix) Rotary-cum-percussion rigs in the consolidated formations of Madhya Pradesh, Bihar, Orissa, Gujarat, etc.

(x) Jetting drill is suitable for unconsolidated formations for holes upto 15 cm diameter and plenty of water is required for the water jet.

(xi) Down-the-hole hammer (DTH) rigs like Ingersoll Rand, Halco-625, Sanderson Cyclone, RMT, etc. for fast drilling of deep borewells in the hard rock areas of peninsular India, i.e. in crystalline areas.

(xii) Down-the-hole hammer rigs like Halco Tiger, Halco Minor, Atlas Copco, etc. for fast drilling of borewells in the Deccan traps of Maharashtra or the hard rock areas of Coimbatore and North Karnataka where the yield of borewells is very low.

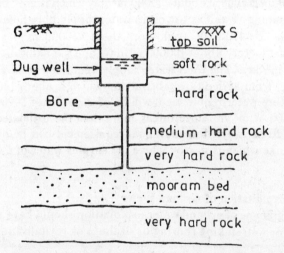

Fig. 1.5 Dug-cum-bore well tapping mooram bed.

(xiii) Calyx drills can drill bore wells in hard rock areas (the rock cores are cut by feeding chilled steel shots) but they are very slow. Calyx drills can be used for drilling bores at the bottom of open wells either for well revitalisation or dug-cum-borewells, Fig. 1.5 (a centrifugal pump can be installed at the bottom of the dug well with the suction pipe in the bore).

Introduction 15

As on March 1979 there are about 4500 hand boring sets, 230 percussion rigs, 330 rotary rigs, 80 reverse circulation rigs, 20 rotary-cum-percussion rigs, 225 down-the-hole hammer rigs, 150 Calyx rigs and 500 pneumatic rigs in the 17 States of the country.

Ground Water Development

Ground water can be developed at a small capital cost and the time taken for development is very small. The chemical quality of ground water is found to be generally good and can be used for drinking, agriculture and industrial purposes. The hard rock drilling programme by UNICEF has significantly lowered the unit cost 'per capita' of making available a safe water supply; water is usually struck at a depth of 30–60 m in hard rock formations and requires no treatment.

The tapping of ground water—location, spacing and yield, in a well field should be so phased that the annual recharge and discharge of the aquifer are almost balanced without causing an overdraft in the area. As the ground water development increases, problems of well field management will become dangerously critical in many places and the studies on optimum well spacing will be required in order to minimise mutual interference between pumped wells.

The recharge depends upon the rock or soil formation and the a.a.r. of the region and is given in Table 1.1.

Table 1.1 Recharge rates for different formations

Formation	Recharge (% of a.a.r.)
Hard rock formations and Deccan traps	10
Consolidated formations (sandstones)	5–10
River alluvium	15–20
Indo-Gangetic alluvium	20
Coastal alluvium	10–15

In several areas the recharge estimates made on the above assumed rates have agreed well with the estimates of recharge based on actual water level rise in the region and the specific yield of the formations.

Ground Water Investigation

The problems facing any ground water investigation programme are the zones of occurrence and recharge. The various phases of a ground water investigation programme are given below.

(a) Hydrometeorological study
(b) Hydrogeological study
  (i) Geological mapping
  (ii) Test drilling, sampling and logging
  (iii) Pumping tests (aquifer tests)
(c) Geophysical survey
  (i) Surface
  (ii) Down-the-hole
(d) Aerial photographic survey
  (i) Black and white
  (ii) Colour
  (iii) Infra-red
  (iv) Radar imagery
(e) Tracer techniques
(f) Geochemical and geothermal surveys
(g) Systems analysis, mathematical modelling and computer applications for ground water basin management
(h) Water balance studies
(i) Intensive irrigation and water management

The objectives of any hydrogeological investigations are:
(i) Define recharge and discharge areas
(ii) Define major water bearing units
(iii) Define location, extent and inter-relationship of aquifers
(iv) Establish physical parameters of aquifers like transmissibility, storage coefficient and specific yield
(v) Estimate total subsurface storage capacity
(vi) Establish geologic factors which affect quality of ground water
(vii) Arrive at the location, probable depth of drilling and yield from the bore well (tubewell)

Aerial and infra-red photography, electrical resistivity surveys and logging techniques can provide valuable information in regard to the zones of occurrence and recharge. The U.S. Geological Surveys (USGS) by the use of an infra-red scanner, has published an atlas of Hawaii's coastal areas, pinpointing the location of underground fresh water flows.

With the advancement of space technology, it has been possible now to resort to the remote sensing technique for the estimation of surface and subsurface waters over large areas. This technique employs the surveying of ultra-violet visible microwave radiations emitted and reflected from the surface of the earth. This method would be extremely useful for rapid hydrogeological mapping of large and inaccessible areas. Areas of dry and wet rocks, aquifers, open water surfaces, springs, etc. can be delineated by skillful interpretation.

Introduction

With the hydrometeorological data combined with geophysical and hydrogeological investigations and pumping tests, it is possible to develop and manage the ground water resources of a basin.

Conjunctive Use of Ground Water

Surface water and ground water may be viewed as two different forms of occurrence of the same total water resources. Tubewell schemes may be integrated with the canal irrigation schemes (composite irrigation) by suitably spacing them along a line in between the distributory and the drainage line and so designing that the subsoil water level is kept steady at a desired level. The tubewells intercept the canal seepage and serve as an anti-water logging measure and enable the benefit of irrigation facilities to be spread to wider areas. Supplemental ground water irrigation is proposed to be introduced in the command areas of a number of major irrigation systems like the Yamuna canal, the Cauvery and the Krishna deltas to enable intensive agricultural development.

Maximising Irrigation Efficiency and Water Management

As old varieties of cereals, pulses and millets are being replaced by new high yielding varieties which respond to chemical fertilisers, fresh studies on soil-water-plant relationship are being carried out to determine the irrigation requirements of different crops at different stages of growth. This has to be carried down to the level of the farmer through extension services for introducing improved and intensive agricultural practices.

By charging for the water supplied (by measuring the tubewell discharge over the V-notch or by noting the power units consumed) and lining the water courses by any cheap material locally available like clay tiles, laterite sheets, cuddapah slabs with joints finished with 1 : 4 cement mortar or soil cement, brick in cement mortar, etc., the economic use of water is achieved.

Hence it is necessary to have an All India Water Resources Council to coordinate, compile and computerise data and apply modern methods of 'systems approach' to irrigation.

Ground Water Legislation and Pollution

There is need for legislation for ground water exploitation and regulation to check indiscriminate draining of ground water resources. Precautions should be taken against pollution of surface and subsoil waters by enacting legislation.

Water is a national asset available in a finite quantity. One cannot afford to forget that if one cannot afford to pollute air which is available in unlimited quantities, much less can one afford to take liberties with the use of the limited asset of water.

QUIZ 1

I. Fill in the blanks choosing the right word(s) given in the brackets:
   1. Ground water is a .... resource of the earth.
      (replenishable, scarce, evenly distributed)
   2. .... is the greatest user of ground water.
      (agriculture, rural water supply, industry)
   3. The present irrigated area in the country is .... M ha of which about .... % is from ground water.
      (25, 45, 55; 25, 40, 50)
   4. The a.a.r. of India is ..... cm.
      (30, 50, 80, 114, 200)
   5. The total ground water potential of India is ...... M ha-m of which ...... M ha-m is the present utilisation and ...... M ha-m is available for further exploitation.
      (46, 31, 13, 6.5, 15.5, 19.5)
   6. Brick-lined dug wells existed in the country as early as ..... (Moghul period, since India became independent, Indus Valley Civilization).
   7. The ETO was set up during ...... and was merged with GSI Ground Water Directorates in ... to form the .... which is now the apex body for ground water at the national level.
      (1934, 1944, 1954, 1970, 1972; CWPB, CGWB, CWPRS, ICID, ICAR)
   8. Since 1970, a large scale well drilling programme has been launched in hard rock areas of India with the assistance of ... for .... schemes which utilises air hammer drilling rigs.
      (UNDP, UNICEF, UK; irrigation, rural water supply, industrial)
   9. 50% of the total land formations in the country comprises ......
      (crystallines, river alluvium, trap)
   10. In arid zones of Rajasthan and Gujarat excess concentration of .... has been observed in the ground water resulting in ....
       (chloride, fluoride, boron, iron, hydrogen sulphide; saline patches, sulphuretted water, mottling of teeth)
   11. Ground water can be developed at a ... capital cost with a ..... irrigation efficiency.
       (small, medium, high, maximum)
   12. Remote sensing technique employs the surveying of .... visible microwave radiations emitted and reflected from the ... of the earth.
       (infra-red, ultra-violet; surface, subsurface; aquifer).
   13. The scale of the hydrogeological map of India released by GSI in 1969 was ... and was updated by CGWB in 1976 to a scale of ....
       (1 cm = 10 km, 1 cm = 20 km, 1 cm = 50 km, 1 cm = 100 km)

II. Select the correct answer(s):
   1. The prime factors to be considered in any ground water investigation programme are:
      (a) zones of occurrence and recharge
      (b) opportunity for recharge
      (c) hydraulic connection between recharge and discharge areas
      (d) balancing annual extraction and recharge
      (e) optimum spacing of wells

Introduction 19

   (f) all the above factors
   (g) none of these factors
2. Conjunctive use of ground and surface water helps
   (a) prevent water logging
   (b) more area to be irrigated
   (c) supplemental irrigation
   (d) all the above factors
   (e) none of these factors
3. Economic use of water can be obtained by
   (a) volumetric sale of water
   (b) charging a crop-wise water rate per hectare of crop
   (c) lining the water courses
   (d) matching the soil-water-plant relationship, the crop pattern and planning
   (e) landshaping and providing adequate drainage
   (f) using improved seeds and fertilisers
   (g) all the above answers
   (h) none of these answers
4. Ground water legislation is needed to
   (a) check indiscriminate draining of ground water
   (b) check pollution of ground water
   (c) maximise irrigation efficiency
   (d) limit the use of water
   (e) all the above answers
   (f) none of these answers
5. Infra-red radiations can be utilised for demarcating the areas of;
   (a) surface water and ground water
   (b) underground fresh water and salt water

III. Match the items in A, B and C (more than one item can fit in):

Part I

	A	B	C
1.	Siwaliks at the foot hills of Himalayas	(i) Cretaceous	(a) Kathmandu Valley
2.	Sandstones of Kathiawar and Kutch	(ii) Archaeans	(b) Chhatisgarh basin and Baghelkhand region of M.P.; North coastal belt of Ceylon
3.	Sandstones of Rajahmundry, Sriperambadore and Cuddalore	(iii) Vindhyan system	(c) 50% of land formation in the country—poor yield
4.	Lathi sandstones of Bikaner and Jaisalmer areas	(iv) Tertiary	(d) Kashmir Valley
5.	Cavernous limestones of Rajasthan	(v) Upper Gondwana	(e) Terai zone
6.	Sediments in Tanjore and South Arcot districts	(vi) Deccan traps	(f) Maharashtra and M.P.
7.	Doon gravels	(vii) Miocene	(g) Gneisses, granites and schists

	A	B	C
8.	Sediments of lacustrine, fluvial and glacial origin	(viii) Eocene	(h) Charnockites
9.	Multilayered lacustrine aquifer	(ix) Precambrian	(i) Vast reservoir of fresh ground water
10.	Karstic limestones	(x) Pleistocene	(j) Laterite deposits in the west coast of Peninsular India
11.	Indo-Gangetic alluvium	(xi) Artesian springs and wells at shallow depth	(k) Principal rock types of Karnataka
12.	Vesicular basalts	(xii) Can support > 200 tubewells of yield ≈150 m³/hr each	(l) Crystallines
13.	Gollapalli and Tirupati sandstones		

Part II

	A	B	C
1.	Augers	(i) Large size holes	(a) Hard rock areas (crystallines)
2.	HB sets	(ii) Small size holes	(b) Semi-consolidated formations
3.	RB units	(iii) Cable tool	(c) Unconsolidated formations
4.	Air compressor	(iv) Atlas Copco-OD	(d) Consolidated formations
5.	Bencher units	(v) Halco Tiger	(e) Boulder formations
6.	Percussion rigs	(vi) Halco Minor, Halco-625	(f) Limestone formations
7.	Rotary rigs	(vii) RMT rig	(g) Soft formations
8.	Air rotary	(viii) Ingersoil Rand	(h) Coastal shallow aquifers
9.	Reverse rotary rigs	(ix) Plenty of water for drilling	(i) Alluvium
10.	Rotary-cum-percussion rigs	(x) Well revitalisation	(j) Terai region
11.	Jetting drill	(xi) Blast holes	(k) Deccan traps
12.	DTH rigs	(xii) Extension drilling	
13.	Calyx drills	(xiii) Well development	
14.	Top hammer	(xiv) Bores at the bottom of shallow dug wells	
		(xv) Large capacity suction pump	

QUESTIONS

I. Give an account of the geohydrological zones of India.

II. Give an account of ground water resources of India with special reference to Karnataka.

Introduction

III. In the map of your State, indicate the following, after collecting information from State Ground Water Directorate/Cell and specialists in the field:
   (i) Potential areas of ground water in the State.
   (ii) Type of formations and their probable yield.
   (iii) Recommended method of drilling (percussion, rotary, DHD, etc.) in the different formations.
   (iv) Recommended depths of drilling and the probable yield.
   (v) Any information regarding the quality of water with reference to any particular area.
   (vi) Any restrictions on the depth of drilling or pumping rates, particularly in coastal aquifers.
   (vii) Extractions already existing, type of drilling rigs operating in the area, trained personnel and drilling crew.
   (viii) Further potential and scope for further ground water development.

2

Hydrometeorology

Hydrometeorological data are required to determine the water balance of a basin for developing and managing its water resources. The most useful hydrometeorological elements are precipitation, evaporation, evapotranspiration, solar radiation (sunshine hours), air temperature and humidity, soil moisture, water levels (surface and underground), stream discharge, water quality, etc.

Rainfall may be measured by a network of non-recording and recording rain gauges. The non-recording rain gauge used in India is the Symon's rain gauge shown in Fig. 2.1, erected on a masonry foundation with the rim 30.5 cm above the ground. The rain water in the gauge should be measured every day at 08.30 hours IST. The rain water falling into the funnel is collected in the receiver kept inside the body and is measured by pouring into the special measure glass graduated in mm of rainfall. During heavy rains, it must be measured three or four times a day so that the receiver may not fill and overflow, but the last reading must be taken at 08.30 hours IST and the sum total of all the measurements during the previous 24 hours entered as the rainfall of the day.

Self-recording Rain Gauge

The natural siphon recording rain gauge gives a continuous record of rainfall, its intensity and duration. Rain water entering the gauge at the top of the cover is led via the funnel to the receiver, consisting of a float chamber and a siphon chamber. The pen is mounted on the stem of the float and as the water level rises in the receiver the float rises and the pen records, on a chart placed on a clock drum, the amount of water in the receiver at any instant. The clock drum revolves once in 24 hours or 7 days, so that a continuous record of the movement of the pen is made on the chart. Siphoning occurs automatically when the pen reaches the top of the chart and as the rain continues, the pen rises again from the zero line

Hydrometeorology

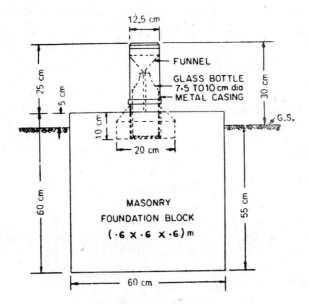

STANDARD RAINGAUGE

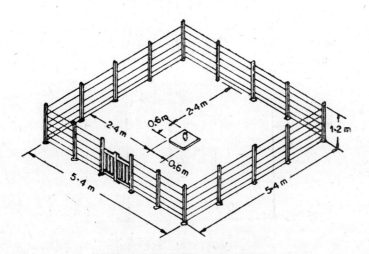

ENCLOSURE FOR RAINGAUGE

Fig. 2.1 Symon's rain gauge

of the chart, Fig. 2.2. If there is no rain the pen traces a horizontal line. Siphoning occurs automatically for every 10 mm of rainfall.

The gauge is installed on a masonry foundation, with its rim 75 cm above the ground, by the side of the ordinary rain gauge (2-3 m away from it) for checking and adjustment if necessary.

The optimum number of rain gauge stations (N) for a desired degree of percentage error (p) in the estimate of the mean rainfall of the basin calculated by the simple arithmetic average is given by

$$N = \left(\frac{C_v}{p}\right)^2 \qquad (2.1)$$

where C_v is the coefficient of variation of the rainfall of the existing rain gauge stations.

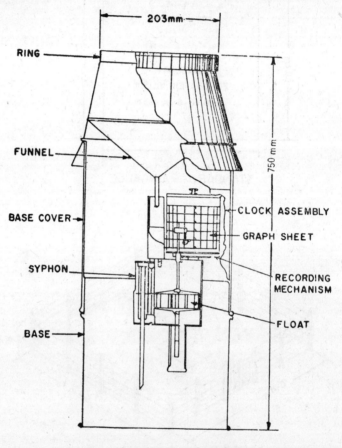

VERTICAL SECTION
Fig. 2.2 Automatic rain gauge.

The rain gauge density may be 1 in 520 km² in plains, 1 in 260-390 km² in elevated regions, and 1 in 130 km² in hilly and very heavy rainfall areas preferably with 10% of the rain gauge station equipped with the self-recording type gauge. The length of record needed to obtain stable frequency distribution of rainfall may be recommended as about 30 years in islands, 40 years in shore and plain areas and 50 years in mountainous regions.

Hydrometeorology

Mean Areal Depth of Rainfall

The average depth of rainfall over a specific area may be determined as:

(i) The arithmetic average of the rainfalls recorded by individual rain gauge stations in that area.

(ii) *The weighted mean*: Thiessen polygons are obtained by drawing perpendicular bisectors to the lines joining adjacent stations on a base map and each polygon area is assumed to be influenced by the rain gauge station inside it. For example, if P_1, P_2 and P_3 are the point rainfalls in a basin and the areas of the polygons surrounding these stations are A_1, A_2 and A_3 respectively, the average depth of rainfall for the entire basin is given by

$$P_{mean} = \frac{\Sigma A_1 P_1}{\Sigma A_1} \qquad (2.2)$$

The results obtained are usually more accurate than those obtained by simple arithmetic averaging. The gauges should be properly located over the catchment to get regular shaped polygons.

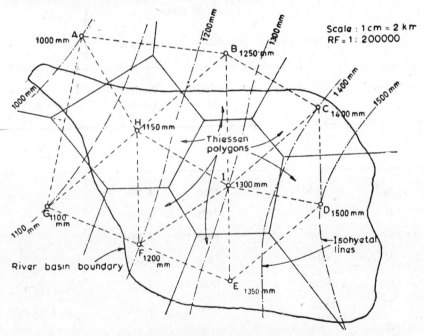

A, B, C, D, E, F, G, H, I: Locations of the rain gauge stations with annual rainfalls recorded

Fig. 2.3 Thiessen polygon and isohyetal map for the river basin.

(iii) *The isohyetal method*: The point rainfalls are plotted on a suitable base map and the lines of equal rainfall (isohyets) are then drawn giving consideration to orographic effects and storm morphology. The average rainfalls between successive isohyets (usually taken as the average of the two isohyetal values) are weighted with the area between the isohyets, added up and divided by the total area which gives the average rainfall for the entire area. This method, if analysed properly, gives the best results.

Example 2.1: An outline of a river basin with the locations of the rain gauge stations and their annual rainfalls recorded is shown in Fig. 2.3. Construct the Thiessen diagram and the isohyetal map, and determine the mean depth of precipitation for the basin.

Solution: The Thiessen diagram and the isohyetal map are constructed as shown in Fig. 2.3 and the results are tabulated in Tables 2.1 and 2.2.

The isohyetal method gives almost the same result as the Thiessen polygon method.

The arithmetic average $= \dfrac{\Sigma P_1}{n} = \dfrac{11{,}250 \text{ mm}}{9}$, from Table 2.1

$= \mathbf{1250 \text{ mm}}$

Table 2.1 Mean depth of precipitation—Thiessen polygon method

Station	Weighted area (units)*	Rainfall (mm)	Product (2)×(3)	Mean Rainfall Σ(4)/Σ(2)
1	2	3	4	5
A	6.5	1000	6,500	
B	5.5	1250	6,875	
C	17.0	1400	23,800	
D	40.0	1500	60,000	
E	29.5	1350	39,800	$= \dfrac{227{,}785}{173}$
F	17.0	1200	20,400	$= \mathbf{1317 \text{ mm}}$
G	2.5	1100	2,750	
H	25.5	1150	29,300	
I	29.5	1300	38,360	
Total 9	173.0	11,250	227,785	

*1 unit=1 cm² in Fig. 2.3.

Hydrometeorology

Table 2.2 Mean depth of precipitation—Isohyetal method

Isohyetal range (mm)	Average rainfall (mm)	Area between isohyets (units)*	Product (3)×(2)	Mean rainfall Σ(4)/Σ(3) (mm)
1	2	3	4	5
1000-1100	1050	12 (approx.)	12,600	
1100-1200	1150	28	32,200	
1200-1300	1250	42	52,500	$= \dfrac{229300}{173}$
1300-1400	1350	30	40,500	$= 1324$ mm
1400-1500	1450	32	46,500	
	1550	29	45,000	
	Total	173	2,29,300	

*1 unit=1 cm² in Fig. 2.3.

Thus the arithmetic mean differs from the more accurate results that can be obtained from other methods.

Solar Radiation

The solar radiation may be measured by the Bellami spherical pyranometer installed as shown in Fig. 2.4. It consists essentially of a glass sphere filled with alcohol, which when warmed by solar radiation falling on the sphere distils into a graduated glass tube below. The radiation received during a given period is directly read on the graduated tube at 12.30 IST and half an hour after sunset and reset both the times.

Sunshine Recorder

The hours of bright sunshine are usually recorded by the Campbell-Stokes pattern sunshine recorder which consists essentially of a glass sphere about 10 cm in diameter, Fig. 2.5. The hours of bright sunshine are measured by the length of the burn of the cards due to the heat of the sun rays focussed on the card by the glass sphere and the cards are changed each day.

Evaporation

Estimates of evaporation from free water surfaces and [the soil, and

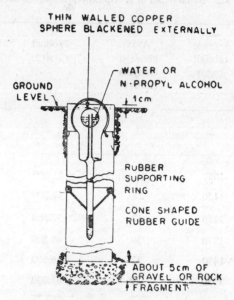

Fig. 2.4 Gunn-Bellani radiation integration.

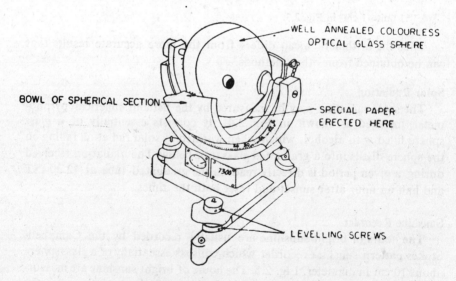

Fig. 2.5 Sunshine recorder (Campbell-Stokes pattern).

transpiration from vegetation are of great importance in hydrometeorological studies. A variety of devices and techniques have been developed for estimating vapour transport from water or land surfaces. Under given conditions, evaporation (E) is proportional to the difference between the

Hydrometeorology

saturated vapour pressure e_w at the temperature of the water and the aqueous vapour pressure of the air e_a (Dalton's law)

$$E = K(e_w - e_a) \tag{2.3}$$

where K is a constant. Dalton's law is the basis for many other equations.

Pan Evaporimeter

The USWB class A land pan is 122 cm diameter and 25.5 cm deep with stilling well, vernier point gauge, thermometer with clip, Fig. 2.6. The pan

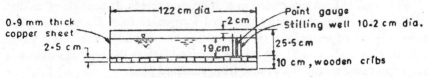

Fig. 2.6 Pan evaporimeter.

is placed on a wooden frame work of 10 cm thickness and is covered with a wire screen. The amount of water lost by evaporation from the pan after any given interval of time is measured by adding known quantities of water to the pan from a graduated cylinder till the water level touches a reference point 19 cm above the base of the pan or can be directly measured by the point gauge. Readings are taken twice daily at 08.30 hours and 17.30 hours IST. The air temperature is determined by reading a dry bulb thermometer kept in the Stevenson screen erected in the same enclosure as the pan. A totalizing anemometer is normally mounted at the level of the instrument to provide the wind speed information required. Rain falling into the pan is allowed for on the assumption that the catch of a nearby rain gauge represents the added depth of water due to the rain. If the rainfall exceeds the water lost by evaporation, water has to be removed from the pan, instead of being added.

Work done in India covering the pan with a GI wire screen shows that, on the average, the unscreened pan reading is 1.144 times that of the screened one, since the screen affects the free air movement.

Soil Evaporation

Evaporation from a wet soil surface immediately after rain or escape of water molecules with more resistance when the water table lies within a metre from the ground surface is called soil evaporation. This expressed as a percentage of evaporation from free water surface is called 'evaporation opportunity'. Soil evaporation will continue at a high rate for some time after the cessation of rainfall, then decreases as the ground surface starts drying, until a constant rate is reached which is dependent on the depth of the water table and nature of the soil in addition to meteorological conditions.

Measurement of soil evaporation can be done with tanks (lysimeters) filled with earth and with the surface almost flush with the ground. To measure the evaporation from a soil whose surface is within the capillary fringe, tanks equipped to maintain the water table at any desired elevation may be used. The soil evaporation is determined by weighing the tanks at stated intervals and knowing the amount of water that was added in the interim.

Piche Evaporimeter

It is usually kept suspended in a Stevenson screen. It consists of a disc of filter paper kept constantly saturated with water from a graduated glass tube. The loss of water from the tube over a known period gives the average rate of evaporation. Though it is a simple instrument the readings obtained are often more erratic than those from standard pans.

Combination of Aerodynamic and Energy Balance Equations

The most widely used method for computing lake evaporation from meteorological factors is based on a combination of aerodynamic and energy balance equations. The Penman's equation in its general form is

$$E_0 = \frac{\Delta H + E_a \gamma}{\Delta + y} \qquad (2.4)$$

$$E_a = b(a + U)(e_a - e_d)$$

$$H = R_a(1-r)\left(0.18 + 0.55 \frac{n}{N}\right) - \sigma\, T_a^4 (0.56 - 0.092 \sqrt{e_d})\left(0.10 + 0.90 \frac{n}{N}\right)$$

where E_o = estimated evaporation from a free water surface, mm of water/day; Δ = slope of the saturation vapour pressure curve at air temperature T_a; γ = a constant in the wet and dry bulb psychrometer equation for °C and mb pressure = 0.65; E_a = a desired parameter; U = mean wind speed, at 2 m above ground, km/day; a, b = constants; e_a = saturation vapour pressure at temperature T_a, mm of mercury; e_d = actual vapour pressure in the air at the temperature T_a, mm of mercury; H = daily heat budget at the surface, mm of water/day; R_a = mean monthly extra-terrestrial radiation expressed in equivalent evaporation, mm of water/day; r = reflection coefficient of water surface (≈ 0.06), albedo; $\frac{n}{N}$ = ratio of actual to possible hours of bright sunshine per day; σ = Stefan-Boltzmann constant, expressed in equivalent evaporation in mm/day (2.01×10^{-9} mm/day); T_a = mean daily air temperature (absolute) and °K (= 273 + °C).

A similar approach was used by Kohlar and others using a graphical representation of relation. Meteorological observations of solar radiation, air temperature, dew point and wind movement at class A pan anemometer height, are required for application at this height.

Evapotranspiration

Transpiration is the amount of water lost by plants mainly through leaves. Evapotranspiration (E_t) or consumptive use (U) is the total water lost from a cropped land due to evaporation from the soil and transpiration by the plants, or used by the plants in building up of plant tissue. Potential evapotranspiration (E_{pt}) is the evapotranspiration from the short green vegetation when the roots are supplied with unlimited water covering the soil. It is usually expressed as a depth (cm) over the area.

Estimation of Evapotranspiration

There are many methods of estimating evapotranspiration. Some of these are briefly described in the following:

1. SOIL MOISTURE SAMPLING

A large number of samples are taken from various depths in the roots zone and the soil moisture depletion studies are made.

2. TANKS AND LYSIMETER EXPERIMENTS

Keeping an account of water added and soil moisture changes by weighing the tanks.

3. INFLOW-OUTFLOW METHOD (by water balance)

For large areas where yearly inflow into the area, annual precipitation, yearly outflow from the area and the change in ground water level are evaluated.

4. FIELD EXPERIMENTAL PLOTS

Small amouuts of water are applied at each irrigation to avoid deep percolation.

5. INSTALLATION OF COLORADO TANKS

The tanks are $90 \times 90 \times 90$ cm made of 1.6 mm mild steel sheets with 2.5×2.5 cm angle stiffeners welded to the edges and rims. The tanks, with bottoms for studies of evapotranspiration and without bottoms to include percolation, with crop and without crop inside, are placed in position (sunk) in a level field with the rim 15 cm above ground level and 7.5 cm standing depth of water inside as shown in Fig. 2.7 (UNDP). After lowering the tanks inside the ground to the desired level the repacking of the soil excavated has to be done layer by layer in the same order and well compacted.

After installation of tanks and preparation of fields for planning, the water level changes inside the tanks are measured accurately by means of a point gauge. The inflow to and outflow from the fields are measured over V-notches installed and the ground water levels are noted in burried piezometers. The separated values of percolation, evaporation and transpiration

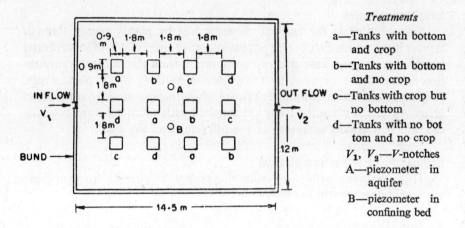

Fig. 2.7 UNDP water use experiment at Marudandanallur, Tamil Nadu.

are obtained and basin-wise decrements are worked out for accurate water balance studies.

The daily and seasonal consumptive use rates are arrived at and the total water requirements calculated for ready adoption in the field by working out suitable irrigation schedules. Thus the evapotranspiration data are obtained by climatological factors with correlation to soil properties and crop factors.

6. Evapotranspiration Equations

(i) *Lowry-Johnson equation used by USBR*: The constants in this equation are based on estimates of consumptive use by the inflow-outflow method for a number of irrigated valleys. The constants can also be determined from the consumptive use data in the immediate region of the project and then adjusted for temperature differences.

(ii) *Penman equation*: Estimation of evaporation is made on the basis of data for wind velocity, vapour pressures for saturated vapour and air and the crop evapotranspiration determined by multiplying the evaporation values by empirical constants depending on latitude and length of day light. This method has been used throughout the world and found to give good results in humid areas. However its use has been limited because of its complexity and the fact that it utilises several parameters which are not available in the published weather data.

Hydrometeorology

(iii) *Thornthwaite equation**

$$E_{pt} = ct^a \quad (2.5)$$

where E_{pt} = potential evapotranspiration, cm/month; t = mean monthly temperature, °C; and c, a = constants to be selected for the climatological conditions of the area latitude and the month of the year for specific crops.

(iv) *P/E Index method*: Based on studies of weather and crop data from the irrigated areas in the western US, Munson found that the following *P/E* (precipitation-evaporation ratios) hold adequately for normal plant growth.

Month	Jan.	Feb.	March	April	May	June	July	Aug.	Sep.	Oct.	Nov.	Dec.
P/E Ratio	1.0	1.8	3.2	4.4	5.8	6.0	6.8	6.1	4.6	3.5	2.3	1.5

The sum of the 12 monthly *P/E* ratios is the *P/E* index equal to 47. For a given month, the corresponding value of the *P/E* ratio and the average monthly temperature t in °F can be substituted in the modified Thornthwaite formula

$$P = 0.014(t-10)\left(\frac{P}{E}\right)^{0.9} \quad (2.6)$$

to determine P the required monthly precipitation or actually the monthly consumptive use. It is necessary to exercise judgement in the use of this method.

(v) *Blaney-Criddle equation*: The consumptive use is given by the empirical formula

$$u = kf \text{ and } U = KF = \Sigma kf; f = \frac{tp}{100}$$

$$U = \frac{\Sigma ktp}{100} \quad (2.7)$$

---

*$E_{pt} = 16\left(\frac{10t}{J}\right)^a$, mm/month

Monthly heat index $j \approx 0.09\, t^{3/2}$

Yearly heat index $J = \sum_{1}^{12} j$, $a = 0.016 J + 0.5$ } (Serra)

$j = \left(\frac{t}{5}\right)^{1.514}, J = \sum_{1}^{12} j$

$a = (675 \times 10^9)J^3 - (771 \times 10^{-7})J^2 + (179 \times 10^{-4})J + 0.492$ } (Thornthwaite) where t = mean monthly temperature, °F; p = monthly percentage of day time hours of the year; k = monthly consumptive use coefficient determined from experimental data (see Table 2.3); u = monthly consumptive use; U = seasonal consumptive use or evapotranspiration; f = monthly consumptive use factor; $F = \Sigma f$, for the growing season; and K = seasonal comsumptive use coefficient for the crop.

In metric units,

$$u = kp\frac{(45.7t + 813)}{100} \text{ mm} \qquad (2.7\text{a})$$

where t = mean monthly temperature, °C and $f = p(0.46t + 8.13)$.

Phelan proposed an empirical modification of the Blaney-Criddle formula

$$k = k_t k_c \qquad (2.7\text{b})$$

where k_t = climatic coefficient related to the mean temperature t as $k_t = 0.0173t - 0.314$ and k_c = coefficient reflecting growth stage of crop which has to be determined experimentally.

Table 2.3 Monthly consumptive use crop coefficient k for use in Blaney-Criddle formula* (after Dastane, 1972)

Crop	Consumption use coefficient k for different months											
	Jan.	Feb.	Mar.	April	May	June	July	Aug.	Sep.	Oct.	Nov.	Dec.
Rice				0.85	1.00	1.15	1.30	1.25	1.10	0.90		
Maize				0.50	0.60	0.70	0.80	0.80	0.60	0.50		
Wheat	0.50	0.70	0.75	0.70								
Sugarcane	0.75	0.80	0.85	0.85	0.90	0.95	1.00	1.00	0.95	0.90	0.85	0.75
Cotton				0.50	0.60	0.75	0.90	0.85	0.75	0.55	0.50	0.50
Vegetables	0.50	0.55	0.60	0.65	0.70	0.75	0.80	0.80	0.70	0.60	0.55	0.50
Berseem	0.50	0.70	0.80	0.90	1.00				0.60	0.65	0.70	0.65
Citrus	0.50	0.55	0.55	0.60	0.60	0.65	0.70	0.70	0.65	0.60	0.55	0.55

*For India lying between 8°4' and 37°6' North latitude (in the absence of local experimental data).

The Blaney-Criddle formula does not take into consideration such factors as humidity, wind velocity, elevation, etc.

The monthly use for rice in the Cauvery delta in Tanjore has been worked out by UNDP as follows.

The Cauvery delta lies between latitudes 9°50' and 11°50' North and the maximum hours of sunshine for 10°N latitude are:

Month	June	July	Aug.	Sep.	Oct.	Nov.	Dec.	Jan.	Feb.
Max. possible sunshine hours	12.7	12.6	12.5	12.2	11.9	11.7	11.6	11.6	11.8

The monthly sunshine hours expressed as a percentage of yearly sunshine hours are

Month	June	July	Aug.	Sep.	Oct.	Nov.	Dec.	Jan.	Feb.
Monthly % sunshine hours	8.60	8.82	8.75	8.26	8.33	7.92	8.12	8.12	7.46

(vi) *Evaporation index method*: Analysis of data on consumptive use indicate a high degree of correlation between pan evaporation values and consumptive use. The relationship between the evapotranspiration E_t and pan evaporation E_p is usually expressed as

$$E_t = kE_p \qquad (2.8)$$

where k is the E_t/E_p ratio. The values of k for different crops at 5% increments of the crop growing season are presented by Hargreaves and has recently introduced several modifying coefficients into his pan evaporation formula. The values of k at 10% increments of the stage of growth for the principal Indian crops are given in Table 2.4. Since these coefficients are average values, care must be taken in their use.

In recent years greater emphasis is being given for determining pan evaporation by using climatological factors. Christiansen and others have developed the pan evaporation equation from the climatological factors as

$$E_p = 0.459\, R\, c_t\, c_w\, c_h\, c_s\, c_e \qquad (2.9)$$

where E_p = computed pan evaporation (equivalent to class A pan evaporation); R = extraterrestrial radiation and c_t, c_w, c_h, c_s, and c_e = coefficients for temperature, wind velocity, relative, humidity, % of possible sunshine and elevation, respectively.

Hargreaves has developed a similar formula for estimating pan evaporation.

Table 2.4 Crop consumptive use coefficients k (to be multiplied by class A pan evaporation or E_p computed from climatological factors) [2]

% growth stage of crop	Rice	Sugarcane and alfalfa	Maize, cotton potato, jowar beans, peas and sugar-beets	Hybrid jowar, tomatoes, walnuts, dates olives and plumes	Wheat, barley, celery, flax and other small grains	Citrus crops—oranges, grape-fruit	Pastures, orchard with cover crop and plantains	Melons, onions, carrots, hops and grapes
0	0.80	0.50	0.20	0.15	0.08	0.60	0.90	0.12
10	0.95	0.60	0.36	0.27	0.15	0.60	0.90	0.22
20	1.05	0.70	0.64	0.48	0.27	0.60	0.90	0.38
30	1.14	0.80	0.84	0.63	0.40	0.60	0.90	0.50
40	1.21	0.90	0.97	0.73	0.52	0.60	0.90	0.58
50	1.30	1.00	1.00	0.75	0.65	0.60	0.90	0.60
60	1.30	1.00	0.99	0.74	0.77	0.60	0.90	0.60
70	1.20	0.90	0.91	0.68	0.88	0.60	0.90	0.55
80	1.10	0.80	0.75	0.56	0.90	0.60	0.90	0.45
90	0.90	0.70	0.46	0.35	0.70	0.60	0.90	0.28
100	0.20	0.50	0.20	0.20	0.20	0.60	0.90	0.17

Hydrometeorology

The monthly consumptive use for rice is worked out by Blaney-Criddle method as shown in Table 2.5 (UNDP).

Table 2.5 Monthly consumptive use for rice

Month	Mean monthly temp. °C	% Sunlight hours p	Monthly consumptive use coefficient for rice, k (adopted)	Consumptive use $u = \dfrac{kp}{100}(4.6t + 81.3)$ cm
June	31.5	8.60	1.15	22.4
July	30.8	8.82	1.30	25.6
Aug.	29.9	8.75	1.25	24.2
Sept.	29.6	8.26	1.10	19.8
Oct.	28.2	8.33	0.90	15.8
Nov.	26.3	7.92	0.90	14.5
Dec.	25.1	8.12	0.90	14.4
Jan.	25.1	8.12	0.90	14.4
Feb.	26.1	7.46	0.90	13.5

Seasonal consumptive use for the paddy crop *kuruvai* from June 16 to October 15 (growing season) is worked out as follows:

June 16-30	11.2 cm
July	25.6
Aug.	24.2
Sept.	19.8
Oct. 1-15	7.9
Seasonal consumptive use	$U = \overline{88.7}$ cm

Example 2.2: For the rice growing area in the Cauvery delta the following data are available:

Growing season for the first crop of pady *kuruvai* June 16-Oct. 15

Month	June	July	Aug.	Sept.	Oct.
Pan evaporation, cm	28.80	26.80	20.10	12.40	6.94
Effective rainfall, cm	9.20	10.20	11.40	9.37	3.45

Calculate the field irrigation requirement for the *kuruvai* rice crop assuming an irrigation efficiency of 70%.

Solution

Growing season = June 16-Oct. 15 = 122 days, from Table 2.6
Seasonal consumptive use $U =$ **85.78 cm**
Total depth of irrigation water for the growing season $d =$ **67.79 cm**

From the above discussion it follows that the main factors governing consumptive use or evapotranspiration are the mean temperatures and daylight hours. Other climatic factors affecting evapotranspiration are humidity and wind velocity. The type of crop, its stage of growth and the moisture available in the root zone, also influence the evapotranspiration.

Soil Moisture

The region between the water table and the ground surface is the zone of aeration in which there is a free exchange of air and moisture. Water in the zone of aeration is divided into three belts—soil water, intermediate suspended or vadose water and capillary fringe. Below the zone of aeration is the zone of saturation or ground water; the surface separating these zones is called the water table or phreatic surface, Fig. 2.8. The level of water table is ascertained by measuring the depth to which water rises in the bore holes.

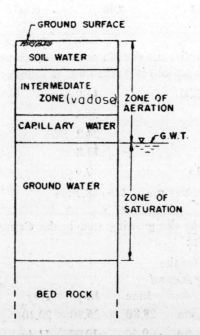

Fig. 2.8 Classification of subsurface water.

Hydrometeorology

Table 2.6 Field irrigation requirement for the *kuruvai* rice crop

Dates	No. of days up to mid-point of interval	% of growing season (122 days)	k from Table 2.4 for rice	Pan evaporation E_p (cm)	$u = k.E_p$ (cm)	P_{eff}[1] (cm)	NIR[2] $= u - P_{eff}$ (cm)	FIR[3] $= NIR/0.70$ (cm)
June 16-30	8	6.5	0.91	14.40	13.10	4.60	8.50	12.14
July	30	24.6	1.10	26.80	29.50	10.20	19.30	27.60
Aug.	61	50.0	1.30	20.10	26.10	11.40	14.70	21.00
Sept.	92	75.5	1.15	12.40	14.30	9.37	4.93	7.05
Oct. 1-15	114	93.5	0.80	3.47	2.78	3.45	—	—
					$U^4 = \overline{85.78}$			$\overline{67.79}$

[1] P_{eff}—effective rainfall: taken from rainfall data of a number of preceding years, which are available in 75-80% of the years. Mean values would be available only in 50% of the years and should not be taken.
[2] NIR—net irrigation requirement, NIR/η_I, η_I = irrigation efficiency = 70%.
[3] FIR—field irrigation requirement.
[4] U—determined from pan evaporation data is very near to that determined by Blaney-Criddle method ($U = 88.7$ cm) by UNDP.

There are three types of soil moisture associated in the zone of aeration:

(i) *Hygroscopic moisture*: It is held so tightly (tension > 31 atm) as a thin film around each soil particle by the force of adsorption (also called adsorbed moisture) that the plants cannot use it. Adsorbed moisture film improves cohesion and therefore the shear strength. The loss of moisture obtained by drying an air dried soil mass in an oven at 110°C to constants weight gives the hygroscopic moisture. Saturation by rain and capillary rise destroy the film causing loss of stability.

(ii) *Capillary moisture*: Below the hygroscopic layer is the capillary fringe held in the pores by force of surface tension or capillarity (tension 1/3 to 31 atm). Plants receive water and most of nutrients from this fringe.

(iii) *Gravitational moisture*: After saturation, the excess water drains through the soil under the force of gravity and builds up the ground water table. When the water table intersects, the land surface springs result. A water table confined from above by an impermeable layer so that the water is under pressure represents artesian conditions, Fig. 2.9. Occassionally a body of ground water will be found above a bed of impervious material. Such an anomalous condition is called 'perched water table'.

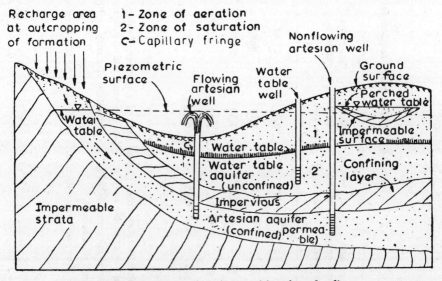

Fig. 2.9 Types of aquifers and location of wells.

Movement of Soil Water (unsaturated flow)

Soil suction and gravity are the two forces causing movement of soil water in the liquid phase. The potential ϕ at a point A relative to a point B is defined as the work done per unit quantity of water to move from B to A. The unit quantity may be a unit volume, unit mass or unit weight and

Hydrometeorology

the corresponding potentials are denoted by ϕ_v, ϕ_m and ϕ_w, respectively.

$$\phi_v = p'' + \rho g z = \gamma_w(h_c + z) \qquad (2.10)$$

$$\phi_m = p''/\rho + gz = g(h_c + z) \qquad (2.10a)$$

$$\phi_w = h_c + z \qquad (2.10b)$$

where p'' (suction pressure) $= -\gamma_w h_c$, $h_c =$ capillary rise above the free water surface called suction head. The negative sign indicates below atmospheric pressure; and $z =$ height of A relative to B.

Capillary movement is always from a lower suction region to a region of higher suction, i.e. from wet soil towards a moist or dry soil. It obeys Darcy's law but with a different permeability than for saturated flow. The velocity of the unsaturated flow is given by

$$v = k_u \frac{\phi_A - \phi_B}{L} \qquad (2.11)$$

where $k_u =$ coefficient of unsaturated permeability and $L =$ distance between points A and B.

The volume of unsaturated flow through a cross-section of soil A in time t is

$$Qt = A v t \qquad (2.12)$$

The unsaturated permeability depends on the average water content (degree of saturation) of soil through which flow occurs, i.e. on the mean value of p''_A and p''_B. It increases rapidly as the water content increases. The curve for k_u/k plotted against the degree of saturation is shown in Fig. 2.10.

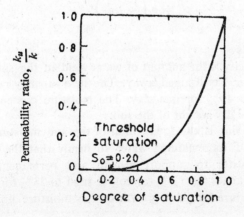

Fig. 2.10 Ratio of k_u/k as a function of saturation (after Irmay, 1954).

It can be seen that k_u is always less than k (coefficient of permeability for saturated flow). k_u near the threshold of saturation approaches zero, indicating no flow. In unsaturated flow the pressure gradient is the sum of the difference in hydrostatic head (gravity) and the difference in soil moisture tension (suction head).

The capillary rise (suction head) is given by

$$h_c = \frac{2\sigma}{\gamma_w r} \tag{2.13}$$

where σ = surface tension of water and r = radius of pores (taken as effective size or one-tenth of the mean diameter of soil grains).

For water at 20°C, $\sigma = 0.073$ N/m and for a soil having an effective size $D_{10} = 0.005$ mm, pore size being related to particle size,

$$h_c = \frac{2(0.073)}{9810(0.005/1000)} = 2.97 \text{ m} = 297 \text{ cm}$$

or $\qquad pF = \log_{10} 296 = 2.4713$

In the moist range unsaturated conductivity is greatest in fine textured soils in the sequence 'sand < loam < clay' while for saturated flow (below ground water table) conductivity increases as the fourth power of the particle size and the rate of flow is in the sequence 'sand > loam > clay'. While in saturated flow no air-water interface exists, meniscus formation and change has much to do with the rate of unsaturated flow. If the meniscus is not of uniform curvature but has r_1 and r_2 as the radii of curvature in two orthogonal principal planes, then the suction pressure is given by

$$p'' = \sigma \left(\frac{1}{r_1} + \frac{1}{r_2} \right) \tag{2.14}$$

Equilibrium Points

The field capacity is the amount of water held in the soil after excess gravitational water has drained away. The field capacity ranges from 8% for coarse sand to 40% for fine clay. The moisture content is expressed as a percent of the dry weight of the soil.

The wilting point is the lower limit of the soil moisture reservoir at which the soil holds its remaining water so firmly that the plants cannot extract enough water to support growth and permanent withering of plants occurs. It ranges from 3% for coarse sand to 23% for fine clay.

Available moisture for plant growth or the moisture holding capacity of the soil

m.h.c. = field capacity − wilting point.

The moisture holding capacities of some soils are given in Table 2.7.

Hydrometeorology

Table 2.7 Soil-water relationships

Soil type	Moisture holding capacity		Infiltration rate (cm/hr)
	$w\%$*	d/D (cm/30 cm of soil)	
Sand	5	2.5	4
Sandy loam	7	3.6	3
Silt	9	4.5	2
Silty loam	11	5.5	1.2
Clay loam	13	6.5	0.5
Clay	15	7.5	0.2

*$d = wG_mD$ (see eq. 2.16).

Moisture Extraction

In general plants extract water from the soil in the following pattern—40% of their requirement from the first quarter (top-most layer) of root zone, 30% from the second quarter, 20% from the third quarter and 10% from the fourth quarter (bottom-most layer) of their effective root zone depth, since a greater part of the roots (60–75%) lie in the top half of the depth of root zone, Fig. 2.11. Usually irrigation is applied when the top half of the root zone depth approaches 50% depletion level even though the lower half of the root zone depth may be at 75% of field capacity. Generally the depth of irrigation is taken as 50% of field capacity.

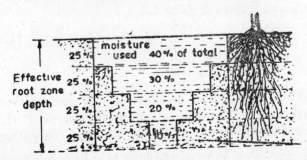

Fig. 2.11 Moisture extraction at different depths in the root zone.

Soil moisture may be determined either from direct weighing of the samples taken from the irrigated fields or by rapid moisture meters commercially developed, embedded in the field.

Soil moisture tension is a measure of the suction required to extract water from the soil. It is low at field capacity (about 1/10 to 1/3 atm) when the soil readily gives up water to plant roots and gradually increases as the moisture is depleted and is about 15 atm at permanent wilting point. Soil moisture tension is often expressed as pF which is the common logarithm (to the base 10) of the numerical value of the negative pressure of the soil moisture expressed in cm of water. Thus a pF of 2 represents a suction of 100 cm of water or a suction pressure of 100 gm/cm² (0.1 kg/cm² $\approx$ 10 kN/m²).

$$pF = \log_{10} h_c \text{ (cm of water)} \qquad (2.15)$$

$pF = 2.5\text{--}4.5$—moist soil

< 2.5—wet soil

≈ 4.2—at permanent wilting point

Soil moisture tension is measured by a tensiometer. It consists of a porous cup connected through a glass tube to a vacuum gauge, Fig. 2.12. The cup and the glass tube are filled with water and is then closed. The cup is then embedded in the moist soil sample. The unsaturated soil in contact with the cup tries to draw water through the pores of the cup creating negative pressure which can be read on the dial of the vacuum gauge.

The moisture equivalent in the soil moisture retained after a wet soil sample is subjected to a centrifugal force of 1000 times the acceleration due to gravity (i.e. 1000 g) for 40 minutes in a soil centrifuge and corresponds to field capacity.

Effective root zone is the depth of soil from which the plant draws water during its water sensitive stage of growth. It should not be confused with the true root zone. Effective root zone depths for crops for moisture extraction is given in Table 2.8.

Depth of irrigation

$$d = \frac{1}{E_i} \sum \frac{w_f - w_i}{100} G_m D \qquad (2.16)$$

where d = depth of irrigation required; w_i = % moisture content just before irrigation; w_f = % moisture content at field capacity; G_m = apparent specific gravity of the soil = dry unit weight of the soil/unit weight of water = $\frac{W_s/V}{\gamma_w}$; D = depth of soil layer to be irrigated; Σ = stands for the summation of the soil layers from the surface up to the effective root zone depth and

E_i = water application efficiency or simply, irrigation efficiency.

Irrigation interval or frequency = $\dfrac{\text{Moisture holding capacity of the soil}}{\text{Daily consumptive use}}$

Hydrometeorology

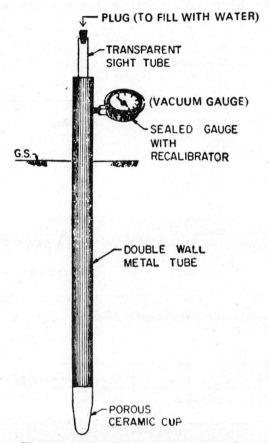

Fig. 2.12 Soil moisture gauge (tensiometer).

The cycle will be short at the period of peak water use. If it rains between consecutive irrigation, the depth and the frequency will have to be corrected according to the moisture contents of the soil. Water in excess of field capacity will be lost to the lower strata and may build up a water table. Over-irrigation causes leaching away of the plant nutrients, waste of valuable water, prevents proper aeration of plant roots and also causes water logging. A certain quantity of air in the soil is essential to satisfy the requirements of crop growth. The growth of crop is stimulated by moderate quantities of soil moisture and retarded by excessive or deficient amounts. Depending on the depth of irrigation and frequency, the average moisture content in the soil may be closer to the lower limit (wilting point) or upper limit (field capacity), Fig. 2.13, the evapotranspiration being more in the latter case than in the case of the former. Most crops react to such variation in soil moisture (and fertilizer dose) and have an optimum soil mosture content, and hence depth of

Table 2.8 Effective root zone depths for crops for moisture extraction*

Shallow rooted 60 cm	Moderately deep rooted 90 cm	Deep rooted 120 cm	Very deep rooted 180 cm
Cauliflower and cabbage, groundnut onion potato, lettuce, rice, vegetables (45 cm) cucumber (45 cm)	Carrots, french bean, garden pea, chillies, potato, muskmelon, wheat tobacco, castor, turnips,	Cotton, watermelon, maize, bajra, jowar, sugar beet, soyabeans pearl millet, citrus & fruit trees, grass (pasture)	Apple, grapevine, lucerene, coffee, sugar cane, safflower, orange, tomato, water melon,

*Usually a 90-100 cm depth is assumed in design.

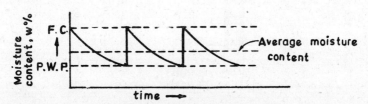

F.C.—Field capacity;
P.W.P.—Permanent wilting point
(a) Average moisture content closer to wilting point

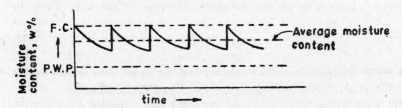

(b) Average moisture content closer to field capacity

Fig. 2.13 Moisture levels in soils for crop growth.

application and frequency, at which maximum yields are obtained. For example wheat is quite sensitive to over-irrigation and has a well defined optimum consumptive use, Fig. 2.14 curve (a), whereas in the case of other crops the increase in yield is not appreciable under increased depths of irrigation, Fig. 2.14 curve (b). In the former type of crops it is desirable to provide the optimum depth of irrigation whereas in the latter case the increase in yield is not worth the cost of additional water supplied, and water can be more economically utilised for growing other crops. Hence the depth

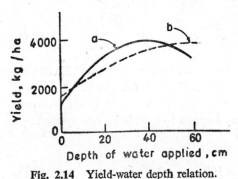

Curve a—Optimum consumptive use well defined
Curve b—No appreciable yield with increased depth

Fig. 2.14 Yield-water depth relation.

and frequency of irrigation should be so selected that the average soil moisture is at the corresponding optimum level.

Leaching Requirement

Leaching requirement is the extra depth of water to be applied to move down the salts to a sufficient depth for the safe crop growth. It depends upon the salt content of irrigation water, depth of reclamation and the soil properties. If the salinity of irrigation water is C and the total quantity of water applied is Q, the total salt applied to the field is CQ. If C_s is the salinity of the soil solution after the consumptive use U is taken from the soil and the effective precipitation during the irrigation period is P_{eff}, then

$$CQ = (Q + P_{eff} - U)C_s$$

$$Q = \frac{C_s(U - P_{eff})}{C_s - C} \tag{2.17}$$

from which the leaching requirement over and above the crop requirements can be calculated to maintain a desired salt concentration in the soil that can be tolerated by the crops. The precipitation falling during the growing season which is available to meet the consumptive water requirements of crops is called effective precipitation. It does not include deep percolation below the root zone or surface runoff.

The total water requirement for raising a crop in a given period includes the consumptive use, leaching requirement if any, other economically unavoidable losses and that applied for special operations such as land preparation, transplanting, etc. and are termed water needs of a project.

Irrigation requirement is the quantity of water exclusive of precipitation, required for successful growth of crops. The percentage irrigation water that is stored in the soil and is available for crop growth is called irrigation efficiency and is designated by the place of measurement of irrigation water like the head of the farm or field, point of diversion etc. The farm irrigation efficiency (E_i) should not fall below 60% and rarely exceeds 80%. Irrigation requirement is given by

$$IR = \frac{U - P_{eff}}{E_i} \tag{2.18}$$

The concept of the consumptive use (or water use) efficiency (E_u) stresses on the availability of the water stored in the soil for use by the crop as water may evaporate from the ground surface or continually move downward beyond the root zone as it may happen in a wide furrow spacing.

Consumptive use efficiency (%)

$$E_u = \frac{U}{d} \times 100 \tag{2.19}$$

where U = water consumptively used by the crop and d = death of water depleted from the root zone of the soil.

This concept of efficiency is useful in explaining the difference in crop response from different methods of irrigation. The water use efficiency* is often expressed as kg per ha/mm, i.e. crop yield per hectare for a water depth of 1 mm.

Example 2.3: The average alkali content of the first 1.3 m of the depth in a given field is 0.7% of the dry weight of the soil whose apparent specific gravity is 1.25. The irrigation water has an alkali content of 200 ppm and the drainage water contains 1,000 ppm.

What annual depth of water over and above the crop requirements must be applied to reduce the alkali content of the soil to the average maximum salt concentration of 0.2% tolerated by the particular crop, in four years? (Neglect the salt content of the water required by the crop.)

Solution: Consider a unit area of 1 m².

Salt to be removed in 4 years,

---

*Crop water use efficiency = $\dfrac{Y}{U}$

Field water use efficiency = $\dfrac{Y}{FIR}$

where Y = crop yield on field.

Hydrometeorology

$$S = \frac{0.7 - 0.2}{100} (1 \times 1.3)(1.25 \times 1000) = 8.12 \text{ kg}$$

If d is the annual depth of water applied (over and above the crop requirements), salt removed in 4 years

$$S' = \frac{1000 - 200}{10^6} (1 \times 4d)\, 1{,}000 = 3.2d \text{ kg}$$

Equating S and S',

$$d = \frac{8.12}{3.2} = 2.54 \text{ m}$$

Infiltration

Water entering the soil at the surface is called infiltration. It replenishes the soil-moisture deficiency and the excess moves downward by the force of gravity called deep seepage or percolation and builds up the groundwater table. The maximum rate at which the soil in any given condition is capable of absorbing water is called its infiltration capacity. Infiltration often begins at a high rate (20 to 25 cm/hr) and decreases to a fairly steady rate as the rain continues, called the ultimate f_p (1.25 cm/hr), Fig. 2.15.

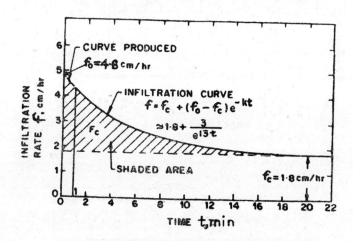

$$F_c \approx 7 \times \frac{1 \text{ cm}}{60 \text{ min}} \times 2 \text{ min} \approx 0.23 \text{ cm}$$

$$K = \frac{f_o - f_c}{F_c} = \frac{(4.8 - 1.8) \text{ cm/hr}}{0.23 \text{ cm}} \approx 13/\text{hr, or } 13 \text{ hr}^{-1}$$

Fig. 2.15 Infiltration rate curve.

The infiltration rate f at any time t is given by the Horton's equation

$$f = f_c + (f_o - f_c)e^{-kt} \tag{2.20}$$

$$k = \frac{f_o - f_c}{F_c}$$

where f_o = initial rate of infiltration capacity; f_c = final constant rate of infiltration (at saturation); k = a constant depending primarily upon soil and vegetation; e = base of the natural logarithm and F_c = shaded area in Fig. 2.15.

Infiltration depends upon the intensity and duration of rainfall, weather

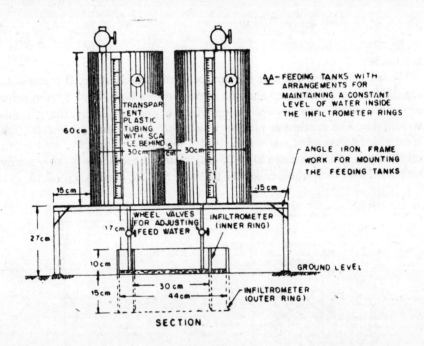

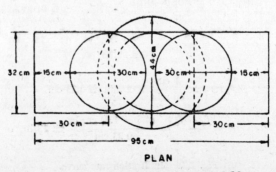

Fig. 2.16 Double ring infiltrometer with constant head tanks.

Hydrometeorology

soil characteristics, vegetal cover, land use, initial soil moisture content, entrapped air and depth of water table.

Measurement of Infiltration

Infiltration data can be obtained as follows:

(i) By analysis of rainfall and runoff records of small drainage basins with homogeneous soils.

(ii) By establishing plots or watersheds and operating them for several years to obtain sufficient data.

(iii) Use of infiltrometers or rate simulators by which water can be applied artificially to small areas.

A double ring infiltrometer is shown in Fig. 2.16 with two feeding tanks to maintain a constant level of water inside the infiltrometer rings. The two rings are driven into the ground by a driving plate and hammer, to penetrate into the soil uniformly without tilt or undue disturbance of the soil surface, to a depth of 15 cm. After the driving is over, any disturbed soil adjacent to the sides is tamped with a metal tamper. Two point gauges are fixed in the centre of the rings and in the annular space between the two rings. A minimum water level of 2.5 cm up to a maximum of 15 cm is usually maintained. The water added to maintain the desired depth after the start of the experiment is noted at regular time intervals, say, 5, 10, 15, 20, 30, 40, 60 min up to a period of at least 6 hours and the results plotted as infiltration in mm versus time. A typical infiltration rate curve is shown in Fig. 2.17.

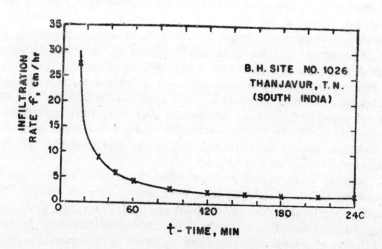

Fig. 2.17 Typical infiltration rate curve.

(iv) Observation from infiltration pits and ponds and deducting the loss due to evaporation.

(v) Placing a catch basin called a lysimeter under a laboratory sample or at some depth below the land surface to measure the infiltrating water at that point.

(vi) Measuring the differences in soil moisture and rise in water table.

(vii) By the general solution of the hydrologic equation for an area.

Infiltration due to rainfall, from irrigated and unirrigated lands, and seepage from reservoirs, lakes, ponds and canals constitute the groundwater accretion.

Ground Water Table Fluctuations

Measurement of water levels or piezometric heads are used in the analysis of ground water with respect to its occurrence, storage, movement, recharge and discharge; to be most valuable, they should be made as continuous records of water level fluctuations.

An indigenous automatic water level recorder (AWLR) manufactured by M/s Hindustan Clock Works, Poona has a 8 day spring clock, 1:1 to 1:10 gears for water level measurement, 12.5 cm float, lead counter weight and 6 m float line with hooks. It costs around Rs. 4,000. The rise and fall of the float with changing water levels turns the drum proportionately, as the clock controlled pen moves across the chart at constant speed. The resulting graph shows water levels against record of time. The range of stage is unlimited, as the chart drum may make any number of revolutions, depending only on the length of the float line.

The number of observation wells may be 1 for every 100-200 km^2. All must tap the same aquifer; wells tapping different aquifers must be separated. Non-pumping wells and wells used for domestic purposes are desirable; disused wells should be avoided. Permanent measuring points should be marked and the bench marks should be connected to mean sea level. A minimum of 3 years data is required for analysis.

Evapotranspiration processes may result in dynamic water level responses seasonally. Confined aquifers respond readily to pressure effects due to natural phenomena such as barometric pressures, tidal movement near oceans and earthquakes. Man's activities also produce water level changes in aquifers, mainly by pumping, but locally also through other factors such as pressure of moving vehicles.

Time lag features and pressure effects in artesian aquifers are to be considered in analysing water level fluctuations. In such analysis, hydrographs, water level profiles and contours are used. From these three, graphic forms of representation estimates may be made on occurrence, movement and change of storage resulting from variation in the natural regimen or from over development.

For computation of changes in storage under water table conditions, data on water levels on a rigid pattern of observation wells located 1 to 1.5 km apart must be collected round the year. Such data preferably

Hydrometeorology

should be supplemented by continuous water table hydrographs recorded on AWLR. A basin water table contour map may be prepared for minimum and maximum changes of water levels. The total change in storage is given by the product of the change in water level, the specific yield and the involved area of the aquifer. For any confined aquifer, specific yield has to be replaced by storage coefficient.

The factor of subsurface inflow or outflow often poses difficulty since comprehensive geological investigations are necessary. Ground water inflow or outflow from one basin to another, particularly when subsurface geological formation trangress physiographic basin limits, the inflow or outflow may often be of great significance. This is true for karstic limestone formations also. However, directions of flow can be obtained from the water table or piezometric surface (for confined aquifers) contour maps. From estimation of slope of formation and permeabilities (determined by pump tests) subsurface flow can be computed.

From a record of water level fluctuations the following interpretations can be made:

(i) Pattern of fluctuation.
(ii) Period of recharge and discharge.
(iii) Correlation with pumpage, rainfall, surface flow, etc.
(iv) Hydraulic gradient.
(v) Direction of ground water flow.
(vi) Quantity of subsurface flow.
(vii) Troughs and mounds.
(viii) Average ground water fluctuation.
(ix) Monthly change in ground water storage.
(x) Cumulative change in ground water storage.
(xi) Quantity of ground water discharge and recharge.
(xii) Identification of effluent and influent stretches of streams.

QUIZ 2

I. Fill in the blanks choosing the correct word(s) given in the brackets:
  1. The rain gauge readings are taken every day at......
     (08.00 and 18.00 hr IST, 08.30 hr IST, 08.30 and 17.30 hr IST, 08.00 and 14.00 hr IST)
  2. The self-recording rain gauge gives a continuous record of....... while the Symon's rain gauge gives......
     (total depth, intensity and duration of rainfall, hourly rainfall, one day's rainfall)
  3. The a.a.r. is the average of......consecutive records of yearly rainfall.
     (40, 50, 35, 15, 60)
  4. The water use efficiency is expressed as......
     (%, cm/crop, kg/cm, qn/ha of crop, kg per ha/mm)

II. Match the items in A and B:

A	B
1. Thiessen polygon	i. Evaporation-erratic
2. Sunshine recorder	ii. Evaporation from water surface
3. Piche evaporimeter	iii. Wind velocity
4. Assman hygrometer	iv. Soil evaporation
5. Land pan	v. Humidity
6. Anemometer	vi. Evapotranspiration
7. Pan coefficient	vii. Hours of bright sunshine
8. Observation wells	viii. Mean areal depth of precipitation
9. Lysimeters	ix. Consumptive use
10. Blaney-Criddle formula	x. Capillary moisture
11. Colarado sunken pans	xi. Gravitational moisture
12. Plants	xii. Soil suction
13. Ground water	xiii. Infiltration equation
14. Tensiometer	xiv. Infiltration
15. Double ring infiltrometer	xv. Ground water table fluctuations
16. Horton	xvi. Ratio of lake evaporation to pan evaporation

[1-viii, 2-vii, 3-i, 4-v, 5-ii, 6-iii, 7-xvi, 8-xv, 9-iv, 10-vi, 11-ix, 12-x, 13-xi, 14-xii, 15-xiv, 16-xiii]

III. State whether 'true' or 'false'; if false give the correct statement:
1. As the area increases the average depth of precipitation increases for a particular storm.
2. The more the number of rain gauge stations the better the estimate of mean rainfall over the area.
3. In hilly and heavy rainfall areas at least 10% of the rain gauge stations should be equipped with the self-recording type rain gauge.
4. For hydrological estimates about 35 years of rainfall records are required.
5. The rain gauge density in plains should be more than that in elevated regions.
6. The isohyetal method of estimation of mean areal depth of rainfall gives the best results if analysed properly.
7. Evapotranspiration cannot be determined from pan evaporation data.
8. Evapotranspiration is the same as consumptive use.
9. The evapotranspiration is higher on a hot windy day.
10. Higher the moisture, higher will be the soil suction.
11. Capillary movement is always from a wet soil towards a moist or dry soil and is given by Darcy's law having the same permeability as that of saturated flow.
12. The permeability of sand is higher than loam while it is the reverse for unsaturated flow.
13. Plants extract most of water requirement from the lower half of their root zone depth and irrigation is applied when the moisture at this depth depletes to 50% of field capacity.
14. Soil suction: $pF\ 2 = 100$ cm of water $= 10$ kN/m^2.

Hydrometeorology

15. Higher the salt concentration in the irrigated soil lesser should be depth of water applied.
16. Penman's equation is usually not used for computation of evaporation from water surfaces because of too many meteorological observations are required.

[False: 1, 5, 7, 10, 11, 13, 15]

IV. Select the correct answer(s):

1. Vegetation tends to
    a. increase the runoff from the catchment
    b. decrease the runoff from the catchment
    c. does not affect the runoff
    d. none of these answers

2. The water level in a well responds to
    a. atmospheric pressure changes
    b. high and low tides
    c. earthquakes
    d. pumping rates
    e. movement of loaded trucks
    f. all the above answers
    g. none of these answers

3. The factors which affect evapotranspiration are
    a. climatic factors like mean temperature, hours of bright sunshine, wind velocity, humidity, etc.
    b. crop factors like type of crop and the stage of its growth
    c. the moisture level in the soil
    d. all the above answers
    e. none of these answers

4. Isohyet is a line joining all places having
    a. the same atmospheric pressure
    b. the same depth of rainfall
    c. the same temperature
    d. the same depth to the ground water table

5. Rain gauges are erected
    a. vertically 30 cm above ground level
    b. perpendicular to the ground surface on which they are installed

6. The initial infiltration rate is at capacity rates
    a. if the intensity of rainfall is less than the average rate of infiltration
    b. if the intensity of rainfall is less than the infiltration capacity of the soil
    c. if the intensity of rainfall is equal to or more than the average rate of infiltration
    d. if the intensity of rainfall is equal to or more than the infiltration capacity of the soil.
    e. none of these answers.

[1-b, 2-f, 3-d, 4-b, 5-a, 6-d]

PROBLEMS

I. Determine the monthly and seasonal consumptive use and total depth of irrigation for the ADT 27 rice crop in the Cauvery delta given the following data:

Growing season—July 1 to October 15

Month	July	Aug.	Sept.	Oct.
% sunlight hours	8.82	8.75	8.26	8.33
Mean monthly temp. °C	30.8	30.3	29.5	28.0
k (Blaney-Criddle)	1.30	1.25	1.10	0.90
Effective rainfall, cm	10.20	11.40	9.37	6.90

Use the Blaney-Criddle formula and assume a field irrigation efficiency of 70%.

(U=76.7 cm, FIR=60.5 cm)

II. Determine the monthly and seasonal consumptive use and total depth of irrigation for jowar at Bellary in Karnataka (India) given the following data:
Growing season—October 16 to February 2, 1976

Month	Oct.	Nov.	Dec.	Jan.	Feb.
Pan evaporation, cm	16.86	15.57	16.59	19.10	22.34
Effective rainfall, cm	6.63	2.19	0.54	0.15	0.29

Use the evaporation-index method (Coefficients k can be obtained from Table 2.4) and assume a field irrigation efficiency of 70%.

(U=51.37 cm, FIR=59.2 cm)

III. Determine the monthly and seasonal consumptive use and total depth of irrigation for wheat in U.P. given the following data:
Growing season—November 1 to March 15

Month	Nov.	Dec.	Jan.	Feb.	March
Pan evaporation, cm	15.0	12.7	9.5	15.5	20.65
Effective rainfall, cm	1.0	2.0	3.2	2.1	—

Use the evaporation index method (Coefficients k can be taken from Table 2.4) and assume a field irrigation efficiency of 70%.

(U=34.85 cm, FIR=37.90 cm)

IV. A sandy loam soil has a dry weight of 1.48 g/cc and wilting point and field capacity of 8 and 16% respectively of the dry weight. If the depth of effective root zone is 1.2 m determine the moisture in the root zone in cm at (a) the wilting point and (b) the field capacity.

If the peak consumptive use of a crop is 5 mm/day, determine the depth and frequency of irrigation in days in the period of peak use, assuming that irrigation water is applied when 50% depletion level in the soil moisture has reached and no rainfall occurs during peak use period.

(14.2, 28.4, 7.1 cm, 14 days)

Hydrometeorology

V. In a project, irrigation water has a salinity of 500 mg/l. The yearly consumptive use is 120 cm and the effective rainfall during the year is 25 cm. If it is desired to maintain the salinity of the soil solution at 2000 mg/l, what annual depth of water should be applied?

(126.7 cm)

VI. What is meant by optimum requirement of a crop in irrigation? Exemplify by means of characteristic curves for crops like sugarcane, rice and potato. List the approximate yield of each crop in terms of tons per hectare (or kg/ha per cm of water applied) or any other equivalent.

VII. The following are the stream flows due to a 6-hour storm commenced at 06 hr on 16 Sept. 1985 producing 71 mm of total rain fall on a basin of 775 km². Assuming a constant base flow of 40 cumec, determine the average infiltration rate for the basin.

Date	Time (hr)	Discharge Q, cumec
16 Sept. 1981	06	40
	12	64
	18	215
	24	360
17 Sept. 1981	06	405
	12	350
	18	270
	24	205
18 Sept. 1981	06	145
	12	100
	18	70
	24	50
19 Sept. 1981	06	40

Hint: Refer 'HYDROLOGY' by the same author and publisher

$$\left[71 - \frac{\sum_{t=1}^{13} (Q_i - 40) 6 \times 60 \times 60, \, m^3}{775 \times 10^6, \, m^2} \times 1000 \right] \div 6 \, hr = 3.5 \, mm/hr$$

3

Hydrogeology and Aerial Photography

Some clues regarding the water bearing properties of rocks are given in the following:

(i) It is the perviousness or permeability and not porosity which is significant in water yielding of rocks. For example the igneous rocks have a porosity of 1% and may yield all water while some clays have a porosity as high as 50% but are practically impervious. The principal factors affecting porosity are grain size, shape, grading or sorting and the amount and distribution of cementing material. Fissures, joints, bedding planes, faults, shear zones and cleavages, vesicles and solution cavities, interstitial or intergranular openings, all contribute to the perviousness of the rock.

(ii) Besides weathering and texture, the presence of numerous sets of joints, fractures, cleavages and fault breccias form the good water bearing zones in igneous and metamorphic rocks.

(iii) Solution openings in limestones and dolomites may yield water. Limestones with cavities are formed due to the action of acid waters.

(iv) In consolidated formations, water may be yielded through cracks called fissures or crevices.

(v) Basalts form a good source of water since they invariably contain vesicles and are easily susceptible for weathering.

(vi) Water in the shale is found in the joints. Shales invariably develop two sets of joints. In such cases open dug wells of large diameter are preferred. Selection of well sites in shaly region must be done carefully, recognising the storage, infiltration and seepage.

(vii) Sandstones form very good aquifers. The porosity of these materials depends mainly on shape and arrangement of their constituent particles, cementation and compaction, degree of assortment, and fractures and joints. Poorly sorted and well cemented sandstones are poor aquifers.

(viii) The most important water yielding formations are the unconsolidated gravels, sands alluvium, lake sediments, glacial deposits, etc.

(ix) Marble with fissures and cracks, weathered gneisses and schists, heavily shattered quartzites, and slates, serve as good aquifers.

(x) Faults generally affect the water table by blocking or diverting the flow, sometimes providing an outlet to underground water that would not otherwise be available. The ground water conditions of Western Utah are different from that of Eastern Utah, on account of great fault zone that extends through the states.

(xi) The most favourable of all sites for a well is provided by a synclinal trough of alternating layers of permeable and impermeable beds of rocks providing artesian conditions, Fig. 3.1.

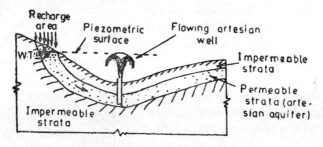

Fig. 3.1 Artesian or flowing well.

(xii) A series of dipping formations which include a pervious or water bearing stratum with impervious beds above and below, the out crop of the pervious bed receiving supply of water from the surface, form the artesian conditions.

(xiii) Most of the water in crystalline rocks is within 90 m and in sedimentary formations wells drilled deeper than 600 m yield little water.

(xiv) The quality of water in the well depends on the various types of rocks encountered.

(xv) Sometimes a small band of impervious strata above the main ground water table (g.w.t.) holds part of the water percolating from above. Such small water bodies of local nature which cannot replenish quickly (and hence cannot provide a sustained yield) are called perched water table, Fig. 3.2. Such water bodies are highly deceptive and misleading in ground water exploration works. In such cases, drilling is to be continued

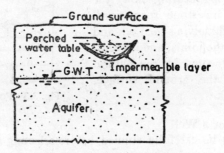

Fig. 3.2 Perched water table.

further to ensure sustained yield (since the perched water table can supply water but the yield is only very short-lived). Swarming of insets in the evenings at certain points on the surface during the dry season and existence of green grass vegetation are indications of a perched or suspended water table.

(xvi) Springs are formed when the ground water starts oozing out from the ground surface. Springs may be formed when:

(a) an impermeable bed, overlain by a permeable bed, intercepts the sloping surface of the natural ground or hill side, Fig. 3.3(a).

(b) a sloping permeable bed is interrupted by a dyke; it is an ideal site for a bore well, Fig. 3.3(b).

(c) a sloping permeable bed is interrupted by an impermeable bed due to the presence of a fault, Fig. 3.3(c).

(d) the water moving along the interconnected joints present in the rock is intercepted by the natural slope of the ground surface, Fig. 3.3(d).

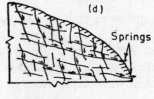

(e) the water permeating along the joints in limestone formations, keeps on dissolving the rock and widening the joints, and is eventually intercepted by the natural slope of the ground surface, Fig. 3.3(e).

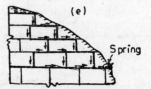

Fig. 3.3 Formation of springs.

Selection of Site for a Well

The factors to be carefully studied before selecting a site for sinking a well are:

Hydrogeology and Aerial Photography

  (i) Topography.
 (ii) Climate.
(iii) Vegetation.
(iv) Geology of the area.
 (v) Porosity, permeability and alteration of rocks.
(vi) Joints and faults in rocks.
(vii) Folded strata.
(viii) Proximity of any tank, river, spring, lake, unlined channels, reservoirs, etc.
(ix) Existing wells in the vicinity.

(i) *Topography*: The valley regions are more favourable than the slopes and the top of the hillock.

(ii) *Climate*: The annual rainfall of region, intensity of sunlight, maximum temperature and humidity are of considerable value. Areas having heavy or moderate rainfall favour more water to percolate in the soil and pervious rock layers, and get stored on impermeable layers to form aquifers. Intensity of summer days evaporates and depletes ground water through direct eveporation from shallow depths and evapotranspiration through plants. Areas of scanty rainfall and severe summer do not favour storage of ground water at shallow depth and the wells do not yield much water. Semi-arid zones are more favourable than the arid zones.

(iii) *Vegetation*: Vegetation can flourish well where the ground water is available at shallow depths. The trees of forests draw their requirements directly from the 'zone of saturation'. Such plants are known as 'phreatophytes'. Some plants can exist under arid conditions by absorbing the soil moisture (intermediate or vadose water) in the zone of aeration and store water in their thick fleshy leaves and stems. Such plants are known as 'xerophytes'. By studying the vegetation of the area, the condition of ground water can be assessed. This vegetation indicates large storage of ground water at shallow depths whereas bald hillocks with large number of xerophytes indicate the scarcity of ground water at shallow depths.

(iv) *Geology of the area*: Areas comprising thick soil or alluvium cover, highly weathered, fractured, jointed or sheared and porous rocks indicate good storage of ground water, whereas bald hillocks of massive igneous and metamorphic rocks or impermeable shales indicate paucity of ground water.

(v) *Porosity, permeability and alteration of rocks*: Highly porous, permeable and altered zones of dense rocks encourage storage of ground water. Massive rocks do not permit the water to sink.

(vi) *Joints and faults in rocks*: The movement of water is through joints, fractures, fissures and cracks which are interconnected. Wells sunk into the highly jointed rocks or along the fault plane yield copious supply of water. Faults in certain areas behave as barriers for the movement of ground water and create artesian conditions where flowing wells and springs are commonly noticed, Fig. 3.3 (c).

(vii) *Folded strata*: When the rocks are folded into anticlines and synclines, the synclines are favourable for storage of ground water in the pervious layers and water is stored under pressure under artesian conditions. Wells sunk from the top of synclined hills so as to reach the pervious layers will be successful. On the contrary, a well sunk in the anticlinal valley will be a failure since it will be at a point of 'ground water divide' and the water flow will be away from the crest of the anticline towards the synclinal basin.

(viii) *Proximity of tank, river, etc.*: These water bodies serve as sources of recharge and the water is stored in the pervious layers. The wells sunk in these areas yield water throughout the area. For example, 'percolation dams' are built in Maharashtra and the wells are sunk in the zones of seepage.

(ix) *Study of existing wells in the vicinity*: The subsurface geology, rock formations, depths, fractures, etc. can be observed in the existing wells in the neighbourhood. The depth of water table and the yield can be assessed by observing the water levels. Care should be taken to see that no dykes, veins or faults are situated in between the existing wells and the proposed well.

In addition to the above factors, aerial photographs and hydrogeological maps are helpful in making rapid reconnaissance of the area, where a large-scale well sinking programme is on hand.

Besides, geophysical methods of exploration of ground water, namely, magnetic, seismic, gravity and electrical methods are also employed, of which electrical resistivity method is found to be more helpful in the selection of well sites. Subsurface exploration by test drilling and studying the various rock formations at different depths and their water bearing properties can be done by more sophisticated methods which are briefly described in the subsequent chapters.

Spacing of Wells

If the wells are situated very closely, the supply of water will be greatly affected, due to interference, when both the wells are pumped simultaneously. As shown in Fig. 3.4 an open well is situated in the land of '*A*'. A borewell is sunk very close to the open well in the land of '*B*', and water is pumped from the borewell. This creates a large 'cone of depression' around the borewell. The water in the open well is depleted rapidly and it can get water only when the pumping is stopped and fast recuperation takes place (unless the open well is further deepened). Otherwise the open well remains a dry well though the water table is at shallow depth. So it is always advisable to space the wells beyond the 'radii of cones of depression' of the adjacent wells. This is roughly estimated to be around 200-300 m in alluvial areas and around 75-150 m in hard rock areas.

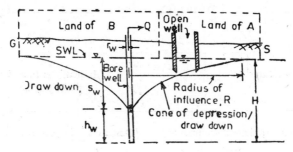

Fig. 3.4 Open well becomes dry due to pumping of new borewell.

Aerial Photography

Aerial photography consists of taking photographs at regular time intervals from an aeroplane flying along definite lines at a certain height (flight altitude) above ground level. Economy of time and cost are the most important advantages in the use of aerial photographs. Practical application of aerial photographs in ground water studies are confined to surface geological mapping and morphological study.

Aerial photo-interpretation is of considerable use in:

(i) Identifying rock types—whether porous, massive, bedded, jointed, fractured, etc. of consolidated material and data for analysis of geomorphological features, nature of sediments of unconsolidated deposits.

(ii) Evapotranspiration studies—demarcation and measurement of lakes, streams, rivers, marshes and swamps that of 100% evaporation opportunity; differentiation of areas with various vegetation densities and land use, selection of sites for meteorological stations.

(iii) Infiltration studies and runoff—area can be devided into various infiltration components on the basis of rock type, soil cover, vegetative cover, land use pattern and intensity, surface drainage network and density, fracture and joint pattern, land form, and study of sedimentation and river regimen.

(iv) Selection of drilling sites, gauging stations, alignment of water mains and power lines; for geophysical survey study of terrain conditions for locating traverse lines, camps, access routes, etc.

Selective field investigations are a must to confirm surfacial geological interpretations and carry out subsurface studies to evaluate the rocks and their water yielding properties at depths. Correlation and refinement in interpretation of geological features may be made in the field or in a follow up photogeologic study.

Aerial photographic investigation is the first phase in planning any ground water exploration programme. Many of the features which cannot be always brought out by ground surveys can be interpreted with the help of aerial photographs. Aerial photographs provide information in respect of geomorphological features, drainage pattern and presence of springs in the area. Some of the features as indicated by aerial photographs are as follows:

Features	Indications in aerial photographs
Igneous, metamorphic and sedimentary rocks	Demarcated by texture and tonal changes
Joints, fissures and faults	Appear as linear features
Shales, clays or clastic rocks	Show dark tones
Coarse grained and permeable rocks	Show light photographic tones
Alluvial, unconsolidated formations	Recognised by fluvial features such as river terraces, alluvial fans, etc.
Extrusive rocks (lava flows)—joint pattern, fractures, fissures and dykes	Geometrical pattern
Sink holes in Karstic topography	Pitted appearance of land surface

PHOTO-INTERPRETATION—ITS PLANNING AND EXECUTION

The factors which should be taken into consideration which will guide photo-interpretation are:

A. The boundaries of the area to be photographed should be marked on a toposheet giving coordinates, and the purpose which determines the specification and type of photography should be stated.

B. For hydrological studies the specifications to ensure best results are as follows:

(i) The photographs should be absolutely vertical with a maximum tilt limit of 3°.

(ii) The photographs should be clear and free from such obscuring factors as haze, cloud, fingerprints and blurred images, as these will affect the interpretation and measurement of details.

(iii) The best season of photography is immediately after the monsoons as the light condition remains favourable and obscuring factors like haze and cloud do not mar the quality of photographs. The time of photography is very important. During morning and evening, long and deep shadows will tend to obscure details unless it is a flat and plain country. In the tropics it is necessary to limit the time of photography to between $1\frac{1}{2}$ to 3 hours after sunrise to avoid the atmospheric haze as soon as the temperature rises.

(iv) For interpretation purpose, photographs should preferably be in glossy prints.

(v) Colour photographs are best suited for hydrological studies but being extremely costly are rather prohibitive. Photographs taken with infra-red films and filters are generally used for hydrographic purposes, particularly in forest areas because the water spots show up very conspicuously dark owing to their absorption of infra-red light. In India at present only black and white photography is used.

(vi) The scale of photography depends on the purpose of investigation and nature of the terrain and must always be specified while ordering for photographs. The following examples serve as a guideline in the selection of scale:

Nature of terrain	Scale
Desert country of low relief, flood plains where measurement of micro relief features, meander characteristics of rivers and study of soil characteristics are important	$\frac{1}{10,000}$ to $\frac{1}{15,000}$
Areas of modetate relief and low elevation differences with open jungle or shrubby forests	around $\frac{1}{20,000}$
Areas of moderate to high relief with dense jungle	$\frac{1}{30,000}$ to $\frac{1}{40,000}$
Areas of high relief and tall trees obscuring geological elements, i.e. mountainous areas with tropical rainforest	$\frac{1}{40,000}$ to $\frac{1}{50,000}$

In flat terrain the flying height is not so important but in rugged country with elevation differences, flying should be done as high as possible using the lens of largest possible focal length for the particular scale required (scale = focal length/flying height). High flight will cause less relief distortion and less scale variation.

Aerial photographs are flown in strips to cover the area with the flight direction usually kept along the length of the area. However in areas of high dipping geological formations, if the strike is along the length of the area, it may be necessary to resort to cross stripping. Hence, the type of stripping should be specified. Minimum overlap required to have a continuous stereoscopic cover is forward overlap 52% and lateral overlap 15%. But in high relief terrain higher percentages are taken so as to cut down the relief displacement.

STEREOSCOPIC PHOTOGRAPHY

A pair of photographs taken from two camera stations but cover a common area constitutes a stereo pair which when viewed under a stereoscope in a certain manner with parallel eye axes give a three dimensional view of the common area.

A schematic diagram of a minor stereoscope is shown in Fig. 3.5. The distance between corresponding points is generally 240 mm large enough

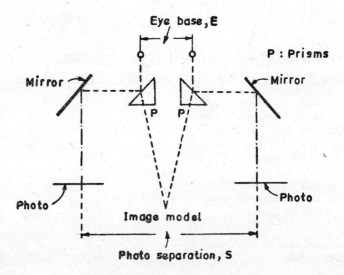

Fig. 3.5 Minor stereoscope.

for 23 × 23 cm photographs. Lenses having a focal length equal to the distance from the lenses to the photographs via the mirror-prism are placed. This distance is usually about 300 mm. Since the normal viewing distance is 250 mm, the magnification is $250/300 \approx 0.8$. Occulars having an enlargement factor of 3 to 8 are placed over the lenses, resulting in a net enlargement of 2 4 ($= 3 \times 0.8$) to 6.4 ($= 8 \times 0.8$).

The pocket stereoscope, Fig. 3.6, usually has plano-convex lenses with a focal length of 100 mm. The parallel rays, entering the eyes converge at infinity. Since the normal viewing distance is 250 mm, the magnification due to closer view is $250/100 = 2.5$. The pocket stereoscope is cheap, portable, has a large field of view and good image quality due to the simple optical system. The limitations are as follows:

(i) They have limited magnification. Pocket stereoscope with more

Hydrogeology and Aerial Photography

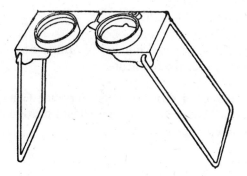

Fig. 3.6 Pocket stereoscope.

than three times magnification cannot be equipped with simple planoconvex lenses due to the large increase in lens aberration.

(ii) The distance between the eye and the photograph is too small for adequate illumination and working space for tracing, etc.

(iii) The distance between the corresponding points on the photographs must be equal to or smaller than the eye base, which is difficult with normal photograph unless they are bent or folded.

Mosaic

An assemblage of aerial photographs whose edges have usually been cut and matched to form a continuous photographic representation of the particular area is called a mosaic. If this assemblage is made without any control, then it is called uncontrolled mosaic. If, before being laid, the prints have been properly rectified, i.e. enlarged or reduced and fitted on adequate ground control, i.e. to fit pre-determined locations of certain important features, the mosaic is said to be a controlled mosaic. The controlled mosaic, though more accurate, retains the changes in scale and displacements due to differences in relief within the individual prints. A contoured mosaic shows the relief by means of contours, and may be either controlled or uncontrolled. A semicontrolled mosaic may be prepared from unrectified photograph assembled to have ground control; alternatively rectified photographs may be used with no ground control. Semi-controlled mosaics are a compromise between economy and accuracy.

The Surveyor General of the Survey of India, Dehradun, is the coordinating authority for all aerial photography in India. A library of the existing photographs is also maintained for restricted use and it is essential to obtain security clearance from the Ministry of Defence, Government of India. While ordering for aerial photography the following details have to be furnished:

(a) Purpose.
(b) Extent and location (Index map).

(c) Scale of photography.
   (d) Camera and focal length.
   (e) Season and time.
   (f) Requirements—type and number of prints, mosaic, etc.
   (g) Any other special requirements.

On receipt of the estimated cost from the Surveyor General of India for the type and scale of photograph, the user should provide for the necessary funds and submit a certificate.

PHOTO-INTERPRETATION

The interpretation from aerial photographs is done in five steps:

(i) Consultation of literature, previous maps and reports and a reconnaissance field trip, if necessary.

(ii) Interpretation, hydrographic measurements and annotation of details on overlay sheets from aerial photographs.

(iii) Combination of annotated overlays to an average scale or transfer of details to a controlled base.

(iv) Field work in key areas with the interpretation map.

(v) Preparation of the final map and report.

In spite of all these steps, the whole project can be completed in a very short time.

A qualitative estimate of the grain size, degree of compaction and permeability of materials composing the land surface is possible from a study of the aerial photographs with the help of recognition elements such as photographic tone, geomorphic texture, drainage factors, shape and size of geomorphic elements, spatial association of features, colour differences, erosion pattern, etc. The drainage pattern and drainage density are indicators of underlying geology. A granitic country develops a dendritic drainage pattern with a low drainage density whereas a sandstone or quartzite country develops deeply incised streams with moderate to low drainage density depending on the degree of permeability. The higher the permeability, the lower is the drainage density. Similarly a shale country develops a close knit shallow drainage and a lime stone country hardly any surface drainage. The drainage lines may be srtucture controlled, i.e. developed along some faults, joints, fractures systems, around domal structures, along the crests or troughs of some fold system, parallel or perpendicular to bedding, etc. Then they develop some definite orientation. Thus a study of drainage features from aerial photographs brings out a wealth of information regarding the water-bearing potentiality of the underlying materials. Thus photo-geological studies are considered very advantageous in tackling various hydroproblems.

Analytical Treatment in Aerial Photography

From Fig. 3.7, Photo-scale

$$S = \frac{ab}{AB} = \frac{f}{H} \text{ or } \frac{1}{H/f} \text{ as it is usually expressed} \qquad (3.1)$$

Let l = length of photograph in the direction of flight
 w = width of photograph normal to direction of flight
 p_f = forward overlap (%)
 p_l = lateral or side overlap (%)
 L = net ground distance corresponding to l
 W = net ground distance corresponding to w
 a = net area of each photograph
 A = total area photographed
 N = number of photographs to cover gross area A
 V = speed of flight in kmph
 B = the distance the plane travels between exposures in km (air base)
 t = time interval between exposures in sec

then

$$L = Sl(1 - p_f) = B, \qquad \text{here } S = \frac{H}{f} \qquad (3.2)$$

$$W = Sw(1 - p_l) \qquad (3.2a)$$

$$a = LW \qquad (3.2b)$$

$$N = \frac{A}{a} \qquad (3.2c)$$

$$t = \frac{3600\, B}{V} \qquad (3.2d)$$

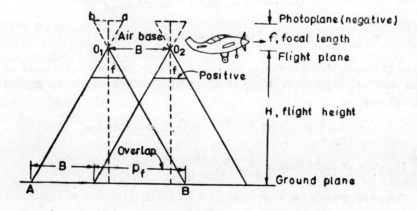

Fig. 3.7 Analytical treatment in aerial photography.

Example 3.1: A project area is 16 km long in east-west direction and 8 km wide in the north-south direction. It is desired to cover the area with vertical aerial photography having a scale of 1:10000. The aeroplane has a speed of 160 kmph. A 150 mm focal length camera with a 23 cm square format is to be used. The average elevation of the ground is 350 m above m.s.l. The forward lap and the lateral lap are to be 60 and 30 per cent, respectively. Prepare the flight map on a base map whose scale is 1:20000, and compute the total number of photographs necessary for the project, the flight altitude and the intervalometer setting.

Solution
(a) Fly east-west to reduce the number of flight lines
(b) Dimension of square ground coverage per photograph

$$G = \frac{23 \times 10000}{100 \times 1000} = 2.30 \text{ km}$$

(c) Lateral advance per strip $W = 2.30 (1 - 0.3) = 1.61$ km
(d) Number of flight lines (align the first and last line with $0.3G$ coverage outside the north and south project boundary lines as shown in Fig. 3.8,

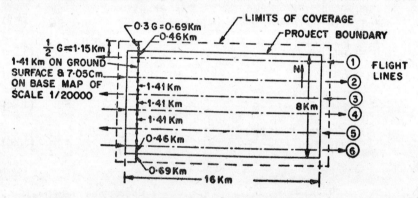

Fig. 3.8 Flight map of the project area.

so as to ensure lateral coverage outside of the project area). Distance of first and last flight lines inside their respective north and south project boundaries

$$= \frac{1}{2} G - 0.3G = \frac{2.30}{2} - 0.3 \times 2.30 = 0.46 \text{ km}$$

Number of flight lines $= \dfrac{8 - 2(0.46)}{1.61 \text{ km/strip}} + 1 = 5.4$, use 6

Adjusted spacing between flight lines for integral number of flight lines

$$W_a = \frac{8 - 2(0.46)}{5} = 1.41 \text{ km}$$

(e) Linear advance per photograph (air base)
$$L = 2.30(1 - 0.6) = 0.92 \text{ km}$$

(f) Number of photograph per strip $= \dfrac{16 \text{ km}}{0.92 \text{ km/photograph}} + 1$

$= 18.4$, use **19**

(g) Total number of photographs $= 19$ photographs/strip $\times 6$ strips
$= \mathbf{114}$

(h) Total number of photograph required theoretically:
Area of ground covered by each photograph
$$a = LW = 0.92 \times 1.61 = 1.48 \text{ km}^2$$
Project area $\quad A = 16 \times 8 = 128 \text{ km}^2$

Total number of photographs required theoretically
$$N = \frac{A}{a} = \frac{128}{1.48} = 86.5, \text{ say } \mathbf{87}$$

(i) Spacing of flight lines on the map $= \dfrac{1.41 \text{ km/strip} \times 1,000 \times 100}{20,000}$

$= 7.05 \text{ cm}$

and the first and the last lines at a distance inside the project boundary
$$= \frac{0.46 \text{ km} \times 1,000 \times 100}{20,000} = 2.3 \text{ cm}$$

(j) Flight altitude
$$S = \frac{f}{H} = \frac{1}{H/f} = \frac{1}{10,000}$$

$$H = 10,000 f = 10,000 \times \frac{150}{10 \times 100} = 1,500 \text{ m}$$

Flight altitude $= 1,500 + 350 = \mathbf{1,850 \text{ m above m.s.l.}}$

(k) Time interval between photographs (intervalometer setting)
$$t = \frac{L}{V} = \frac{0.92 \text{ km} \times 60 \times 60}{160 \text{ km/hr}} = 20.7 \text{ sec, use } \mathbf{20 \text{ sec}}$$

RELIEF DISPLACEMENT

In Fig. 3.9, $r_1 = a_1 a_2$ is the displacement of the image point of A_1 due to its relief h above the datum

$$\frac{r_1}{A_1 A_2} = \frac{f}{H}, \quad A_1 A_2 = h \tan \alpha, \qquad \tan \alpha = \frac{n_1 a_2}{f}$$

Therefore, $\quad r_1 = n_1 a_2 \dfrac{h}{H}$

Similarly $\quad r_2 = n_2 a_1 \dfrac{h}{H}$

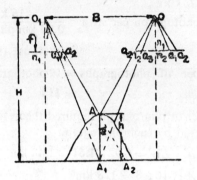

Fig. 3.9 Relief displacement.

Total displacement on the two photographs $r = r_1 + r_2$ is the parallax displacement due to relief and is called parallax difference.

$$r = \frac{h}{H}(n_1 a_2 + n_2 a_4)$$

$$r = \frac{h}{H}(a_1 a_3 + r)$$

$$\frac{B}{H} = \frac{a_1 a_3}{f}$$

Therefore $\quad a_1 a_3 = B\dfrac{f}{H} = \dfrac{B}{S}$

$\qquad\qquad\quad = B_m$, the stereoscopic base of the photographs

$$h = H \frac{r}{B_m + r} \qquad (3.3)$$

From this parallax equation the elevation and ground coordinates of a point can be determined.

The parallax difference can be measured with an ordinary ruler but more accurately by a parallax bar, which consists of two glasses engraved with marks known as floating marks connected by a bar whose length can be varied by a micrometer screw, Fig. 3.10. By adjusting the micrometer, looking through the stereoscope, the two marks are fused to land on the top of the object and on the foot of the object the difference between the corresponding micrometer readings gives the parallax difference r.

Example 3.2: A photographic surveying is carried out to a scale of 1:20000. A camera with a wide angle lens of $f = 150$ mm was cased with 23×23 cm plate for a net of 60% overlap along the flight. Find the error

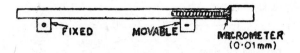

Fig. 3.10 Parallax bar.

in height given by an error of 0.1 mm in measuring the parallax of the point.

Solution

$$H = fS = \frac{150}{1000} \times 20000 = 15 \times 200 = 3{,}000 \text{ m}$$

$$L = 23 \times 200\,(1 - 0.6) = 1{,}840 \text{ m} = B$$

$$B_m = \frac{B}{S} = \frac{1{,}840 \times 100}{20{,}000} = 9.20 \text{ cm}$$

$$r = 0.1 \text{ mm or } 0.01 \text{ cm}$$

Therefore, $h = 3000 \times \dfrac{0.01}{9.20 + 0.01} = 3.26$ m

Example 3.3: Relief $h = 60$ m, $f = 150$ mm, $H = 3000$ m, $n_1 a_2 = 10.8$ cm. Find the displacement of the pictured location of the point A due to elevation h.

Solution

$$r_1 = n_1 a_2 \frac{h}{H} = \frac{(10.8)60}{3000} = 0.216 \text{ cm, or } 2.16 \text{ mm}$$

Note: Focal length need not be known for computation.

The aerial photographic flight mission is fairly expensive. Since a great number of photographs have to be taken in rapid succession while moving in an aircraft at high speed, the aerial cameras should have short cycling times, fast lenses and efficient shutters. They must be capable of faithful functioning under the most extreme weather conditions and inspite of aircraft vibrations. Aerial cameras generally use roll film and have magazine capacities of 60 to 120 m or more. The four main types of aerial cameras are single-lens frame cameras, multi-lens frame cameras, strip cameras, and panoramic cameras.

The camera mount has devices which prevent aircraft vibrations from being transmitted to the camera and also for rotating the camera in azimuth to correct for crab. Crab is the deviation in the aircraft's actual travel direction from its direction of heading. It is usually caused by side winds, Fig. 3.11.

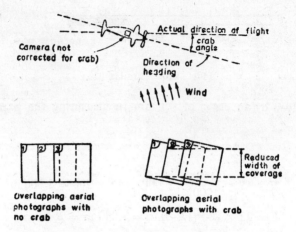

Fig. 3.11 Crab in aerial photography (after P.R. Wolf, 1974).

For topographic mapping, photoscale is usually dictated by the required map scale, required contour interval and capabilities of the instruments that will be used in compiling the map. The minimum possible contour interval that can be reliably traced with a stereoscopic plotting instrument is generally given in terms of a C-factor,* which is a ratio of flying height to the contour interval that can be reliably plotted

$$C\text{-factor} = \frac{H}{\text{C.I.}} \tag{3.4}$$

If a plotter has a C-factor of 1000 and a map with a contour interval of 2 m, then a flying height above ground of 2000 m or less is required. In largescale mapping, the compatible scale and contour intervals (C.I.) are given below:

Scale	Contour interval (C.I.)
1 cm = 6 m	30 cm
1 cm = 12 m	60 cm
1 cm = 25 m	1.5 cm
1 cm = 60 m	3 m

* C-factor values range from 500 to 2000; = 500 for low precision instrument.

Hydrogeology and Aerial Photography

Unavoidable aircraft tilts cause photographs to be exposed with the camera axis tilted slightly from vertical. This is usually less than 1° and rarely exceeds 3° in vertical photography.

A systematic study of aerial photographs involves a consideration of the basic characteristics of photographic images like shape, size, pattern, shadow, tone, texture and site.

SPECIFICATIONS

Most flight plans include a set of detailed specifications which outline the materials, equipment, and procedures to be used on the project. These specifications include requirements and tolerances pertaining to photographic scale (including camera focal length and flying height), end lap, side lap, tilt, crab, and photographic quality.

QUIZ 3

I. State whether 'true' or 'false'; if false give the correct statement:
   1. Porosity is significant in the storage function of rocks while permeability indicates the water yield property of rocks, hence the yield from wells.
   2. Wells are spaced for apart in hard rocks than in alluvial areas.
   3. Tubewells in alluvium should be spaced at least 200 m apart.
   4. Crystalline rocks have low porosity but if fractured can yield water because fractures and fissures increase permeability.
   5. Wells sunk in a synclinal valley are always successful and may be naturally flowing.
   6. An open well may become dry if a borewell is drilled very close to it and pumped continuously.

   [√, ×, ×, √, √, √

II. Select the correct answer(s):
   1. Favourable sites for well sinking are
      (a) valley region rather than the top of the hillock
      (b) bald hillocks with xerophytes
      (c) thick vegetation
      (d) presence of phreatophytes
      (e) joints and faults in rocks
      (f) presence of dykes
      (g) anticlinal folds
      (h) synclinal folds
      (i) areas below tank bunds
      (j) none of the above locations
   2. A perched water table
      (a) may lie below the ground water table
      (b) may lie above the ground water table
      (c) gives a fair amount of sustained yield
      (d) is deceptive and the yield from them is short-lived

3. Artesian conditions may be formed
    (a) due to the presence of a broadly synclinal water bearing formation
    (b) when the well is sunk at the exposed ground surface at the higher level called the recharge area
    (c) when the well is sunk in the trough of the synclinal valley
    (d) usually at the crest of an anticlinal water bearing formation
    (e) at all the above locations
4. A spring may be formed when a sloping permeable bed is
    (a) interrupted by a dyke
    (b) interrupted by an impermeable bed due to the presence of a fault
    (c) interrupted by the naturally sloping ground surface or hill side
    (d) always well below ground level with no interruptions

[1—a, c, d, e, f, h, i; 2—b, d; 3—c; 4—a, b, c]

III. Give the rating as aquifer (Good—G, Medium—M, Poor—P):
Vesicular basalts
Quartzites—heavily shattered
Sandstones
Granite with fissures and joints
Gneisses and schists—weathered
Karstic limestones
Dolomites
Clays and shales
Laterites
Slates
River alluvium
Coarse sand
Marble with fissures and cracks
Crystallines
Tertiary sediments
Quaternary alluvium
Basalts—not weathered

IV. State whether 'true' or 'false':
   (i) Moisture content of a soil can exceed porosity.
   (ii) The wider the range in grain size, the lower the porosity.
   (iii) In general the higher the porosity, the higher the permeability.
   (iv) Clay cannot hold as much water as sand can.
   (v) In consolidated rock formations, water comes through fissures and crevices.
   (vi) Dolomite and limestones yield water from solution openings.
   (vii) Sand should be washed before running sieve analysis to remove the organic matter and soluble salts which may add to the weight.
   (viii) Angularity in shape of grains tends to decrease porosity.
   (ix) If the diameter of the tubewell is doubled, twice the yield can be obtained.

(×, √, ×, ×, √, √, √, ×, ×)

Hydrogeology and Aerial Photography

QUESTIONS

1. Discuss the hydrogeological properties and the well yield in volcanic rocks, sandstones and carbonate rocks.
2. Explain:
   (a) Stereoscopic photography.
   (b) Controlled and uncontrolled mosaics.
   (c) Parallax difference.
3. (a) Mention the salient factors you include while drawing up specifications for conducting a flight mission or a project.
   (b) It is desired to compute a flight mission for a project area 10×20 km. The aeroplane has a speed of 140 kmph. A 200 mm camera with a 20 cm square format is to be used. The approximate scale is 1 : 10,000; the average elevation of the ground is 400 m. The desired overlaps are 60% forward and 30% lateral.

Prepare the flight map on a base map whose scale is 1 : 20,000 and compute the total number of photographs to be taken, the flight altitude and the intervalometer setting.

(2 cm, 6.57 cm, 208 (179) no., 2400 m, 20 sec)

4

Aquifer Properties and Ground Water Flow

Two properties of an aquifer material related to its storage function are its porosity and specific yield. Porosity is the ratio of the volume of voids or pores in a soil mass to its total volume. Shape, size and packing of the grains affect the porosity of granular material. The grain size distribution of the formation or aquifer material can be determined by conducting a mechanical analysis test in a nest of standard sieves with the coarsest on the top and finest at the bottom, covered with lid on the top and bottom pan as receiver. A representative sample, about 150–400 gm, is taken in the laboratory by quartering, oven dried and an exact weight poured into the top sieve and covered with lid. The whole nest is shaken through a mechanical sieve shaker (hand operated or electrical) for about 5 min. and the material retained on each sieve and bottom pan is accurately weighed. Percentage of material passing through each sieve gives a point on the grading curve.

Example 4.1: The results of a mechanical analysis test on a soil sample are given in the following:

Weight of sample taken = 380 gm

IS Sieve designation	2.8 mm	2 mm	1.40 mm	1.00 mm	710 μm	500 μm	300 μm	Bottom pan
Weight retained (gm)	nil	32.3	76.0	97.0	79.8	49.5	33.5	11.5

Plot the grading curve and report the results qualitatively.

Data processed: Weight of sample taken = 380 g

(for results see Table 4.1)

Table 4.1 Results of mechanical analysis test

IS sieve aperture dimension, D	Weight retained (g)	% Retained	Cumulative % retained	Cumulative % passing, p
2.80 mm	nil	nil	nil	100.0
2.00 mm	32.3	8.5	8.5	91.5
1.40 mm	76.0	20.0	28.5	71.5
1.00 mm	97.0	25.5	54.0	46.0
0.71 mm	79.8	21.0	75.0	25.0
0.50 mm	49.5	13.2	88.2	11.8
0.30 mm	33.5	8.8	97.0	3.0
Bottom pan	11.5	3.0	100.0	0
Total	379.6	100.0		

Note: $1 \,\mu\text{m} = \dfrac{1}{1000}$ mm

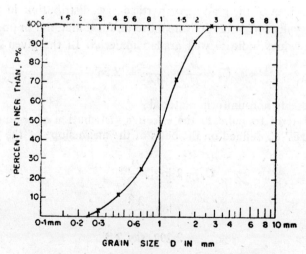

Fig. 4.1 Grading curve of aquifer sample (Ex. 4.1).

Test report: The grading curve is plotted on a semilog paper, Fig. 4.1.

(a) According to IS grain size scale given in the table, below, the soil may be classified as medium to coarse sand with the following percentages:

Grave	Sand			Silt			Clay
	Coarse	Medium	Fine	Coarse	Medium	Fine	
2 mm	0.6 mm	0.2 mm	0.06 mm	0.02 mm	0.006 mm	0.002 mm	

Coarse sand mixed with some fine gravel (from Fig. 4.1, 18% is finer than 0.6 mm size) = 100 − 18 = **82.0%**

Medium sand (between size range 0.6–0.2 mm) = 18 − 0 = **18.0%**

(b) The effective size D_{10} is the size corresponding to 10% of the material being finer and 90% coarser (i.e., the size corresponding to $p=10\%$) and is an index of fineness of the soil. In the given sample

$$D_{10} = \mathbf{0.45 \; mm}$$

(c) The uniformity coefficient C_u is the average slope of the grading curve between 10 and 60% sizes and is given by

$$C_u = \frac{D_{60}}{D_{10}} \tag{4.1}$$

and gives an idea of the grading or particle size distribution in the material. Lower values ($C_u < 2$) indicate more uniform material or poor grading and higher values indicate well graded material. In the given sample

$$C_u = \frac{D_{60}}{D_{10}} = \frac{1.15}{0.45} = \mathbf{2.56}$$

which represents non-uniform material.

Another term to indicate the effective distribution of grain size is the 'range of sizes' C_r defined on the basis of the mean slope of the grain size curve.

$$C_r = 2 \log_{10} \frac{D_{100}}{D_0} \tag{4.2}$$

$$= 2 \log_{10} \frac{2.8}{0.2}$$

$$= 2.292, \text{ say } \mathbf{2.3}$$

A uniformly graded sand has a higher porosity than a less uniform, fine and coarse mixture because in the latter the fines occupy the voids in the coarse material. Wider the range in size, lower is the porosity. The

Aquifer Properties and Ground Water Flow

arrangement of grains or the type of packing also affects porosity—in square packing the porosity is as high as 48% while in rhombic packing it is as low as 26%, Fig. 4.2. Regarding the shape of grains, it is seen that angularity tends to increase porosity.

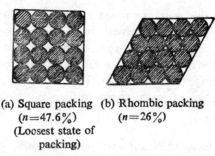

(a) Square packing (b) Rhombic packing
($n=47.6\%$) ($n=26\%$)
(Loosest state of
packing)

Fig. 4.2 Arrangement of grains in square and rhombic packing.

Specific Yield

While porosity is a measure of the water-bearing capacity of the formation, all this water cannot be drained by gravity or by pumping from wells, as a portion of the water is held in the void spaces by molecular and surface tension forces. The volume of water, expressed as a percentage of the total volume of the saturated aquifer, that can be drained by gravity is called the specific yield S_y and the volume of water retained by molecular and surface tension forces, against the force of gravity, expressed as a

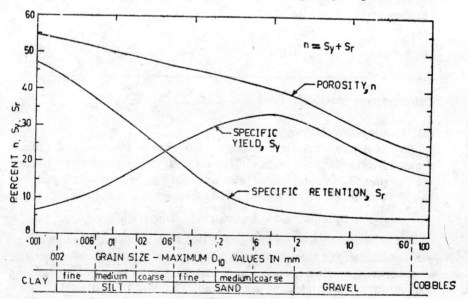

Fig. 4.3 Variations of n, S_y and S_r with grain size.

percentage of the total volume of the saturated aquifer, is called specific retention S_r and corresponds to field capacity.

$$\text{Porosity} = \text{Specific yield} + \text{Specific retention}$$
$$n = S_y + S_r \qquad (4.3)$$

Specific yield is the water removed from unit volume of aquifer by pumping or drainage and is expressed as percentage volume of aquifer. Specific yield depends upon grain size, shape and distribution of pores and compaction of the formation. The values of specific yields for alluvial aquifers are in the range of 10 to 20% and for uniform sands about 30%, Fig. 4.3. Characteristics of some common formation materials are given in Table 4.2.

Table 4.2 Characteristics of some common formation materials

Formation material	Porosity (%)	Specific yield (%)	Permeability lpd/m^2
Clay	45–55	1–10	0.05—100 (for silt also)
Sand	35–40	10–30	5×10^3—15×10^4
Gravel	30–40	15–30	5×10^4—7.5×10^5
Sand and gravel	20–35	15–25	10^3—2.5×10^5
Sandstone	10–20	5–15	5—2.5×10^3
Shale	1–10	0.5–5	10^{-5}—0.1
Limestone	1–10	0.5–5	—

Determination of Specific Yield

(a) LABORATORY METHODS

(i) *Simple saturation and drainage*: Columns of saturated material are drained by gravity and volumes of the material drained and water yielded are determined. The volume of water yielded can be measured directly or can be computed from the porosity and the moisture content after draining.

$$\text{Specific yield} = \text{Porosity} - \text{Specific retention}$$

(ii) *Correlation with particle size*: By determining the effective size or median diameter of the sample and referring to the curve showing the relation between specific retention and effective size or median diameter, the approximate specific retention can be read. The porosity is determined then

$$\text{Specific yield} = \text{Porosity} - \text{Specific retention}$$

Aquifer Properties and Ground Water Flow

(iii) *Centrifuge moisture equivalent*: Considerable experimental work has indicated that for at least some medium textured materials, the moisture equivalent approximates specific retention.

(b) FIELD METHODS

(i) *Field saturation and drainage*: Similar in principle as the laboratory method.

(ii) *Sampling after lowering of water table*: After appreciable lowering of the water table, samples are taken from the zone immediately above the capillary fringe. The moisture content and porosity of the sample are determined.

$$\text{Specific yield} = \text{Porosity} - \text{Specific retention}$$

(iii) *Pumping method*: A known volume of water is pumped out and the volume of sediments drained is determined by observing the depth of the water table lowered.

$$\text{Specific yield} = \frac{\text{Volume of water pumped out}}{\text{Volume of sediments drained}} \times 100$$

Specific yield could also be estimated from moisture content determined by sampling or nuclear meter logging in the cone of depression around a pumped well.

(iv) *Recharge method*: This method is the reverse of the pumping method. The volume of the sediment saturated by the measured recharge is determined and the specific yield is computed.

The specific yield of the soil in the zone of water table fluctuation must be determined in order to estimate the available water supply due to an increment of rise in the water table during the period of recharge, as well as the water supply obtainable for each incremental lowering of the water table.

Example 4.2: The following data is obtained from the difficult rocky areas of Southern U.P. (India):

Area (rocky)	1 km^2
Normal rainfall	700 mm
Normal fluctuation of water table before and after rains	3.2 m
Specific yield of the rock	2 %
Population	154/km^2

Examine how far the drinking water needs of the local population can be met.

Solution

Ground water storage available annually

$$Q = \text{Area} \times \text{Depth of fluctuation of g.w.t.} \times \text{Specific yield}$$

$$= 10^6 \times 3.2 \times \frac{2}{100} = 64{,}000 \text{ m}^3$$

which can be replenished by normal rainfall whose volume, assuming an infiltration rate of $10\% = 10^6 \times \frac{700}{1000} \times \frac{10}{100} = 70{,}000$ m³ and also as observed by the normal fluctuation of water table. Assuming a per capita consumption of 180 lpd annual drinking water supply required $= 154 \times 180 \times 365 = 10{,}120{,}000$ litres or $10{,}120$ m³.

The annual drinking water supply required is 10,120 m³ against an availability of 64,000 m³. Thus there is enough replenishable ground water resource available in the area to meet the drinking water needs of the local population. The only problem is location and economic construction of potential wells.

Example 4.3: In a phreatic aquifer extending over 1 km² the water table was initially at 25 m below ground level. Sometime after irrigation with a depth of 20 cm of water, the water table rose to a depth of 24 m b.g.l. Later 3×10^5 m³ of water was pumped out and the water table dropped to 26.2 m b.g.l. Determine (i) specific yield of the aquifer, (ii) deficit in soil moisture (below field capacity) before irrigation.

Solution
Volume of water pumped out $= \text{Area of aquifer} \times \text{drop in g.w.t.} \times \text{specific yield}$

$$3 \times 10^5 = 10^6 \times 2.2 \times S_y$$

$$S_y = 0.136, \text{ or } 13.6\%$$

Volume of irrigation water recharging the aquifer $= \text{Area of aquifer} \times \text{rise in g.w.t} \times S_y$

Considering an area of 1 m² of aquifer,

$$1 \times y = 1 \times 1 \times 0.136$$

Recharge volume (depth) $y = 0.136$ m, or 136 mm

Soil moisture deficit (below field capacity) before irrigation $= 200 - 136 = 64$ mm

Example 4.4: In an area of 100 ha, the water table dropped by 4.5 m. If the porosity is 30% and the specific retention is 10% determine (i) the specific yield of the aquifer, (ii) change in ground water storage.

Aquifer Properties and Ground Water Flow

Solution

$$\text{Porosity} = S_y + S_r$$
$$30\% = S_y + 10\%$$
$$S_y = 30 - 10$$
$$= 20\% \text{ or } 0.2$$

Change in ground water storage = Area of aquifer × drop in g.w.t. × S_y
$$= 100 \times 4.5 \times 0.2$$
$$= 90 \text{ ha-m, or } 90 \times 10^4 \text{ m}^3$$

Storage Coefficient

Storage coefficient of an aquifer is the volume of water discharged from a unit prism, i.e., a vertical column of aquifer standing on a unit area (1 m²) as water level (piezometric level in confined aquifer—artesian conditions) falls by a unit depth (1 m). For unconfined aquifers (water table conditions) the storage coefficient is the same as specific yield, Fig. 4.4. The storage coefficient for confined aquifers ranges from 0.00005 to 0.005 and for water table aquifers 0.05 to 0.30.

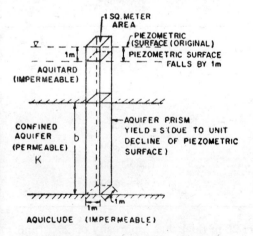
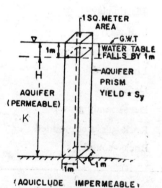

a. Storage coefficient S of a confined or artesian aquifer

b. Storage coefficient or specific yield (S_y) of a water table or unconfined aquifer

Fig. 4.4 Diagrammatic representation of coefficient of storage.

Under artesian conditions, when the piezometric surface is lowered by pumping, water is released from storage by the compression of the water bearing material (aquifer) and by expansion of the water itself*. Thus the

---
*As the water comes out of the pores into the well under atmospheric pressure.

coefficient of storage is a function of the elasticity of water and the aquifer skeleton and is given by (Jacob, 1950) as

$$S = \gamma_w b(\alpha + n\beta) \tag{4.4}$$

where S = coefficient of storage, fraction; n = porosity of aquifer, fraction; b = saturated thickness of aquifer (m); γ_w = units weight of water (9810 N/m³); $\beta = \dfrac{1}{K_w}$, reciprocal of the bulk modulus of elasticity of water $K_w = 2.1$ GN/m² $= 2.1 \times 10^9$ N/m²; and $\alpha = \dfrac{1}{E_s}$, reciprocal of the bulk modulus of elasticity of aquifer skeleton.

The fraction of storage attributable to expansibility water

$$S_w = 4.7 \times 10^{-6} \, nb \tag{4.5}$$

The bulk modulus of compression of some formation material are given in Table 4.3.

Table 4.3 Bulk modulus of compression of formation materials

Material	Bulk modulus of compression E_s, N/m²
Plastic clay	5-40 × 10⁵
Stiff clay	40-80
Medium-hard clay	80-150
Loose sand	100-200
Dense sand	500-800
Dense sandy gravel	1,000-2,000
Rock—fissured, jointed	1,500-30,000
Rock—sound	>30,000

Example 4.5: An artesian aquifer 20 m thick has a porosity of 20% and bulk modulus of compression 10^8 N/m². Estimate the storage coefficient of the aquifer. What fraction of this is attributable to the expansibility of water.

Solution

$$\begin{aligned}
S &= \gamma_w b \, (\alpha + n\beta) \\
&= 9810 \times 20 \left(\frac{1}{10^8} + 0.20 \times \frac{1}{2.1 \times 10^9} \right) \\
&= (1.962 + 0.0187) \, 10^{-3} \\
&= 1.98 \times 10^{-3}, \qquad \text{say, } 2 \times 10^{-3}
\end{aligned}$$

Aquifer Properties and Ground Water Flow

The fraction of storage attributable to the expansibility of water (taking only the second term within the brackets)

$$S_w = 0.0187 \times 10^{-3}$$
$$= \frac{1.87 \times 10^{-5}}{1.98 \times 10^{-3}} \text{ of } S$$
$$\approx \frac{1}{100} \text{ of } S, \text{ or } \mathbf{1\% \text{ of } S}$$

Example 4.6: In a certain place in Andhra Pradesh, the average thickness of the confined aquifer is 30 m and extends over an area of 800 km². The piezometric surface fluctuates annually from 19 m to 9 m above the top of the aquifer. Assuming a storage coefficient of 0.0008, what ground water storage can be expected annually?

Assuming an average well yield of 30 m³/hr and about 200 days of pumping in a year, how many wells can be drilled in the area?

Solution:

$$\Delta GWS = A_{aq} \times \Delta\text{piezo. Surface} \times S = (800 \times 10^6)(19-9)\, 0.0008$$

or 6.4×10^6 m³, or **6.4 M m³**

Annual draft $= (30 \times 24)\, 200 = 0.144 \times 10^6$ m³

or $\quad$ 0.114 M m³

Number of wells that can be drilled in the area $= \dfrac{6.4}{0.144}$

$= 44.5$, say **44 wells**

Of course, the well sites have to be investigated and there should be sufficient spacing for the wells.

Example 4.7: An aquifer has an average thickness of 60 m and an aerial extent of 100 ha. Estimate the available ground water storage if

(a) the aquifer is unconfined and the fluctuation in GWT is observed as 15 m,
(b) the aquifer is confined, and the piezometric head is lowered by 50 m, which drains half the thickness of the aquifer.

Assume a storage coefficient of 2×10^{-4} and a specific yield of 16%.

Solution: (a) $\Delta GWS = A_{aq}.\, \Delta GWT.\, S_y = 100 \text{ ha} \times 15 \text{ m } (0.16)$

$= \mathbf{240\ ha\text{-}m}$

(b) $\Delta GWS = A_{aq}$ [Δ piezo. head $\times S$ + $\Delta GWT \times S_y$]
  (as confined)   (as unconfined)

$= 100$ ha $[20(2 \times 10^{-4}) + 30(0.16)]$

$= \mathbf{480.4\ ha\text{-}m}$

Land Subsidence Due to Ground Water Withdrawals

In areas where unconsolidated and semi-consolidated alluvial aquifers are confined or partially confined by thick fine-grained beds, subsidence of the land surface is likely to result as artesian pressure is reduced by ground water pumpage. Subsidence can be detected by careful relevelling of surface bench marks. Compaction of sediments may extend to depths more than 300 m and is directly related to declines in artesian pressure.

Reduction of artesian pressure results in land subsidence that may cause failure of wells, settlement of buildings. In Southern California and in larger areas near Mexico city, subsidence has exceeded 3 m.

The elastic subsidence of the land surface can be computed from the equation (Lohman, 1961)

$$\Delta b = \Delta p \left(\frac{S}{\gamma_w} - nb\beta \right) \tag{4.6}$$

where $\Delta b =$ land subsidence (m); $\Delta p =$ reduction in artesian pressure (N/m^2) and $\gamma_w =$ specific weight of water.

Example 4.8: Estimate the probable land subsidence when the piezometric head drops by 70 m in an artesian aquifer 30 m thick, having a porosity of 30% and storage coefficient 2×10^{-4}.

Solution

$$\Delta b = \Delta p \left(\frac{S}{\gamma_w} - nb\beta \right)$$

$$= 9810 \times 70 \left[\frac{2 \times 10^{-4}}{9810} - 0.30 \times 30 \times \frac{1}{2.1 \times 10^9} \right]$$

$\simeq 0.011$ m, or **11 mm**

which may cause damage to canal linings under water table conditions. When the water table is lowered by pumping, ground water is obtained from storage by gravity drainage of the interstices in the portion of the aquifer unwatered by pumping.

Specific capacity: Specific capacity of a well is its discharge per unit drawdown usually expressed in 1pm/m. This is a measure of the effectiveness of a well.

Safe yield: Safe yield from a well is the amount of water that can be economically withdrawn from the well in the foreseeable future without causing depletion of the aquifer. Tapping of ground water (location, spacing and yield) in a well field should be so phased that the recharge and discharge of the aquifer are almost balanced without causing an overdraft in the area.

Aquifers: The water bearing geologic formations or strata which yield significant quantity of water for economic extraction from wells are called aquifers.

Rock formations that serve as good aquifers are:

Gravel, sand and sandstone, alluvium
Limestone with cavities formed by the action of ucid waters
Marble with fissures and cracks
Granite rocks with fissures and joints
Weathered gneisses and schists
Heavily shattered quartzites
Vesicular basalts
Slate (better than shale owing to its jointed conditions)

Aquiclude: A geologic formation which can only store water but cannot transmit significant amounts; e.g. clay lenses, shale, etc.

Aquifuge: A geologic formation with no interconnected pores and hence can neither absorb nor transmit water; e.g. basalt, granite, etc.

Aquitard: A geologic formation of a rather impervious nature which transmits water at a slow rate compared to an aquifer but insufficient to supply individual wells; e.g. clay lenses interbedded with sand.

Transmissibility coefficient: The coefficient of transmissibility (T) is the discharge through unit width of aquifer for the fully saturated depth under a unit hydraulic gradient and is usually expressed as lpd/m or m^2/sec. It is the product of field permeability (K) and saturated thickness of the aquifer (b); $T = Kb$ and has the dimensions L^2/T.

Permeability: Permeability (K) is the ability of a formation to transmit water through its pores when subjected to a difference in head. It can be defined as the flow per unit cross-sectional area of the formation when subjected to a unit hydraulic head per unit length of flow (i.e. per unit hydraulic gradient) and has the dimension of velocity, i.e. L/T.

Movement of Ground Water

Ground water moves from levels of higher energy to levels of lower energy, its energy being essentially the result of elevation and pressure, the velocity heads being neglected since the flow is essentially laminar. The velocity is very small in the laminar range—of the order of 1 cm/sec—and the Reynolds number (Re) for ground water flow (Hantush, 1964) varies from 1 to 10 and is given by

$$\text{Re} = \rho v \frac{d_m}{\mu} \tag{4.7}$$

where v = velocity (seepage or bulk) of ground water flow; d_m = mean diameter of the soil grains (usually taken as D_{10}); ρ = density of ground water and μ = dynamic viscosity of ground water.

Velocity of ground water flow which is entirely laminar is given by Darcy's law which states that 'the velocity of flow in a porous medium is proportional to the hydraulic gradient', Fig. 4.5.

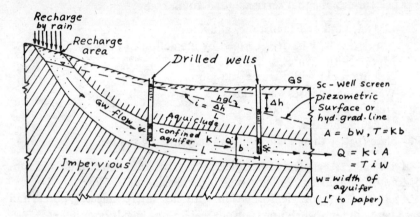

Fig. 4.5 Ground water flow.

$$v = Ki \qquad (4.8)$$

where K = coefficient permeability and i = hydraulic gradient $\left(= \dfrac{h}{L} \text{ if a head } h \text{ is lost in a length } L \right)$.

Ground water flow $Q = Av$
$$= AKi, \qquad A = wb, \qquad T = kb$$

Therefore
$$Q = Tiw \qquad (4.9)$$

where A = cross-sectional area of the aquifer; w = width of the aquifer; b = saturated thickness of the aquifer and T = transmissibility of the aquifer, i.e. capacity of a unit prism of aquifer to transmit water.

The actual velocity (v_a) at which the water is moving through an aquifer, i.e. on an average, the velocity at which a tracer would move through a permeable medium, is given by

$$v_a = \frac{Q}{A_{act}} = \frac{Q}{nA} = \frac{v}{n} \qquad (4.10)$$

where $v (= Q/A)$ is the apparent or seepage velocity given by Darcy's law $v = Ki$.

Aquifer Properties and Ground Water Flow

Flow in coarse-grained aquifer under high drawdown, especially in the area adjacent to the pumping well is given by the non-Darcy regime of flow generally described by the Forchheimer equation (Ahmed, 1969)

$$i = av + bv^2 \qquad (4.10a)$$

where a and b are constants depending upon the properties of porous media and the fluid, and have the dimensions

$$[a] = \frac{1}{[v]} = \frac{T}{L}, \qquad [b] = \frac{1}{[v^2]} = \frac{T^2}{L^2},$$

where [ ] means 'dimensions of'. That is, the flow is no longer laminar, particularly as it arrives at the well face due to high gradients and exhibits non-linear relationship between the velocity and hydraulic gradient. While the use of Darcy's law is valid for low Reynolds number, typically $R_e < 1$, various flow situations have been observed where the Reynolds number of flow is likely to be greater than unity.* For example in a gravel packed well (mean size of gravel $\approx$ 5 mm), $R_e \approx$ 45 and the flow would be transitional at a distance of about 5 to 10 times the well radius.

Example 4.9: It was observed in a field test that 3 hr 20 min was required for a tracer to travel from one well to another 20 m apart, and the difference in their water surface elevations was 0.5 m. Samples of the aquifer between the wells indicated a porosity of 15%. Determine the permeability of the aquifer, seepage velocity, and the Reynolds number for the flow, assuming an average grain size of 1 mm and v_{water} at 27°C = 0.008 Stoke.

Solution
Actual velocity of flow through the aquifer as indicated by the tracer

$$v_a = \frac{x}{t} = \frac{20 \text{ m}}{3\frac{1}{3} \text{ hr}} = 6 \text{ m/hr}$$

Seepage velocity as given by Darcy's law,

$$v = Ki = K\frac{h}{L} = K\frac{0.5 \text{ m}}{20 \text{ m}} = \frac{K}{40}$$

From Eq. 4.10, $v_a = v/n$

$$6 = \frac{K}{40 (0.15)}$$

$K = 36$ m/hr at the field temperature
   $= 864$ m/day or m³/day/m²
   $= 8.64 \times 10^5$ lpd/m²

---

*In aquifers containing large diameter solution openings, coarse gravels, and also in the immediate vicinity of a gravel packed well, flow is no longer laminar.

Seepage velocity, $v = \dfrac{K}{40} = \dfrac{864}{40}$

$= 21.6$ m/day, or 0.025 cm/sec

Reynolds number,

$$\text{Re} = \dfrac{vd_m}{\nu} = \dfrac{0.025}{100} \times \dfrac{1}{1000} \times \dfrac{1}{0.008 \times 10^{-4}} = 0.3$$

Example 4.10: A 30 cm gravel packed well is pumped at a constant rate of 5000 m³/hr from a confined aquifer of thickness 50 m, having $D_{10} = 0.23$ mm, $D_{50} = 0.6$ mm.

(i) What is the domain around the well for which Darcy's law is applicable, assuming that the law is valid up to $R_e = 6$? $\nu_{\text{water}} = 1$ c St

(ii) If the gravel pack is 14 cm thick and has $D_{10} = 1.5$ mm and $D_{50} = 3$ mm, estimate the Reynold's number and the seepage gradient at mid-pack.

Solution:

(i) $R_e = \dfrac{vD}{\nu}$; $D = D_{50}$ (mean size)

$6 = \dfrac{v(0.6 \times 10^{-3})}{1 \times 10^{-6}}$ $v = 0.01$ m/s

$Q = Av$

$\dfrac{5000}{60 \times 60} = 2\pi\, r.50\,(0.01)$

$r = 0.442$ m, say **44 cm**

Hence Darcy's law is valid beyond $r = $ **44 cm**

(ii) At $r = \dfrac{30}{2} + \dfrac{14}{2} = 22$ cm (mid-gravel pack),

$\dfrac{5000}{60 \times 60} = 2\pi\,(0.22)\,50\,v$, $v = 0.02$ m/s

$R_e = \dfrac{vD}{\nu} = \dfrac{0.02\,(3 \times 10^{-3})}{1 \times 10^{-6}} = 60$

Actually, $R_e > 60$, since the depth of flow near the well is < 50m, due to the cone of depression.

$K = C\,D_{10}^2$, Allen Hazen

$\dfrac{K_g}{K_a} = \left(\dfrac{1.5}{0.23}\right)^2 = 36$

Aquifer Properties and Ground Water Flow

where K_g and K_a are the permeabilities of the gravel pack and aquifer, respectively, $C = 100$ where D_{10} is in cm, K in cm/s,

$$K_a = 100\,(0.023)^2 = 0.053 \text{ cm/s}$$
$$K_g = 0.053 \times 36 = 1.91 \text{ cm/s} \approx 0.02 \text{ m/s}$$

$$v = K_g i, \qquad i = \frac{v}{K_g} = \frac{0.02}{0.02} = 1$$

i.e. the cone of depression has a slope of 45° which is critical.

Factors Affecting Permeability

The coefficient of permeability (K) is the rate of flow per unit cross-sectional area under unit hydraulic gradient (at a specified temperature) and is usually expressed as lpd/m² or m/sec. It has the dimensions of velocity (L/T). The permeability depends upon the grain size distribution, porosity, shape and arrangement of pores, properties of the pore fluid and entrapped air or gas and can be expressed as

$$K = CD^2 \frac{\gamma_w}{\mu} \frac{e^3}{1+e} \qquad (4.11)$$

where C = a constant; D = effective size of the formation material (aquifer); e = void ratio; γ_w = unit weight of water at the flow temperature and μ = viscosity of water at the flow temperature.

The intrinsic or specific permeability (k) of a water bearing medium is given by

$$k = CD^2 \qquad (4.12)$$

where the constant C summarises the geometrical properties of the porous medium and k has the dimensions of L^2. k when expressed in cm² is usually extremely small, so that Darcy has been adopted as a more practical unit.

$$1 \text{ darcy } (k) = 0.987 \times 10^{-8} \text{ cm}^2$$

Coefficient of permeability, $K = k \dfrac{\gamma_w}{\mu}$ \qquad (4.13)

1 darcy $(K) = 0.966 \times 10^{-3}$ cm/sec (for water at 20°C)

Since $\gamma_w = \rho g$, where ρ is the density of water. $K \propto \dfrac{1}{\mu/\rho}$ or $\dfrac{1}{\nu}$, i.e. the permeability at different temperatures varies inversely as the respective kinematic viscosities. Hence if permeability at laboratory temperature K_L (temperature of determination) is known, the permeability at any other temperature K_t can be determined from the equation

$$K_t = \frac{K_L \nu_L}{\nu_t} \qquad (4.14)$$

While determining permeability (K), the water temperature has to be noted and K has to be reduced to a standard temperature of 20°C or 27°C and usually expressed as cm/sec, m/sec, or lpd/m² at the specified temperature. The permeability conversion factors are given in Table 4.4.

For filter sands, $K = CD_{10}^2$, Allen Hazen

K is in cm/sec, if D_{10} is in cm and $C \approx 100$

K is in m/day, if D_{10} is in mm and $C \approx 850$

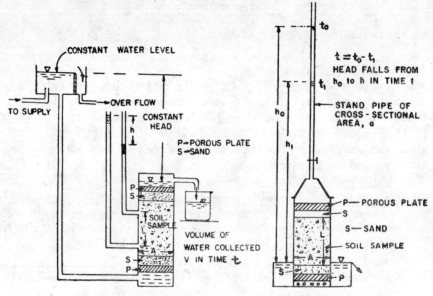

(a) Constant head permeameter   (b) Falling head permeameter

Fig. 4.6 Constant head and falling head permeameters.

Laboratory Permeability

In the laboratory the permeability of relatively coarse-grained soils can be determined by a constant head permeameter, Fig. 4.6a, by measuring the volume of water percolated (V) through the soil sample of cross-sectional area A and length L in a given time t under a constant head h. From Darcy's law,

$$V = Qt = KA \frac{h}{L} t \qquad (4.15)$$

$$K = \frac{VL}{hAt}$$

at the laboratory temperature.

In relatively fine-grained soils, it is not possible to collect an appreciable volume of water to be accurately measured; the permeability is determined by a falling head permeameter, Fig. 4.6b. If h_0 and h_1 are the initial and final readings respectively in the stand pipe of cross-sectional area a, in a time t, above the soil specimen of length L and cross-sectional area A, the **permeability is given by**

Aquifer Properties and Ground Water Flow

Table 4.4 Permeability conversion factors

Specific permeability k, darcy	10^5	10^3	1	10^{-4}
Soil class	Clean gravel	Clean sands, mixtures of clean sands and gravels	Very fine sands, silts, mixtures of sand, silt and clay, glacial till, stratified and highly organic clays	Unweathered clays, shales intact igneous, metamorphic rocks and sound limestone
Flow characteristics		Good aquifers (high permeability)	Poor aquifers (low permeability)	Impervious
cm/sec Permeability (lab. values) K at 20°C	10^2	1	10^{-3}	10^{-7}
lpd/m²	8.64×10^7	8.64×10^5	8.64×10^2	8.64×10^{-2}
m/day	8.64×10^4	8.64×10^2	8.64×10^{-1}	8.64×10^{-5} (m³ per day/m²)

The table shows the values of permeability coefficients (k, K) in different units at 20°C

Note: 1 darcy = 9.66×10^{-4} cm/sec (1 cm/sec = 1.033×10^3 darcy)
1 lpd/m² = 1.16×10^{-6} cm/sec (1 cm/sec = 8.64×10^5 lpd/m²)
1 m/day = 1.16×10^{-3} cm/sec (1 cm/sec = 8.64×10^2 m/day)
= 1,000 lpd/m² (1 lpd/m² = 10^{-3} m/day)

$K = k\, \gamma/\mu$

$$K = 2.3 \frac{aL}{At} \log_{10} \frac{h_0}{h_1} \qquad (4.16)$$

at the laboratory temperature. The soil samples should be well saturated before readings are taken.

The permeability is determined in the field by pumping test. By supplementing a few field tests with laboratory permeability data on samples collected, representative of at least all major textural units within the area of study and geologic interpretation, the permeability or transmissibility of the aquifer can be obtained.

Aquifers usually consist of lithologic units or layers having different permeabilities (anisotropic aquifer). To determine the average permeability, samples are obtained from each layer and their permeability determined. The average permeabilities K_x, K_y, in the horizontal and vertical directions, respectively, are calculated from the equations

$$K_x = \frac{1}{b}(K_1 b_1 + K_2 b_2 + K_3 b_3 + ...) \qquad (4.17)$$

$$K_y = b \bigg/ \left(\frac{b_1}{K_1} + \frac{b_2}{K_2} + \frac{b_3}{K_3} + ...\right) \qquad (4.18)$$

where

b = total thickness of the aquifer ($= b_1 + b_2 + b_3 + ...$)
$b_1, b_2, b_3, ...$ = thickness of each layer
$K_1, K_2, K_3, ...$ = permeability of each layer

Example 4.11: An aquifer of aerial extent of 100 km² is overlain by four strata as given below:

Strata	Thickness (m)	K_x (m/day)	K_y m/day
1 (top)	1	1	0.25
2	3	2	0.3
3	2	1.5	0.2
4	4	0.025	0.005

(a) If a 4 hr storm occurs producing a total rainfall 100 mm, estimate the recharge into the aquifer, assuming that the piezometric surface in the aquifer is at the bottom of layer 4 and that all layers are saturated.

Aquifer Properties and Ground Water Flow

(b) If the four layers are underlain by an impermeable strata instead of the aquifer, estimate the lateral flow per unit width through the layers, assuming that the layer dip by 0.1%.

Solution

(a) $Q_v = K_v i A$

$$K_v = \frac{b}{\Sigma \frac{b_1}{K_{y1}}} = \frac{1+3+2+4}{\frac{1}{0.25} + \frac{3}{0.3} + \frac{2}{0.2} + \frac{1}{0.005}} = \frac{10}{224} = 0.0446 \text{ m/day}$$

Assuming that no water is standing over the ground surface,

$$i = \frac{h}{L} = \frac{10}{10} = 1$$

$$Q_v = 0.0446 \times 1 (100 \times 10^6) = 4.46 \times 10^6 \text{ m}^3/\text{day}$$

Recharge into the aquifer during the 4 hr storm $= (4.46 \times 10^6) \frac{4}{24}$

$$= 0.743 \text{ M.m}^3$$

Infiltration $\quad F = \frac{0.743 \times 10^6 \text{ m}^3}{100 \times 10^6 \text{ m}^2} = 0.00743 \text{ m or } 7.43 \text{ mm}.$

The infiltration is very small since the aquifer is semi-confined.

(b) $Q_h = K_h i A$

$$K_h = \frac{1}{b} \Sigma K_{x1} b_1$$

$$= \frac{1}{10} [1 \times 1 + 2 \times 3 + 1.5 \times 2 + 0.025 \times 4] = 1.01 \text{ m/day}$$

$$Q_h = 1.01 \times \frac{0.1}{100} (10 \times 1) = \mathbf{0.0101 \text{ m}^3/\text{day}}$$

per m width which is the lateral flow rate during the storm.

Example 4.12: During hydrogeological investigation two potential aquifers 32 km apart, were located, one being 5,000 years and the other 25,000 years old. They were found to be connected by a water bearing stratum of 30 m thickness running inclined at 20 m/km. From a few observation wells, the hydraulic gradient was found to be 0.2 m/km. Determine the transmissibility of the water bearing stratum.

Solution

It has taken 20,000 years for the ground water movement through the inclined water bearing stratum, Fig. 4.7, to form a recent potential ground water storage.

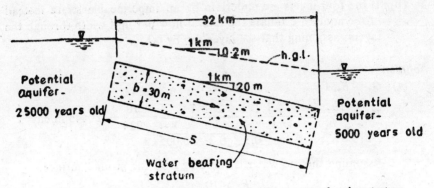

Fig. 4.7 Two potential aquifers connected by a water bearing stratum.

$$v = \frac{s}{t} = \frac{\sqrt{32^2 + 0.64^2} \times 1{,}000}{20{,}000 \times 365} = 0.00438 \text{ m/day}$$

$$v = Ki = K\frac{h}{L} \text{ (Darcy's law)}$$

$$0.00438 = K\frac{0.2}{1000}$$

$$K = 21.90 \text{ m/day}$$

Transmissibility of the water bearing stratum,

$$T = K.\, b = 21.90 \times 30$$

$$= 657 \text{ m}^2/\text{day or m}^3/\text{day/m}$$

$$= 657{,}000 \text{ lpd/m, or } \mathbf{657 \text{ m}^2/\text{day}}$$

Example 4.13: During ground water investigation in the Cauvery basin by UNDP the following data were collected.

Recharge area identified	19×13 km
Annual rainfall	1070 mm
Infiltration	20% of rainfall (approx.)
Transmissibility of the aquifer (from pump tests in the discharge area)	6×10^6 lpd/m
Width of the aquifer	21 km
Hydraulic gradient (towards the discharge area from observation wells)	1.14 m/km

It has to be ascertained whether all the pumpage comes from the recharge area.

Solution

Annual recharge $= (19 \times 13)\ 10^6\ \dfrac{20}{100} \times 1.07 = 5.29 \times 10^7\ m^3$

$$Q = T\ i\ w = (6 \times 10^3)\ \dfrac{1.14}{1000} \times 21{,}000$$

$$= 144 \times 10^3\ m^3/day$$

Annual pumpage $= (144 \times 10^3)\ 365 = 5.25 \times 10^7\ m^3$

Thus the entire pumpage comes from the recharge area.

Water Table Contour Maps and Flow Net Analysis

For given boundary conditions, flow lines (stream lines) and equipotential lines (having the same head) can be constructed so that the flow fields are roughly square (flow net). Contour maps of water tables or piezometric surfaces can be prepared from observation of water levels in bore holes and flow lines can be drawn to form an orthogonal system of small squares, Fig. 4.8. Flow lines must be parallel to an impermeable boundary and in an unconfined aquifer the water table itself becomes one bounding flow surface. Flow lines are drawn perpendicular to water table contours.

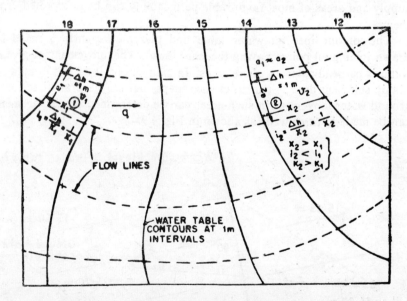

SECTION 2 IS A MORE PROMISING SITE FOR A WELL

Fig. 4.8 Water table (piezometric surface) contour maps.

Since there is no flow across the flow lines

$$q = a_1 v_1 = a_2 v_2 = a_1 . K_1 i_1 = a_2 . K_2 i_2 \ldots \text{from Darcy's law}$$

Therefore $\dfrac{K_1}{K_2} = \dfrac{a_2 i_2}{a_1 i_1}$

$\dfrac{K_1}{K_2} = \dfrac{i_2}{i_1}$

when flow lines are nearly parallel.

If the spacing between the water table contours (contour interval Δh) at section 1 and 2 are x_1 and x_2, respectively, then

$$i_1 = \dfrac{\Delta h}{x_1}, \; i_2 = \dfrac{\Delta h}{x_2}$$

when $\dfrac{K_1}{K_2} = \dfrac{x_1}{x_2}$

which indicates that for areas with uniform ground water flow, the portions having wide water table contour spacing having flat gardients and higher permeabilities. Similarly a_2/a_1 can be estimated from the distances between the flow lines at the two sections. Thus section 2 indicates a better and more promising site for a well than section 1.

Such water table contour maps with flow lines drawn provide useful data for locating new wells as the best possible sources of ground water supply and areas of most favourable permeability can be interpreted from them.

The contour lines on water table and piezometric surface maps are drawn joining all points having the same head. The movement of ground water is perpendicular to these lines. In many areas it is not possible to obtain sufficient data to construct such maps, but the general direction of ground water movement in such areas can be determined if measurements can be made in three wells as shown in Fig. 4.9.

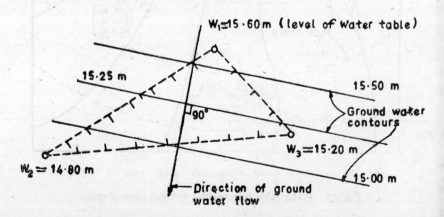

Fig. 4.9 Direction of ground water flow and contours from water level observation in three wells.

Analysis of Watertable Maps

If the flow rates (q_1 and q^2) across successive watertable or piezometric surface contours between two limiting flow lines are known, Fig. 4.10 the following deductions can be obtained:

(i) $q_1 = T_1 i_1 w_1$, $i_1 = \dfrac{\Delta h}{l_1}$, $l_1 = MN$, $w_1 = AA'$ (scaled from map)

$q_2 = T_2 i_2 w_2$, $i_2 = \dfrac{\Delta h}{l_2}$, $l_2 = PQ$, $w_2 = BB'$ (scaled from map)

Δh = water level difference between two successive contours

T_1 and T_2 can be estimated. Repeating this procedure spatial distribution of T can be obtained.

(ii) If there is no leakage from above or below into the aquifer the storage coefficient S can be estimated from

$$q_2 - q_1 = S \Delta h_t A_L \qquad (4.19)$$

where Δh_t = average rate of water level decline or rise, in the area between the two limiting flowlines and successive contours, A_L.

(iii) Recharge by vertical leakage of water through an aquitard can be obtained from

$$P A_L = (q_2 - q_1) \pm S \Delta h_t A_L \qquad (4.20)$$

where P = rate of vertical leakage per unit area;
+ sign is used for rise in water level and
− sign is used for decline in water level.

Example 4.14: The coefficients of transmissibility of the aquifer at cross sections 1 and 2 in Fig. 4.10 are 200 and 180 m²/day, respectively.

(i) Compute the flow rates at the cross sections.

(ii) If the area between the cross sections AA' and BB' is about 450 km² for which the average coefficient of storage is 4×10^{-4} and the average water level decline in the area is about 2 mm/day, compute the recharge into the area.

(iii) If the total withdrawal from the pumping centre P is 30,000 m³/day, determine the average transmissibility of the aquifer in the vicinity of the pumping centre.

Scaled from map: $AA' = 15$ km, $BB' = 70$ km, $CC' = 50$ km

$MN = 18$ km, $PQ = 9$ km, $RS = 4$ km

Solution

(i) $q_1 = T_1 i_1 w_1 = 200 \dfrac{15 \text{ m}}{18 \text{ km}} \times 15 \text{ km} = \mathbf{2500 \text{ m}^3/day}$

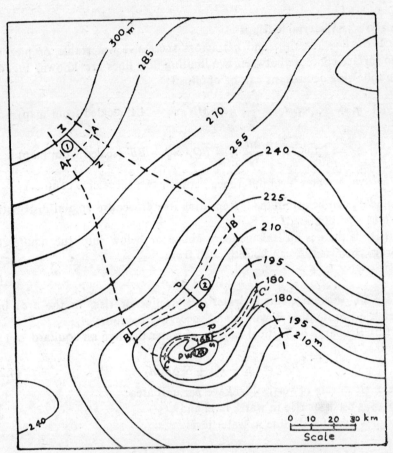

Fig. 4.10 Flow analysis from the contour map of GWT or piezometric surface.

$$q_2 = T_2 i_2 w_2 = 180 \frac{15 \text{ m}}{9 \text{ km}} \times 70 \text{ km} = \mathbf{21000 \text{ m}^3/\text{day}}$$

(ii) $PA_L = (q_2 - q_1) - S \Delta h_t A_L$

$P(450 \times 10^6) = (21000 - 2500) - (4 \times 10^{-4})(0.002)(450 \times 10^6)$

Recharge rate $P = 40 \times 10^{-6}$ m³/day per m² or $\mathbf{40 \text{ m}^3/\text{day per km}^2}$

(iii) $q = Tiw \quad 30000 = T \dfrac{15 \text{m}}{4 \text{km}} \times 50 \text{ km}$

$T = \mathbf{160 \text{ m}^2/\text{day}}$

Flownet—Isotropic Aquifer, Fig. 4.11

If the total head h between sections 1 and 2 divided into n_d number of potential drops, n_f is the number of flow channels, and b is the saturated thickness of the aquifer (perpendicular to paper), the total ground water

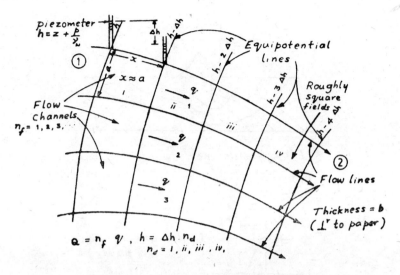

Fig. 4.11 Flownet in an isotropic aquifer.

flow can be obtained from the flownet as follows:

Total head $h = h_1 - h_2$

Head drop between successive equipotential lines $\Delta h = \dfrac{h}{n_d}$ ...(i)

Flow through one flow channel by Darcy's law

$$q = K \cdot \dfrac{\Delta h}{x} \cdot (a \cdot b), \quad a \approx x \text{ for roughly square fields}$$

$$= K \, \Delta h \cdot b$$

Total flow through n_f flow channels

$$Q = q \cdot n_f = n_f (K \, \Delta h \, b), \qquad T = K b \text{... (ii)}$$

$$Q = n_f \, \Delta h \, T \tag{4.21}$$

From (i) and (ii),

$$Q = \dfrac{n_f}{n_d} K \, h \, b \tag{4.21a}$$

Flownet—Anisotropic Aquifer

For an anisotropic system, $K_h > K_v$, Fig. 4.12, the horizontal dimensions must be multiplied by $\sqrt{\dfrac{k_v}{k_h}}$ to obtain a transformed section with an isotropic medium. The flownet is drawn and the flow rate is obtained substituting for $K = \sqrt{K_h \, k_v}$. The flownet is then transferred back to the original

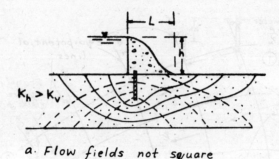

a. Flow fields not square

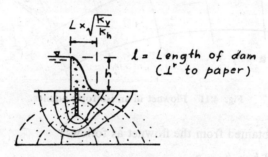

b. orthogonal network
$Q = \dfrac{n_f}{n_d}\sqrt{K_h K_v}\, h$

Fig. 4.12 Flownet in anisotropic soil.

anisotropic section by multiplying the horizontal dimensions by $\sqrt{\dfrac{K_h}{K_v}}$.

Example 4.15: From flownet analysis of a well pumping at 150,000 lpm at Neyveli Lignite Project the following data were obtained:

Number of flow channels	30
Head drop between successive piezometric contours	6 m
Thickness of aquifer	30 m

Determine the transmissibility and permeability of the aquifer.

Solution

$$T = \frac{Q}{n_f \Delta h} = \frac{150{,}000}{30 \times 6}(60 \times 24)$$

$\qquad = 12 \times 10^5$ lpd/m, $\qquad$ or **1200 m²/day**

$$K = \frac{T}{b} = \frac{12 \times 10^5}{30}$$

$\qquad = 4 \times 10^4$ lpd/m², $\qquad$ or **40 m/day**

Aquifer Properties and Ground Water Flow

Ground Water Flow Potential

The total energy (or head, h) at any point in the ground water flow field per unit weight of water is given by

$$h = z + \frac{P}{\gamma_w} + \frac{V^2}{2g}$$

where $z =$ elevation of the point (above a chosen datum); $\frac{P}{\gamma_w} =$ pressure head and $\frac{V^2}{2g} =$ velocity head.

Since the ground water flow velocities are usually very small, $\frac{V^2}{2g}$ is neglected, and

$$h = z + \frac{P}{\gamma_w} = \text{piezometric head, at the point} \qquad (4.22)$$

From Darcy's law,

$$V = Ki$$

where $K =$ coefficient of permeability of the formation; and $i =$ hydraulic gradient $= \text{grad}\left(z + \frac{P}{w}\right) = \frac{dh}{ds}$.

Therefore

$$V = -K\frac{dh}{ds} \qquad (4.23)$$

the negative sign indicates the ground water flow in the direction of decreasing head, i.e. in the direction s.

It is convenient to introduce a velocity potential ϕ, defined as a scalar function of time and space, such that the velocity components in the x-, y- and z- directions are given by

$$u = -\frac{\partial \phi}{\partial x}, \quad v = -\frac{\partial \phi}{\partial y}, \quad w = -\frac{\partial \phi}{\partial z} \qquad (4.24)$$

and generally,

$$V = -\frac{\partial \phi}{\partial s} \qquad (4.25)$$

where V is the velocity in the flow direction s.

From Eqs. (4.23) and (4.25),

$$\frac{d\phi}{ds} = K\frac{dh}{ds}$$

or $\qquad \phi = Kh + C$

where $C =$ a constant which disappears during differentiation and is usually taken as zero (since it does not alter the flow pattern in any manner). Therefore

$$\phi = Kh \tag{4.26}$$

Since ground water can be considered incompressible for all practical purposes, i.e. $\rho =$ constant, the equation of continuity for *steady groundwater flow* is obtained by putting

$$\frac{\partial \rho}{\partial t} = 0 \text{ in Eq. (4.61), as}$$

$$\frac{\partial u}{\partial x} + \frac{\partial v}{\partial y} + \frac{\partial w}{\partial z} = 0 \tag{4.27}$$

From Eq. (4.24),

$$\frac{\partial^2 \phi}{\partial x^2} + \frac{\partial^2 \phi}{\partial y^2} + \frac{\partial^2 \phi}{\partial z^2} = 0 \tag{4.28}$$

$$\text{or,} \quad \Delta^2 \phi = 0 \tag{4.29}$$

which is the Laplace equation.

For a homogeneous and isotropic medium (formation), from Eqs. (4.24) and (4.26),

$$u = -K\frac{\partial h}{\partial x}, \quad v = -K\frac{\partial h}{\partial y}, \quad w = -K\frac{\partial h}{\partial w}$$

From Eq. (4.27),

$$\frac{\partial^2 h}{\partial x^2} + \frac{\partial^2 h}{\partial y^2} + \frac{\partial^2 h}{\partial z^2} = 0 \tag{4.30}$$

$$\text{or} \quad \nabla^2 h = 0 \tag{4.31}$$

i.e. for steady ground water flow in homogeneous isotropic medium, Laplace equation is satisfied, and a flow net can be constructed.

In anisotropic formation, if the permeabilities in the three principal directions are K_x, K_y and K_z, the velocity components are given by Darcy's law as

$$u = -K_x\frac{\partial h}{\partial x}, \quad v = -K_y\frac{\partial h}{\partial y}, \quad w = -K_z\frac{\partial h}{\partial z} \tag{4.32}$$

From the continuity Eq. (4.27), for steady flow,

$$K_x \frac{\partial^2 h}{\partial x^2} + K_y \frac{\partial^2 h}{\partial y^2} + K_z \frac{\partial^2 h}{\partial z^2} = 0 \tag{4.33}$$

If a transformation could be made such that

$$X = \frac{x}{\sqrt{K_x}}, \quad Y = \frac{y}{\sqrt{K_y}}, \quad Z = \frac{z}{\sqrt{K_z}}$$

Eq. (4.33) becomes

$$\frac{\partial^2 h}{\partial X^2} + \frac{\partial^2 h}{\partial Y^2} + \frac{\partial^2 h}{\partial Z^2} = 0 \tag{4.34}$$

Aquifer Properties and Ground Water Flow

and hence Laplace equation is satisfied.

For a two-dimensional anisotropic system $K_x > K_y$, the horizontal dimensions are reduced by multiplying by the factor $\sqrt{K_y/K_x}$, and a flownet is drawn in the transformed section. This creates an isotropic medium having an equivalent $K = \sqrt{K_x K_y}$. The ground water flow is computed by substituting $K = \sqrt{K_x K_y}$ in the discharge equation and then the flownet is transferred back to the original section, which of course will not be an orthogonal network, Fig. 4.12.

Ground Water Flow Problems

Potential flow theory may be applied to obtain discharge and drawdown for various conditions of ground water flow, which are illustrated in the following examples.

Steady One-dimensional Flow:

(a) *Steady flow in a homogeneous aquifer*, Fig. 4.13

Discharge per unit width, assuming flow to be horizontal and uniform everywhere in a vertical section

$$q = V(h.1) = -\frac{\partial \phi}{\partial s} h \approx - Kh \frac{dh}{dx}$$

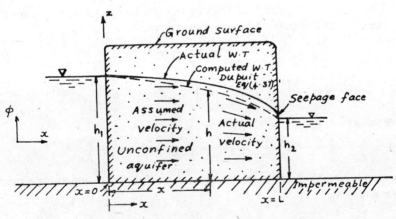

Fig. 4.13 Steady flow in an unconfined aquifer.

Integrating,
$$qx = -\frac{Kh^2}{2} + c$$

$$h = h_1 \text{ when } x = 0, \ c = \frac{Kh_1^2}{2}$$

Therefore
$$q = \frac{K}{2x}(h_1^2 - h^2) \tag{4.35}$$

which is the Dupuit's equation and indicates that for a given discharge the water table is parabolic in form.

Applying the boundary condition $h = h_2$, when $x = L$,

$$q = \frac{K}{2L}(h_1^2 - h_2^2) \tag{4.36}$$

From Eqs. (4.35) and (4.36), the equation for the phreatic line is obtained as

$$h = \sqrt{h_1^2 - (h_1^2 - h_2^2)x/L} \tag{4.37}$$

which is shown in Fig. 4.13 and also the water table actually observed in piezometers or determined by drawing a flownet. The actual water table is higher than that computed by the Dupuits equation, the departure being large in the neighbourhood of the well. The descrepancy is due to the fact that Dupuit flows are all assumed horizontal, where as the actual velocities of the same magnitude have a downward vertical component so that a greater saturated thickness is required for the same discharge; the assumption that $dh/ds \approx dh/dx$ is not correct in the vicinity of the well. The watertable approaches the well face tangentially and forms a seepage face because no consistent flow pattern can connect a watertable directly to a downstream free water surface. However, Eq. (4.36) yields correct values of q or K.

The vertical component of velocity (w) can be determined by considering the continuity equation in two dimensions,

$$\frac{\partial u}{\partial x} + \frac{\partial w}{\partial z} = 0$$

$$w = -\int_0^z \frac{\partial u}{\partial x} dz = -K \int_0^z \frac{\partial^2 h}{\partial x^2} dz = -Kz\frac{d^2h}{dx^2}$$

Because

$$q = -Kh\frac{dh}{dx}, \frac{d^2h}{dx^2} = -\frac{q^2}{K^2h^3}, \text{ and } w = -\frac{q^2 z}{Kh^3}$$

The vertical velocity component increases with elevation and at the phreatic surface, $z = h$,

$$w = -\frac{q^2}{Kh^2} \tag{4.38}$$

The negative sign indicates the downward direction.

(b) *Flow into horizontal galleries dug down to the impervious soil layer,* Fig. 4.14

Depth of GWT above impervious soil layer = H

Depth of water table in the gallery = h_1

Aquifer Properties and Ground Water Flow

The phreatic surface is parabolic dropping from H to h in a distance L.
The quantity of water flowing into the trench from one side per unit length of shoreline is

$$q = VA = Ki(h \times 1) = Kh\frac{dh}{dx}$$

x being measured from the face of the gallery.

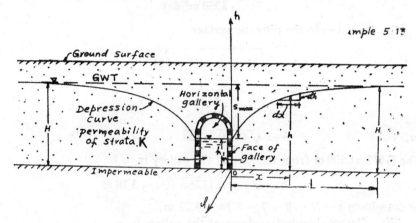

Fig. 4.14 Flow into horizontal galleries.

Integrating, $$qx = \frac{Kh^2}{2} + c_1$$

when $x = 0$, $h = h_1$; $c_1 = -\dfrac{Kh_1^2}{2}$

$$q = \frac{K}{2x}(h^2 - h_1^2) \tag{4.39}$$

which is parabolic. Putting $h = H$ when $x = L$

$$q = \frac{K}{2L}(H^2 - h_1^2) \tag{4.40}$$

From Eqs. (4.39) and (4.40), the equation to the phreatic line is

$$h = \sqrt{h_1^2 + \frac{x}{L}(H^2 - h_1^2)} \tag{4.41}$$

The quantity of water flowing into the gallery of length l from both sides

$$Q = 2ql = \frac{Kl}{L}(H^2 - h_1^2) \tag{4.42}$$

When the watertable in the gallery drops, i.e. h_1 decreases, L increases upon increasing h_1, L decreaes.

As an example, given $H = 7$ m, $h_1 = 2$ m, $L = 400$ m, $K = 60$ m/day, $l = 200$ m.

Flow from one side

$$q = \frac{K}{2L}(H^2 - h_1^2) = \frac{60}{2 \times 400}(7^2 - 2^2) = 3.375 \text{ m}^3/\text{day per m}$$

(i) Flow into the gallery $Q = 2ql = 2 \times 3.375 \times 200$
$$= \mathbf{1350 \text{ m}^3/\text{day}}$$

(ii) Equation to the phreatic surface

$$h = \sqrt{h_1^2 + \frac{x}{L}(H^2 - h_1^2)}$$

$$h = \sqrt{2^2 + \frac{x}{400}(7^2 - 2^2)}$$

or, $\qquad h = \sqrt{4 + 0.1125x}$

The GWT at 100 m from the face of the gallery is

$$h = \sqrt{4 + 0.1125 \times 100} = \mathbf{3.78 \text{ m}}$$

or drawdown $s = H - h = 7 - 3.78 = 3.22$ m.

(iii) The maximum drawdown at the gallery is

$$s_{max} = H - h_1 = 7 - 2 = \mathbf{5 \text{ m}}$$

(c) *Aquifer with recharge*

If the aquifer is being recharged by rainfall and if the steady infiltration rate be P (in suitable units), Fig. 4.15,

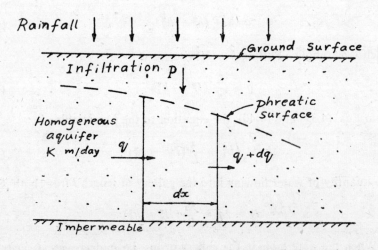

Fig. 4.15 Flow in an unconfined aquifer with rainfall.

Aquifer Properties and Ground Water Flow

$$dq = P \cdot dx \quad \ldots(i)$$

$$q = -Kh\frac{dh}{dx} = -\frac{1}{2}K\frac{d(h^2)}{dx}$$

$$\frac{dq}{dx} = -\frac{1}{2}K\frac{d^2(h^2)}{dx^2} = P, \text{ from (i)}$$

Therefore

$$\frac{d^2(h^2)}{dx^2} = -\frac{2P}{K} \qquad (4.43)$$

Further, $\dfrac{dq}{dx} = -K\dfrac{d}{dx}\left(h\dfrac{dh}{dx}\right) = P$ $\qquad (4.44)$

The solution is $Kh^2 + Px^2 = c_1 x + c_2$.
From the boundary conditions at $x = 0$, $h = h_1$ and at $x = L$, $h = h_2$, the equation for the phreatic surface is

$$h = \sqrt{h_1^2 - \frac{(h_1^2 - h_2^2)}{L}x + \frac{P}{K}(L-x)x} \qquad (4.45)$$

$$q_x = q_0 + Px, \qquad (4.46)$$

where q_0 is at $x = 0$.
Integration of Eq. (4.45) yields

$$q_0 = K\frac{h_1^2 - h_2^2}{2L} - \frac{PL}{2}$$

From Eq. (4.46)

$$q_x = K\frac{h_1^2 - h_2^2}{2L} - P\left(\frac{L}{2} - x\right) \qquad (4.47)$$

If there is no recharge by rain, then

$$\frac{dq}{dx} = 0, \quad \frac{d^2(h^2)}{dx^2} = 0$$

and q, h are given by Eqs. (4.36) and (4.37), respectively.

Example 4.16: Rainfall at the rate 10 mm/hr falls on a strip of land 1 km wide laying between two parallel canals with 2 m difference in their water surface levels, Fig. 4.16. It is underlain by a horizontal impermeable stratum at 10 m below the water surface of the lower canal. Assuming a permeability of 12 m/day with vertical boundaries and all the rainfall infilters into the soil, compute the discharge per metre length into both the canals.

Solution

$$q = AKi = -(h \cdot 1)K\frac{dh}{dx} = -\frac{1}{2}K\frac{d(h^2)}{dx}$$

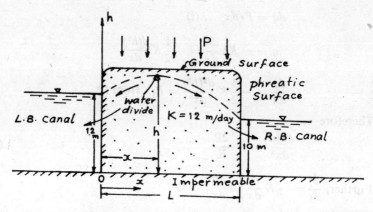

Fig. 4.16 Aquifer between parallel canals recharged by rainfall, Example 4.16.

From Eq. (4.43), $\dfrac{d^2(h^2)}{dx^2} = -\dfrac{2P}{K}$

$$\dfrac{d(h^2)}{dx} = -\dfrac{2P}{K}x + c_1$$

$$h^2 = -\dfrac{P}{K}x^2 + c_1 x + c_2$$

Taking a reference origin as 0, Fig. 4.16,

when $x = 0$, $h = 12$: $\quad 12^2 = -0 + 0 + c_2, \quad c_2 = 144$

when $x = 1000$, $h = 10$: $\quad 10^2 = -\dfrac{0.01 \times 24}{12} \times 10^6 + 1000\, c_1 + 144$

$$c_1 = 20$$

$$q = -\dfrac{1}{2}K\dfrac{d(h^2)}{dx} = -\dfrac{1}{2}K\left(-\dfrac{2P}{K}x + c_1\right)$$

At $x = 0$, $\quad q_{\text{L.B. Canal}} = -\dfrac{12}{2}(-0 + 20) = -\mathbf{120\ m^3/day/m}$

the negative sign indicates flow towards left.

At $x = 1000$,

$$q_{\text{R.B. Canal}} = -\dfrac{12}{2}\left(-\dfrac{2(0.01 \times 24)}{12} \times 1000 + 20\right) = \mathbf{120\ m^3/day/m}$$

So, there is a discharge of 120 m³/day into both the canals, for each metre length of the strip of land.

Note (i): Recharge into land = Discharge into both the canals
$$= (0.01 \times 24)(1000 \times 1) = 120 + 120 = 240\ \text{m}^3/\text{d}/\text{m}$$
$$= 240\ \text{m}^3/\text{d}/\text{m}$$

Note (ii): If there were no recharge by rain, seepage from the upper canal to the lower canal, Eq. (4.36),

$$q = \frac{K}{2L}(h_1^2 - h_2^2) = \frac{12}{2 \times 1000}(12^2 - 10^2) = 0.264 \text{ m}^3/\text{day/m}$$

Note (iii): With recharge, the equation to phreatic surface, Eq. (4.45),
$$h = \sqrt{144 + 19.956x - 0.02x^2}$$
and the water table divide is obtained by putting $dh/dx = 0$,
$$x = 498.9 \text{ m}$$
where $\qquad h = 71.6 \text{ m}$

(d) *Steady flow in a confined aquifer of constant thickness*, Fig. 4.17
Discharge per unit width
$$q = (b \cdot 1) V = b\left(-\frac{d\phi}{ds}\right) = -b \cdot K \frac{dh}{dx}$$
For steady flow,
$$\frac{dq}{dx} = -bK \frac{d^2h}{dx^2} = 0$$
since b and K have finite values,
$$\frac{d^2h}{dx^2} = 0$$
which has the solution
$$h = c_1 x + c_2$$

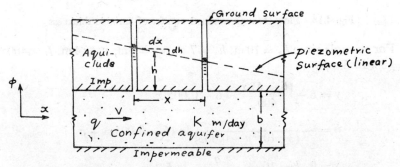

Fig. 4.17 Unidirectional flow in a confined aquifer of constant thickness.

Selecting the datum such that $h = 0$ when $x = 0$, and also from
$$\frac{dh}{dx} = -\frac{q}{bK} = -\frac{V}{K}$$
or
$$h = -\frac{V}{K} x \qquad (4.48)$$
which indicates that the head decreases linearly with x, i.e. the direction of flow. Knowing the piezometric levels h_1 and h_2, X apart, the discharge per unit width of the aquifer,
$$q = b \cdot K \frac{h_1 - h_2}{X} \qquad (4.49)$$

(e) *Steady flow in a confined aquifer of variable thickenss*, Fig. 4.18
Discharge per unit width

$$q = V(y \cdot 1) = -\frac{\partial \phi}{\partial s} y = -K \frac{dh}{dx} y$$

$$dh = -\frac{q}{Ky} dx, \qquad y = a + bx, \text{ say}$$

$$h = -\frac{q}{Kb} \ln(a + bx) + c \qquad (4.50)$$

q and c can be determined by the boundary conditions.

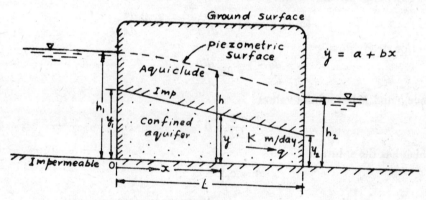

Fig. 4.18 Flow through confined aquifer of variable thickness.

For example given $h_1 = 10$ m, $h_2 = 7$ m, $y_1 = 6$ m, $y_2 = 4$ m, $L = 400$ m, $K = 3 \times 10^{-6}$ m/s,

$$y = 6 - \frac{(6-4)x}{400} = 6 - 0.005 x$$

$$h = -\frac{q}{(3 \times 10^{-6})(-0.005)} \ln(6 - 0.005x) + c$$

$$x = 0, \ h = 10 = \frac{q}{1.5 \times 10^{-8}} \ln 6 + c$$

$$x = 400, \ h = 7 = \frac{q}{1.5 \times 10^{-8}} \ln 4 + c$$

subtracting, $3 = \dfrac{q}{1.5 \times 10^{-8}} \ln \dfrac{6}{4}$

or $\qquad q = 1.11 \times 10^{-7}$ m³/s per width, or metre m²/s.

Alternatively, by taking average values

$$q = KiA = (3 \times 10^{-6}) \frac{10-7}{400} \left(\frac{(6+4)}{2} \times 1\right) = 1.125 \times 10^{-7} \text{ m}^2/\text{s}$$

which is only 1.35% higher than the previous result.

Ground Water Theory

Consider the flow through an elemental prism of volume $\Delta x, \Delta y, \Delta z$ as shown in Fig. 4.19.

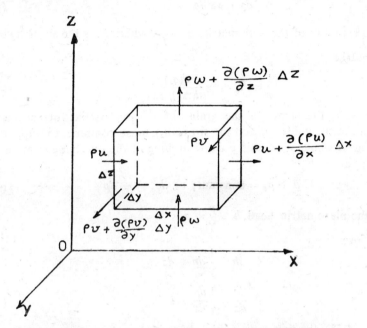

Fig. 4.19 Flow of ground water through an elemental prism.

The rate of change of mass flow rate through the three pairs of faces (net inward flux) is given by

$$-\left[\frac{\partial(\rho u)}{\partial x} + \frac{\partial(\rho v)}{\partial y} + \frac{\partial(\rho w)}{\partial z}\right]\Delta x \Delta y \Delta z = \frac{\partial(\Delta M)}{\partial t} \quad (4.51)$$

The mass of water in the elemental prism

$$\Delta M = \rho n \Delta x \Delta y \Delta z \quad (4.52)$$

where n is the porosity of the aquifer; and

$$\frac{\partial(\Delta M)}{\partial t} = \frac{\partial \rho}{\partial t} n \Delta x \Delta y \Delta z + \frac{\partial(n \Delta z)}{\partial t} \rho \Delta x \Delta y \quad (4.53)$$

assuming no lateral strain.

If ρ is the density of water,

$$\frac{d(\Delta V_w)}{\Delta V_w} = -\frac{d\rho}{\rho} \quad (4.54)$$

Compressibility of water,

$$\beta = -\frac{d(\Delta V_w)/\Delta V_w}{dp} \quad (4.55)$$

Therefore
$$\beta = \frac{d\rho}{\rho\, dp}$$

Therefore
$$d\rho = \rho\beta\, dp \qquad (4.56)$$

Compressibility of the soil matrix, $\alpha = \dfrac{1}{E_s}$, where E_s is the elasticity of the soil matrix

$$\frac{1}{\alpha} = E_s = -\frac{d(\sigma_z)}{d(\Delta z)/\Delta z} \qquad (4.57)$$

where σ_z is the stress in the grains of the soil matrix; further the fluid pressure in the voids (p) and the stress (σ_z) must combine to support the total mass above the unit area if arching of the overlying strata is neglected, i.e.

$$p + \sigma_z = \text{constant}; \quad \text{or} \quad d\sigma_z = -\,dp \qquad (4.58)$$

Also the piezometric head $h = \dfrac{p}{\gamma} + z$

Therefore
$$dh = \frac{1}{\gamma}\, dp + dz$$

$$\frac{dh}{dz} = \frac{1}{\gamma}\frac{dp}{dz} + 1 \qquad (4.59)$$

In Eq. (4.53), the first term on the right side

$$\frac{\partial \rho}{\partial t} n\, \Delta x\, \Delta y\, \Delta z = n\rho\beta \frac{\partial p}{\partial t} \Delta x \Delta y \Delta z \qquad \text{from Eq. (4.56)}$$

Also volume of the soil grains

$$V_s = (1 - n)\Delta x\, \Delta y\, \Delta z$$

$$dV_s = [d(\Delta z) - d(n\Delta z)]\Delta x\, \Delta y = 0$$

since the compressibility of the soil grains is small compared to the compressibility of the water and the change in porosity.

Therefore
$$d(\Delta z) = d(n\Delta z)$$

and the second term on the right side in Eq. (4.53)

$$\frac{\partial (n\Delta z)}{\partial t} \rho \Delta x \Delta y = \frac{\partial (\Delta z)/\Delta z}{\partial t} \rho \Delta x\, \Delta y\, \Delta z$$

$$= \rho\alpha \frac{\partial p}{\partial t} \Delta x\, \Delta y\, \Delta z \qquad \text{from Eq. (4.57) and (4.58)}$$

Therefore, Eq. (4.53) becomes

$$\frac{\partial (\Delta M)}{\partial t} = (\alpha + n\beta)\, \rho\, \frac{\partial p}{\partial t} \Delta x\, \Delta y\, \Delta z \qquad (4.60)$$

Aquifer Properties and Ground Water Flow

from Eqs. (4.51) and (4.60),

$$-\left[\frac{\partial(\rho u)}{\partial x} + \frac{\partial(\rho v)}{\partial y} + \frac{\partial(\rho w)}{\partial z}\right] = (\alpha + n\beta)\,\rho\,\frac{\partial p}{\partial t} \qquad (4.61)$$

From Darcy's law,

$$u = -K_x\frac{\partial h}{\partial x}, \quad v = -K_y\frac{dh}{\partial y}, \quad w = -K_z\frac{\partial h}{\partial z}$$

where K_x, K_y and K_z are the coefficients of permeability in the x, y and z directions, respectively. Substituting in Eq. (4.61)

$$K_x\left(\frac{\partial \rho}{\partial x}\cdot\frac{\partial h}{\partial x} + \rho\frac{\partial^2 h}{\partial x^2}\right) + K_y\left(\frac{\partial \rho}{\partial y}\cdot\frac{\partial h}{\partial y} + \rho\frac{\partial^2 h}{\partial y^2}\right)$$

$$+ K_z\left(\frac{\partial \rho}{\partial z}\cdot\frac{\partial h}{\partial z} + \rho\frac{\partial^2 h}{\partial z^2}\right) = (\alpha + n\beta)\rho\frac{\partial p}{\partial t} \qquad (4.62)$$

The term $K_z\dfrac{\partial \rho}{\partial z}\cdot\dfrac{\partial h}{\partial z} = K_z\,\rho\beta\,\dfrac{\partial p}{\partial z}\cdot\dfrac{\partial h}{\partial z}$ from Eq. (4.56)

$$= K_z\,\rho\beta\gamma\left(\frac{\partial h}{\partial z} - 1\right)\frac{\partial h}{\partial z} \qquad \text{from Eq. (4.59)}$$

$$= K_z\,\rho^2\beta g\left[\left(\frac{\partial h}{\partial z}\right)^2 - \frac{\partial h}{\partial z}\right]$$

Eq. (4.62) can be written as

$$\rho^2 \beta g\left[K_x\left(\frac{\partial h}{\partial x}\right)^2 + K_y\left(\frac{\partial h}{\partial y}\right)^2 + K_z\left\{\left(\frac{\partial h}{\partial z}\right)^2 - \frac{\partial h}{\partial z}\right\}\right]$$

$$+ \rho\left(K_x\frac{\partial^2 h}{\partial x^2} + K_y\frac{\partial^2 h}{\partial y^2} + K_z\frac{\partial h^2}{\partial z^2}\right) = (\alpha + n\beta)\rho^2 g\frac{\partial h}{\partial t}$$

Neglecting the terms which are products of the differentials as of higher order and also $\dfrac{\partial h}{\partial z} \ll 1$ (hence the term $K_z\,\rho^2\beta g\,\dfrac{\partial h}{\partial z}$ becoming very small relative to the remaining terms), Eq. (4.62) becomes

$$K_x\frac{\partial^2 h}{\partial x^2} + K_y\frac{\partial^2 h}{\partial y^2} + K_z\frac{\partial^2 h}{\partial z^2} = (\alpha + n\beta)\,\gamma\,\frac{\partial h}{\partial t} \qquad (4.63)$$

for anisotropic aquifer.

If the aquifer is isotropic with respect to permeability, i.e.

$$K_x = K_y = K_z = K,$$

than Eq. (4.63) can be written as

$$K\left(\frac{\partial^2 h}{\partial x^2} + \frac{\partial^2 h}{\partial y^2} + \frac{\partial^2 h}{\partial z^2}\right) = (\alpha + n\beta)\,\gamma\,\frac{\partial h}{\partial t}$$

or

$$\frac{\partial^2 h}{\partial x^2} + \frac{\partial^2 h}{\partial y^2} + \frac{\partial^2 h}{\partial z^2} = (\alpha + n\beta)\,\frac{\gamma b}{Kb}\,\frac{\partial h}{\partial t}$$

Defining the coefficient of storage as

$$S = (\alpha + n\beta)\gamma b \qquad (4.64)$$

Transmissibility of the aquifer $T = Kb$ and

$$\nabla^2 h = \frac{\partial^2 h}{\partial x^2} + \frac{\partial^2 h}{\partial y^2} = \frac{\partial^2 h}{\partial z^2}$$

Eq. (4.51) finally becomes

$$\nabla^2 h = \frac{S}{T}\frac{\partial h}{\partial t} \qquad (4.65)$$

which is the equation for the piezometric head for unsteady flow in a saturated, confined, isotropic aquifer. For steady flow, the equation becomes the familiar Laplace equation [see Eq. (4.31)]

$$\nabla^2 h = 0 \qquad (4.66)$$

The coefficient of storage (S) is defined as the volume of water released from storage from a vertical column of aquifer of unit cross-sectional area under a unit decline of piezometric head. The first term ($\alpha\gamma b$) gives the fraction attributable to the compressibility of the aquifer skeleton and the second term ($n\beta\gamma b$) that attributable to the expansibility of water. In water table aquifers the coefficient of storage is given by (Hantush, 1964)

$$S = S_y + (\alpha + n\beta)\gamma b \qquad (4.67)$$

where S_y is the specific yield of the aquifer and b is the saturated thickness of the aquifer. Usually $(\alpha + n\beta)\gamma b$ is negligible in comparison to S_y, and S under watertable conditions for all practical purposes is the specific yield.

Relationship between Storage Coefficient, Barometric and Tidal Efficiencies

$$\frac{d(\Delta V_w)/\Delta V_w}{dp} = -\beta. \quad \text{Therefore} \quad \frac{d(\Delta V)_w}{\Delta V_w} = -\beta\, dp$$

$$\frac{d(\Delta V_w)}{\Delta V_w} = \frac{d(\Delta V)}{n\Delta V}, \quad \frac{d(\Delta V)}{\Delta V} = \frac{d(\Delta z)}{\Delta z} = -\alpha\, d\,(\sigma_z) \qquad \text{from Eq. (4.57)}$$

Therefore

$$-\beta\, dp = -\frac{\alpha}{n} d\,(\sigma_z)$$

The negative sign indicates a decrease in volume accompanying an increase in stress.

Therefore

$$\frac{dp}{d\sigma_z} = \frac{\alpha}{n\beta} \qquad (4.68)$$

BAROMETRIC EFFICIENCY (BE)

$$p + \sigma_z = p_a + \text{constant}$$

Aquifer Properties and Ground Water Flow

where p_a is the force of atmospheric pressure acting on the aquitard overlying the aquifer

$$dp + d(\sigma_z) = dp_a$$

$$dp - dp_a = - d(\sigma_z)$$

The pressure due to the column of water in the well above the top of the artesian aquifer plus the pressure of atmosphere is balanced by the pressure of water in the aquifer. Therefore, the change of water pressure is given by

$$dp = dp_a + \gamma\, dh$$

where h is the elevation of the water surface in the well. The barometric efficiency (BE) is the ratio of the change in water level in the well to the change in the atmospheric pressure head, i.e.*

$$BE = \frac{dh}{dp_a/\gamma} = \frac{(dp - dp_a)/\gamma}{dp_a/\gamma} = \frac{-d(\sigma_z)}{dp + d(\sigma_z)} = \frac{-1}{1 + \dfrac{dp}{d(\sigma_z)}}$$

$$= \frac{-1}{1 + \dfrac{\alpha}{n\beta}} \qquad \text{from Eq. (4.68)}$$

Therefore

$$BE = \frac{-1}{1 + \dfrac{\alpha}{n\beta}} \qquad (4.69)$$

The negative sign indicates the opposite nature of the changes in atmospheric pressure and the water level in the well. Unconfined aquifers have practically no barometric efficiency.

From Eq. (4.64), $\qquad S = (\alpha + n\beta)\gamma\, b$

Therefore

$$S = n\beta\gamma\, b \cdot \frac{-1}{BE} \qquad (4.70)$$

TIDAL EFFICIENCY (TE)

Water levels in artesian aquifers respond to tidal fluctuations, i.e.

$$p + \sigma_z = \gamma H + \text{constant}$$

where H is the stage of the tide (corrected for density if necessary).

---

*BE $= \dfrac{\Delta s_w}{\Delta h_b} \times 100\%$. $h_b =$ water barometer reading, $s_w =$ drawdown in the well $\approx 10-75\%$

Therefore
$$dp + d(\sigma_z) = \gamma\, dH$$

The change of pressure in the aquifer
$$dp = \gamma\, dh$$

The tidal efficiency is the ratio of a change of water level in the well to a change in the stage of the tide, i.e.

$$\mathrm{TE} = \frac{dh}{dH} = \frac{\gamma dh}{\gamma dH} = \frac{dp}{dp + d(\sigma_z)} = \frac{dp/d(\sigma_z)}{1 + \dfrac{dp}{d(\sigma_z)}} = \frac{\alpha/n\beta}{1 + \dfrac{\alpha}{n\beta}} \qquad \text{from Eq. (4.68)}$$

Therefore
$$\mathrm{TE} = \frac{\alpha/n\beta}{1 + \dfrac{\alpha}{n\beta}} \tag{4.71}$$

The sum of the barometric and tidal efficiencies equals unity, i.e.
$$\mathrm{BE} + \mathrm{TE} = 1 \tag{4.72}$$

Solutions of the Differential Equation (4.65) for Unsteady Confined Flow

Radial flow to a well in a confined aquifer, Fig. 4.20, permits the transformation of Eq. (4.65) to the polar coordinate form

$$\frac{\partial^2 h}{\partial r^2} + \frac{1}{r}\frac{\partial h}{\partial r} = \frac{S}{T}\frac{\partial h}{\partial t} \tag{4.73}$$

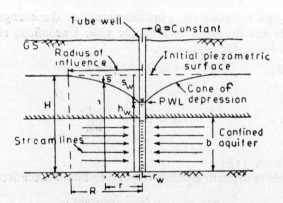

Fig. 4.20 Well in a confined aquifer.

The solution of this equation, when referred to an aquifer of infinite thickness, is given by

$$h = H - \frac{Q}{4\pi T}\int_{\frac{r^2 S}{4Tt}}^{\infty} \frac{e^{-u}}{u}\, du \tag{4.74}$$

Aquifer Properties and Ground Water Flow

The dummy variable u is often written as

$$u = \frac{r^2 S}{4Tt} \qquad (4.75)$$

Q is the constant well discharge and $H - h = s$. The integral is a function of the lower limit and is known as the exponential integral and is termed as 'well function' $W(u)$

Therefore

$$s = \frac{Q}{4\pi T} W(u) \qquad (4.76)$$

where u is the argument of the well function. Equation (4.76) is known as the non-equilibrium or Theis equation (Theis, 1935). The integral $W(u)$ can be expanded as a convergent series as

$$W(u) = -0.577216 - \log_e u + u - \frac{u^2}{2.2!} + \frac{u^3}{3.3!} - \frac{u^4}{4.4!} + \cdots \qquad (4.77)$$

It can be seen from Eq. (4.76) that the drawdown varies with the well function as the argument of the well function varies with the variables r, t and the aquifer constants S and T. The relationships above mean that a plot of s versus r^2/t is similar in shape and slope to a plot of $W(u)$ versus u, called the 'type curve', Fig. 4.21. The region of similarity can be found by plotting the pump test data of s versus r^2/t on log-log graph paper of the same type and physical size as the type cure and superimposing by keeping the axes of the graphs parallel. Any common point (match point)* on the two graphs, Fig. 4.22, gives corresponding values of s, r^2/t, $W(u)$ and u that can be used in Eqs. (4.75) and (4.76) to compute the values of T and S.

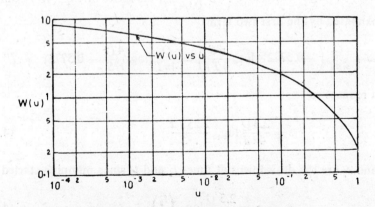

Fig. 4.21 Type curve for the well function.

---

*Match point may be P (on the overlapping portion) or any convenient point like A for which $W(u) = 1$, $u = 0.1$ (for easy calculation).

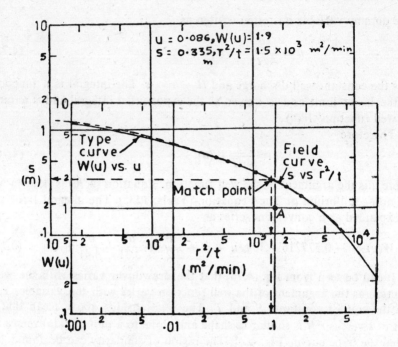

Fig. 4.22 Superposition for match point coordinates (Theis).

Table 1.5 gives the values of $W(u)$ versus u.

Approximate solution (Jacob, 1946, 1950): For small values of u ($u \leqslant 0.01$) or large values of t $\left(t \geqslant \dfrac{r^2 S}{0.04 T} \right)$, the series of Eq. (4.77) can be approximated by the first two terms

$$s = \frac{Q}{4\pi T}\left(-0.5772 - \log_e \frac{r^2 S}{4Tt}\right) = \frac{Q}{4\pi T}\left(\log_e \frac{4Tt}{r^2 S} - 0.5772\right)$$

which reduces to

$$s = \frac{2.3\,Q}{4\pi\,T} \log_{10} \frac{2.25\,T\,t}{r^2 S} \tag{4.78}$$

If s_1 and s_2 are the drawdowns at times t_1 and t_2 since pumping started

$$s_2 - s_1 = \frac{2.3\,Q}{4\pi\,T} \log_{10}\left(\frac{t_2}{t_1}\right) \tag{4.79}$$

If the time-drawdown data on a pumping well is plotted on a semi-log paper, Fig. 4.23 and for convenience t_1 and t_2 are chosen one log cycle

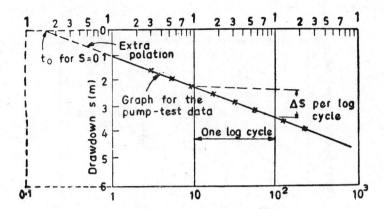

Fig. 4.23 Time-drawdown graph (Jacob's).

apart, $\log_{10} \frac{t_2}{t_1} = 1$ and if $s_2 - s_1 = \Delta s$, i.e. the drawdown difference per log cycle of t, then from Eq. (4.79)

$$T = \frac{2.3\, Q}{4\pi\, \Delta s} \qquad (4.80)$$

From Eq. (4.78), $s = 0$, when $\log_{10} \frac{2.25\, T t}{r^2 S} = 0$, i.e. when

$$\frac{2.25\, T t}{r^2 S} = 1$$

By extrapolating the straight line of the semi-log plot to intersect the zero-drawdown axis, t_0, the time for $s = 0$ can be noted and S can be computed as

$$S = \frac{2.25\, T t_0}{r^2} \qquad (4.81)$$

Similarly from the semi-log plot of distance-drawdown data (when there are many observation wells), Fig. 4.24,

$$s_2 - s_1 = \frac{2.3\, Q}{4\pi\, T} \log_{10}\left(\frac{r_1}{r_2}\right)^2 = \frac{2.3\, Q}{2\pi\, T} \log_{10}\left(\frac{r_1}{r_2}\right)$$

$$T = \frac{2.3\, Q}{2\pi\, \Delta s} \qquad (4.82)$$

where Δs is the drawdown difference log cycle of r. By extrapolating the straight line plot to intersect the zero-drawdown axis, the distance r_0 for $s = 0$ can be noted and S can be computed as

$$S = \frac{2.25\, T t}{r_0^2} \qquad (4.83)$$

On the other hand if the drawdown in several observation wells at

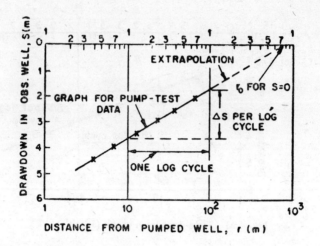

Fig. 4.24 Distance-drawdown graph (Jacob's).

different times are known, a semi-log graph of time-distance-drawdown data, i.e. s versus t/r^2 (composite graph) can be drawn, Fig. 5.7 and s obtained per log cycle of t/r^2. Then

$$S = \frac{2.3}{4\pi} \frac{Q}{\Delta s} \tag{4.84}$$

By extrapolating the straight line plot to intersect the zero-drawdown axis.

$$S = 2.25 \, T \left(\frac{t}{r^2}\right)_0 \tag{4.85}$$

where $(t/r^2)_0$ is the value when $s = 0$.

Correction for very thin aquifers under water table conditions (Jacob, 1944): Under water table conditions the drawdown of the water level by a discharging well reduces the saturated thickness of the aquifer. If this reduction in thickness is appreciable, the transmissibility is not constant but decreases with time. From Fig. 4.25,

$$h = H - s$$

$$h_2^2 - h_1^2 = (H - s_2)^2 - (H - s_1)^2$$

$$= H^2 - 2Hs_2 + s_2^2 - H^2 + 2Hs_1 - s_1^2$$

$$= 2H\left[(s_1 - \frac{s_1^2}{2H}) - (s_2 - \frac{s_2^2}{2H})\right]$$

$$Q = \frac{\pi K (h_2^2 - h_1^2)}{2.303 \log_{10} \frac{r_2}{r_1}}$$

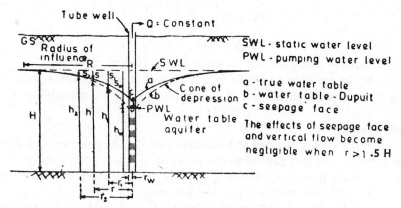

Fig. 4.25 Water table aquifer.

which is the well known Thiem or Theis formula for steady flow conditions (see Eq. 5.3)

$$Q = \frac{\pi K 2H\left[(s_1 - \frac{s_1^2}{2H}) - (s_2 - \frac{s_2^2}{2H})\right]}{2.303 \log_{10} \frac{r_2}{r_1}}, \qquad T = KH$$

$$Q = \frac{2\pi T (s_1' - s_2')}{2.303 \log_{10} \frac{r_2}{r_1}} = \frac{2\pi T \Delta s'}{2.303} = 1.72 T \Delta s'$$

or
$$T = \frac{Q}{1.72 \Delta s'} \qquad (4.86)$$

where $\Delta s'$ is the drawdown difference (corrected) per log cycle of r. See Example 5.4, Fig. 5.5.

Therefore

$$T = \frac{2.303 \, Q \, \log_{10} \frac{r_2}{r_1}}{2\pi \left[(s_1 - \frac{s_1^2}{2H}) - (s_2 - \frac{s_2^2}{2H})\right]} \qquad (4.87)$$

If the drawdowns s_1 and s_2 are very small compared to the original saturated thickness H, the correction fraction may be omitted, and Eq. (4.87) reduces to

$$T = \frac{2.303 \, Q \, \log_{10} \frac{r_2}{r_1}}{2\pi (s_1 - s_2)}, \qquad s_1 - s_2 = h_2 - h_1$$

$$= \frac{2.303 \, Q \, \log_{10} \frac{r_2}{r_1}}{2\pi (h_2 - h_1)}$$

or
$$= \frac{2.303 \, Q \, \log_{10} \frac{R}{r_w}}{2\pi \, (H - h_w)}$$

which compares with Eq. (5.5) for small s_w, i.e. $H + h_w \approx 2H$.

One of the basic assumptions in the derivation of the Thiem and the Theis formula is the stipulation of a constant value of transmissibility.

Theis Recovery

Suppose a well is pumped at a constant rate Q. The pumping is stopped after a time t_1 and it is desired to compute the residual drawdown s' after a time t' since pumping stopped, Fig. 4.26. The recovery of peizometric head can be determined by considering a negative discharge (recharge, in a sense) for the time t'. For small r and large t', Jacob's approximation can be made, and the residual drawdown s' after time $t_1 + t' = t$ since pumping started, can be obtained from Eq. (4.78) as

$$s' = \frac{2.303 \, Q}{4\pi \, T} \log_{10} \frac{2.25 \, T \, t}{r^2 s} - \frac{2.303 \, Q}{4\pi \, T} \log_{10} \frac{2.25 \, T \, t'}{r^2 s}$$

or,
$$s' = \frac{2.303 \, Q}{4\pi \, T} \log_{10} \left(\frac{t}{t'}\right)$$

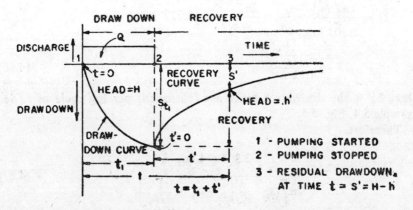

Fig. 4.26 Drawdown and recovery of G.W.L. in the vicinity of a pumping well.

In Theis recovery method the residual drawdowns in a pumped well or a nearby observation well are measured at different time intervals after pumping is stopped. From a semi-log plot of s' versus t/t', the transmissibility T can be obtained similar to Jacob's method as

$$T = \frac{2.303 \, Q}{4\pi \, \Delta s'} \tag{4.88}$$

where $\Delta s'$ is the drawdown difference per log cycle of t/t', see Example 5.9.

The storage coefficient S can be determined from the value of drawdown

Aquifer Properties and Ground Water Flow

at the time the pumping was stopped, from Eq. (4.78). S can also be determined from the time-drawdown plot (Jacob's) of the data during pumping, Eq. (4.81).

The drawdowns resulting from variable pumping rates can be determined by superposition. For example if a well is pumped at the rate Q_1 for time t_1 and then pumping rate is increased to Q_2 for a further time t_2, the drawdown at the end of time $t_1 + t_2$ is given approximately by

$$s \simeq \frac{2.303\,Q_1}{4\pi\,T} \log_{10} \frac{2.25\,T(t_1 + t_2)}{r^2 S} + \frac{2.303\,(Q_2 - Q_1)}{4\pi\,T} \log_{10} \frac{2.25\,T\,t_2}{r^2 S}$$

or accurately by

$$s = \frac{Q_1}{4\pi\,T} \cdot W\left(\frac{r^2 S}{4T(t_1 + t_2)}\right) + \frac{Q_2 - Q_1}{4\pi\,T} \cdot W\left(\frac{r^2 S}{4T\,t_2}\right) \qquad (4.89)$$

for which values of $W(u)$ for corresponding values of u can be obtained from Table 5.1.

Barrier and Recharge Boundaries—Image Wells

Assume two observation wells at distances r_1 and r_2 from a pumping well. From Eq. (4.78) the drawdowns in the two wells are

$$s_1 = \frac{2.303\,Q}{4\pi\,T} \log_{10} \frac{2.25\,T\,t_1}{r_1^2 S}$$

$$s_2 = \frac{2.303\,Q}{4\pi\,T} \log_{10} \frac{2.25\,T\,t_2}{r_2^2 S}$$

If the drawdowns were to be equal, i.e. $s_1 = s_2$

$$\log_{10} \frac{2.25\,T\,t_1}{r_1^2 S} = \log_{10} \frac{2.25\,T\,t_2}{r_2^2 S}$$

or
$$\frac{t_1}{r_1^2} = \frac{t_2}{r_2^2} \qquad (4.90)$$

For the drawdowns to be zero for the two observation wells

$$\frac{2.25\,S\,t_1}{r_1^2 S} = \frac{2.25\,T\,t_2}{r_2^2 S} = 1$$

or
$$\frac{t_1}{r_1^2} = \frac{t_2}{r_2^2}$$

which is the same as Eq. (4.90). Hence it follows that for a given aquifer the times of occurrence of zero drawdown or of equal drawdown vary directly as the squares of the distances of the observation wells from the pumping well, and are independent of the rate of pumping. This principle is also known as the 'law of times', i.e.

$$\frac{t_1}{r_1^2} = \frac{t_2}{r_2^2} = \ldots = \frac{t_n}{r_n^2} \qquad (4.91)$$

If the time intercept (t_p) for a given drawdown in an observation well at a distance r from the pumping well, Fig. 4.27, is known and if the time intercept (t_i) of an equal amount of divergence of the time-drawdown curve caused by the effect of the image well (which simulates the barrier and recharge boundaries) is also known, Fig. 5.24, then from the law of times it follows that

$$\frac{t_p}{r^2} = \frac{t_i}{r_i^2}$$

or
$$r_i = r\sqrt{\frac{t_i}{t_p}} \tag{4.92}$$

where r_i is the distance from image well to the observation well.

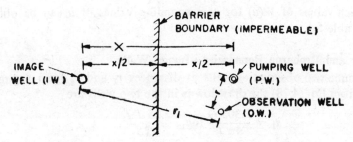

Fig. 4.27 Image well and aquifer boundary location.

Alternatively, if the time (t_0) for zero drawdown is obtained from the Jacob's time-drawdown graph and t is the time at which a charge of slope is indicated (due to the image well coming into effect), then from the law of times, Fig. 5.24,

$$\frac{t_0}{r^2} = \frac{t}{r_i^2} \quad \text{or} \quad r_i = r\sqrt{\frac{t}{t_0}} \tag{4.92a}$$

The distance of the aquifer boundary from the pumping well in both cases is taken as nearly equal to $r_i/2$. If the drawdowns are measured in the pumping well itself, $r = r_w$, $r_i = x =$ distance between the pumping well, and image well, and the distance of the aquifer boundary from the pumping well is exactly equal to $x/2$ (Fig. 5.10 b).

The practical application of the ground water theory and equations derived above to actual field problems are discussed in great detail in the next chapter on 'Well Hydraulics'.

Aquifer Properties and Ground Water Flow

QUIZ 4

I. State whether 'true' or 'false'; if false give the correct statement:
   1. The specific yield of a formation may be more than its porosity
   2. Land subsidence may be caused by excessive pumpage of artesian aquifers
   3. The transmissibility of a confined aquifer depends upon the depth of the water table while that of the water table aquifer does not
   4. Flow through a coarse gravelly sand aquifer is entirely laminar
   5. Closer the water table contour spacings, higher the gradients of subsoil flow and higher the permeabilities, and hence more promising are the well sites
   6. Storage coefficient is the same as the specific yield for water table aquifer

II. Select the correct answer(s):   [×, √, ×, ×, ×, √]
   1. The annual ground water storage in an area is equal to
      (a) land area × drop in ground water table
      (b) land area × rise in ground water table × porosity of formation
      (c) involved area of aquifer × maximum seasonal fluctuation in ground water table × specific yield of aquifer
   2. The velocity at which a tracer would move is
      (a) the same as the seepage velocity given by Darcy's law
      (b) $1/n$th times the seepage velocity, where n is the porosity (fraction) of the formation
   3. The seepage (ground water flow) is calculated as
      (a) cross-sectional area of aquifer × slope of ground water table × permeability of the aquifer
      (b) cross-sectional area of aquifer × slope of ground water table × transmissibility of the aquifer
      (c) width of aquifer × slope of piezometric surface × transmissibility of aquifer (artesian)
      (d) width of aquifer × slope of ground water table × permeability of aquifer
      (e) none of the above methods
   4. For laminar flow in a medium sand aquifer, the Reynolds number is
      (a) <2000
      (b) <1
      (c) 1 to 10
      (d) <50.
   5. The storage coefficient of an aquifer is
      (a) attributable to the compressibility of the aquifer skeleton and the pore water
      (b) attributable to the compressibility of the aquifer skeleton and expansibility of the pore water
      (c) attributable to the expansibility of the aquifer skeleton and the pore water
      (d) the volume of water released from a vertical column of aquifer of unit cross-sectional area due to a unit decline of piezometric head
      (e) the volume of water taken into storage in a vertical column of aquifer of unit cross-sectional area due to a unit increase in piezometric head

(f) the volume of water released from storage from the entire aquifer due to a unit decline of piezometric head

[1—c; 2—b; 3—b, c; 4—c; 5—b, d, e]

QUESTIONS

1. In Chandrapalem basin, AP, India, consisting of 22 km² of plains, the maximum fluctuation of ground water table is 3 m. Assuming a specific yield of 16% what is the probable ground water storage? (10.56 M.m²)

2. Calculate the fresh water flow in a coastal aquifer extending to a length of 40 km along the coast, assuming an average permeability of 40 m³/day/m², average thickness of aquifer of 20 m, and the piezometric gradient at 5 m/km.
   (160,000 m³/day)

3. An aquifer averages 50 m in thickness and is 100 ha in area. Determine the volume in ha-m of water available if
   (a) the aquifer is unconfined and is completely drained
   (b) the aquifer is confined and the piezometric head is lowered from 30 m to 10 m above the aquifer
   (c) the aquifer is confined and the piezometric head is lowered 55 m which brings the water table 25 m below the confining layer.
   Assume $S_y = 15\%$, $S = 2 \times 10^{-4}$
   (750, 0.4, 375.6 ha.m)

4. (a) State Darcy's law and its limitations.
   (b) Discuss briefly the three regions that may be distinguished in flow through porous media. Write down the corresponding equations that are applicable.
   (c) A fully penetrating well is pumped at a constant rate of 1020 m³/hr from a confined aquifer of thickness 30 m and average grain diameter 1 mm. What is the domain around the well for which Darcy's law is applicable? Assume that Darcy's law is valid up to $R_e = 6$ and $v_{water} = 1$ c St ($r > 25$ cm)

5. In a homogeneous isotropic confined aquifer of constant thickness of 20 m. effective porosity of 20% and permeability of 15 m/day, two observation wells 1200 m apart indicate piezometric heads of 5.4 m and 3.0 m, respectively, above m.s.l. Assuming uniform flow, average grain diameter of sand 1 mm and v water $= 0.01$ cm²/sec, state
   (a) whether Darcy's law is applicable?
   (b) what is the average flow velocity in pores?
   (Yes, because Re <1, 15 cm/day)

6. Distinguish between 'ground surface contours' and 'water table contours'. Explain how the water table contours are prepared and state their uses.

7. During a falling head permeability test on a soil sample of 10 cm diameter and 20 cm length, the head in the stand pipe of 2cm diameter dropped from 50 cm to 25 cm in 2 min. Determine the premeability of the sample.
   (4.56×10^{-4} cm/sec)

8. During a constant head permeability test on a soil sample of 7 cm diameter and 15 cm length, 90 cc of water was collected in 2 min under a constant head of 50 cm. Determine the permeability of the sample.
   (5.86×10^{-3} cm/sec)

Monthly water levels (below MP in metres) in five observation wells located in an area of 260 km²

Obs. well No.	Polygon area (km²)	Jan.	Feb.	March	April	May	June	July	August	Sept.	Oct.	Nov.	Dec.	Jan.
1.	39	3.60	4.05	4.12	4.57	4.80	4.95	5.02	4.80	4.42	4.20	3.90	3.30	3.00
2.	65	1.80	1.95	2.10	2.32	2.77	2.92	3.00	2.62	2.47	2.40	2.25	2.10	2.25
3.	52	2.70	2.92	3.22	3.37	3.52	3.60	4.20	3.60	3.37	3.15	2.77	2.40	2.55
4.	39	1.50	1.80	1.92	2.17	2.40	2.70	3.15	2.85	2.62	2.32	2.02	1.80	1.95
5.	65	1.35	1.65	1.87	1.80	2.40	2.77	3.22	3.00	2.62	2.18	1.80	1.42	1.50

(Ans: $-14, -23.4, -32.4, -51.2, -62.3, -79.5, -61.40, -47.4, -34.4, -18.9, -1.8, -5.2$M. m³, cumulative)

9. (a) Define the terms:
   (i) Transmissibility
   (ii) Specific yield
   (iii) Storage coefficient.

   (b) 'The coefficient of storage of an artesian aquifer represents the entire thickness of the aquifer, whereas the coefficient of storage of a free aquifer does not'—Explain.

10. From the piezometric-surface map of a non-leaky artesian aquifer with an average coefficient of transmissibility of 6×10^5 lpd/m in the vicinity of a pumping centre the following data are noted:

    Number of flow channels = 12
    Head drop between successive piezometric contours = 3 m

    What is the quantity of water moving through the aquifer into the pumping center?

    (21,600 m^3/day)

11. In an area of 260 km^2, there are five observation wells and monthly water level readings are recorded. The area of the polygon in which each observation well is located and the monthly ground water levels are given in the table (p. 132). Ground water occurs under water table conditions. The average specific yield for the area works out to 20%. Determine the monthly change in ground water storage in the area and plot the cumulative change in storage against the months (Hint: take the ground water storage in January as the datum)

12. Under 30 ha fallow tract of comparatively level sandy land, the water table is kept at an elevation of 1 metre. The value of 'k' for the unsaturated soil above the water table is 9×10^{-8} cm/sec. If the capillary potential in the surface soil is 686 N.m/kg, compute the loss of water by upward capillary flow in cubic metres per square metre per month.

13. The coefficient of storage of an artesian aquifer is 3×10^{-4}. If the thickness of the aquifer is 50 m and the porosity 30%, estimate the fraction of the coefficient of storage attributable to expansibility of water and that attributable to compressibility of the aquifer skeleton. $K_w = 2.1$ GN/m^2.

    (23.3, 76.7%)

14. If the laboratory coefficient of permeability of a sample of soil is 3.2×10^4 lpd/m^2 at 20°C, what would be the permeability value at 30°C?

    (4×10^4 lpd/m^2)

15. An undisturbed soil sample has an oven-dry weight of 524.6 gm. After saturation with kerosene its weight is 628.2 gm. The saturated sample is then immersed in kerosene and displaces 256.3 gm. What is the porosity of the soil sample?

    (40%)

16. In a field test it was observed that a time of 5 hr was required for a tracer to travel from one observation well to another. The wells are 30 m apart and the difference in their water table elevations is 50 cm. Samples of the aquifer between the wells indicate that the porosity is 15%. Compute:
    (a) the coefficient of permeability of the aquifer assuming it to be homogeneous.
    (b) the actual velocity of flow as indicated by the tracer.
    (c) the seepage velocity.
    (d) Reynolds number for the flow assuming an average grain size of 1 mm and ν_{water} at 27°C = 0.008 stoke.

    (12.96×10^5 lpd/m^2, 144 m/day, 21.6 m/day, 0.3)

Aquifer Properties and Ground Water Flow

17. Describe various hazards consequential to augmentation and depletion of ground water.
18. Determine the storage coefficient of an aquifer from the following data:
    porosity $= 30\%$
    thickness of aquifer $= 25$ m
    bulk modulus of water, $K_w = 2.1$ GN/m^2
    Modulus of elasticity of the soil skeleton, $E_s = 3 \times 10^8$ N/m^2

    (8.55×10^{-4})
19. Determine the effective size, mean size, and the uniformity coefficient of the aquifer material from the following data from a sieve analysis:

Sieve opening (mm)	Percentage finer by weight
4	95.5
2	92.3
1.4	86.0
1.0	78.0
0.5	57.8
0.25	25.5
0.125	12.4
0.075	5.4

$(0.11$ mm, 0.41 mm, $5)$

20. Is it possible to apply potential theory to flow through porous media? Substantiate your answer.
21. Explain: (a) Anisotropy in soil strata and computation of horizontal and vertical flows,

    (b) Non-Darcy regime giving actual instances,

    (c) What are the parameters which characterise a porous medium?
22. An artisian aquifer 30 m thick has a porosity of 26% and elastic modulus of 0.26 GN/m^2. Estimate the storage coefficient of the aquifer. What fraction of this is attributable to the expansibility of water? Bulk modulus of elasticity of water $= 2.1$ GN/m^2.

    $(1.165 \times 10^{-3}, 3.12\%)$
23. Rainfall at the rate of 12 mm/hr falls on a strip of land 1200 m wide lying between two parallel canals. It is underlain by a horizontal impervious stratum. The depth water above this stratum in the two canals are 10 m and 8 m. Assuming a permeability of 10 m/day with vertical boundaries, and all the rainfall infilters into the soil, compute the discharge per metre length into both the canals.

    Determine the equation to the phreatic surface and the water table divide.

    If there is no rainfall, what would be the seepage to the lower canal?

    $[172.8$ m^3/day/m; $h = \sqrt{100 + 34.53x - 0.0288x^2}$; $x = 598$ m, $h = 102$ m; 0.15 m^3/day/m$]$

24. What is the flow rate in the confined aquifer shown in Fig 4.28

(≈ 0.008 m^3/day/m)

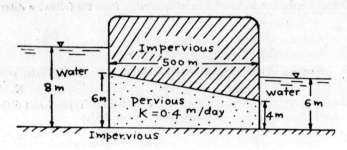

Fig. 4.28

25. Determine the flow into a horizontal infiltration gallery 200 m long resting on an impervious strata 10 m below ground surface. The GWT in the area is 3 m bgl and drops at the face of the gallery to 8 m bgl in a length of 400 m. Permeability of the flow strata can be taken as 45 m/day. Give the equation of the phreatic surface.

(1010 m^3/day, $h = \sqrt{4+0.1125\,x}$)

5

Well Hydraulics

The principle objective of ground water studies is to determine how much ground water can be safely withdrawn perennially from the aquifers in the area under study. This determination involves

(i) Transmissibility and storage coefficients of the aquifers.

(ii) The lateral extent of each aquifer and the hydraulic nature of its boundaries.

(iii) Vertical leakage if any, i.e. vertical seepage of water from either above or below the aquifer being tested, for which the hydraulic characteristics of overlying and underlying beds should be known.

(iv) The effect of proposed developments on recharge and discharge conditions.

Most of the above information can be obtained by conducting aquifer or pumping tests. In all these studies two different flow conditions are assumed:

(a) a steady state condition, i.e. when the flow is steady and the water levels have ceased to decline.

(b) a non-steady state condition, i.e. when the rate of flow through the aquifer is changing and the water levels are declining; water is taken from storage within the aquifer and the water level or piezometric head (in confined aquifer) is gradually reduced.

For steady flow of water in homogeneous and isotropic media

$$\nabla^2 h = 0 \qquad (5.1)$$

where $\nabla = \frac{\partial}{\partial x} + \frac{\partial}{\partial y} + \frac{\partial}{\partial z}$, and h = head causing flow.

For unsteady flow

$$\nabla^2 h = \frac{S}{T}\frac{\partial h}{\partial t} \qquad (5.2)$$

Steady Radial Flow into a Well
(Dupuit 1863, later modified by Thiem, 1906)

After prolonged pumping from the well, the drawdowns stabilise, when the cone of depression spreads to natural discharge and recharge areas, i.e. the steady state and steady shape conditions have been developed, Fig. 5.1 (a), (b).

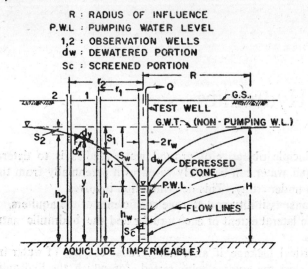

(a) Well in a water table aquifer

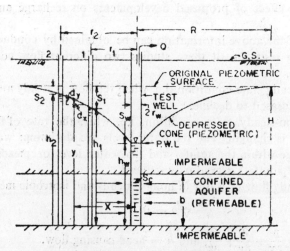

(b) Well in a confined aquifer

Fig. 5.1 Steady radial flow into a well.

a. Water table conditions (unconfined aquifer), Fig. 5.1 (a)
The yield from the well is

Well Hydraulics

$$Q = KiA \text{ (Darcy's law)} = K\frac{dy}{dx}(2\pi xy)$$

$$Q\int_{r_1}^{r_2}\frac{dx}{x} = 2\pi K\int_{h_1}^{h_2} y\,dy$$

$$Q = \frac{\pi K(h_2^2 - h_1^2)}{2.303 \log_{10}\frac{r_2}{r_1}} \quad (5.3)$$

Applying Eq. 5.3 between the points of zero-drawdown (radius of influence R) and face of the well (r_w) on the drawdown curve

$$Q = \frac{\pi K(H^2 - h_w^2)}{2.303 \log_{10}\frac{R}{r_w}} \quad (5.4)$$

If the drawdown in the pumping well $s_w (= H - h_w)$ is small, then

$$H^2 - h_w^2 = (H + h_w)(H - h_w)$$
$$= 2H(H - h_w), \qquad T = KH,$$

when
$$Q = \frac{2.72\,T(H - h_w)}{\log_{10}\frac{R}{r_w}} \quad (5.5)$$

where T = transmissibility of the water table aquifer ($= KH$); Q = yield from the well; H = saturated thickness of the aquifer; h_w = depth of water in the well during pumping; r_w = radius of the well; R = radius of influence (usually assumed as 300 m) and $H - h_w = s_w$ the resultig drawdown in the pumped well.

b. *Confined aquifer (artesian conditions)*, Fig. 5.1 (b)
The yield from the well is

$$Q = KiA, \text{ (Darcy's law)} = K\frac{dy}{dx}(2\pi xb)$$

$$Q\int_{r_1}^{r_2}\frac{dx}{x} = 2\pi Kb\int_{h_1}^{h_2} dy$$

$$Q = \frac{2.72\,Kb(h_2 - h_1)}{\log_{10}\frac{r_2}{r_1}} \quad (5.6)$$

If the drawdowns in the observation wells are s_1 and s_2, then $h_2 - h_1 = s_1 - s_2$; also $Kb = T$, the transmissibility of the confined aquifer. Applying Eq. (5.6) to the points of zero-drawdown (radius of influence R) and the face of the well (r_w) on the drawdown curve

$$Q = \frac{2.72\,T(H - h_w)}{\log_{10}\frac{R}{r_w}} \quad (5.7)$$

which is similar to Eq. (5.5). From Eq. (5.5), it follows that for the same drawdown in the pumping well, the yield is inversely proportional to $\log_{10} \frac{R}{r_w}$.

The assumptions in the Thiem's equations are:
(i) Stabilised drawdown.
(ii) Constant thickness of the aquifer with constant permeability (isotropic).
(iii) Complete penetration of the well with 100% efficiency.
(iv) Radial flow into the well.

Equations (5.3) and (5.6) can be used with reasonable accuracy even when the water table or piezometric surface has an initial slope, provided that h_1 and h_2 are taken from wells lying in a straight line through the well being tested, and in the direction of the initial slope of the water table or piezometric surface, and that h_1 and h_2 are taken as the average values at a distance r_1 and r_2 on the upslope and downslope of the test well respectively (Wenzel, 1936). From Fig. 5.2 it is apparent that the circular area of

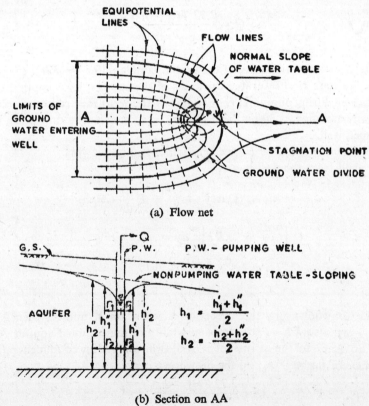

(a) Flow net

(b) Section on AA

Fig. 5.2 Flow into a well with sloping water table (after Wenzel, 1936).

Well Hydraulics

influence associated with a radial flow pattern becomes distorted. The distance r_1 should be great enough to extend beyond the immediate distortion of streamlines near the test well.

Example 5.1: A 30 cm well fully penetrates a confined aquifer 30 m deep. After a long period of pumping at a rate of 1,200 lpm, the drawdowns in the wells at 20 and 45 m from the pumping well are found to be 2.2 and 1.8 m, respectively. Determine the transmissibility of the aquifer. What is the drawdown in the pumped well?

Solution
From Eq. (5.6) it follows that

$$Q = \frac{2.72\, T(s_1-s_2)}{\log_{10} \frac{r_2}{r_1}}$$

$$1.2 = \frac{2.72\, T(2.2-1.8)}{\log_{10} \frac{45}{20}}$$

$$T = 0.388 \text{ m}^3/\text{min/m} = 5.59 \times 10^5 \text{ lpd/m or } \mathbf{559 \text{ m}^2/\text{day}}$$

Applying Eq. (5.6) between the face of the well (r_w) and r_2, it follows that

$$Q = \frac{2.72\, T(s_w-s_2)}{\log_{10} \frac{r_2}{r_w}}$$

$$1.2 = \frac{2.72\, (0.388)(s_w-1.8)}{\log_{10} \frac{45}{0.15}}$$

$$s_w - 1.80 = 2.82 \text{ m}$$

$$s_w = \mathbf{4.62 \text{ m}}$$

Example 5.2: A 30 cm well penetrates 50 m below the static water table. After a long period of pumping at a rate of 1800 lpm, the drawdowns in the wells at 15 and 45 m from the pumped well were 1.7 and 0.8 m, respectively. Determine the transmissibility of the aquifer. What is the drawdown in the pumped well?

Solution

$$Q = \frac{\pi k\, (h_2^2-h_1^2)}{2.303 \log_{10} \frac{r_2}{r_1}}$$

$$\frac{1.8}{60} = \frac{\pi K (49.2^2 - 48.3^2)}{2.303 \log_{10} \frac{45}{15}}$$

$$K = 1.165 \times 10^{-4} \text{ m/sec} = 10.07 \text{ m/day}$$

$$T = KH = 10.07 \times 50$$

$$= 503 \text{ m}^3/\text{day/m or m}^2/\text{day}$$

$$= 5.03 \times 10^5 \text{ lpd/m}$$

The drawdown in the pumped well can be determined by applying Eq. (5.3) to the face of the well (r_w) and $r_2 = 45$ m, finding (h_2-h_w) and then adding s_2 (0.80 m) to this, i.e.

$$s_w = (h_2 - h_w) + s_2$$

or by Eq. (5.4), assuming $R = 300$ m, when

$$s_w = H - h_w$$

From Eq. (5.5), assuming $R = 300$ m, $s_w = H - h_w = 6.25$ m.

A near quick approach is to find the specific capacity, i.e. yield per metre drawdown (see Section on Specific Capacity and safe yield, Eq. (5.93)).

$$\text{Specific capacity } \frac{Q}{s_w} = \frac{T}{1.4} = \frac{(1.165 \times 10^{-4}) \, 50}{1.4}$$

$$= 41.6 \times 10^{-4} \text{ m}^3/\text{sec/m-drawdown}$$

Drawdown in the pumped well,

$$s_w = \frac{1.8}{60 \, (41.6 \times 10^{-4})} = 7.22 \text{ m}$$

Unsteady Radial Flow into a Well

Confined aquifer: A well pumping at a constant rate from an extensive confined aquifer produces an area of influence which expands with time. Water is taken from storage within the aquifer as the piezometric head is reduced. In 1935, C.V. Theis developed the non-equilibrium equation sometimes referred to as the nonleaky-artesian formula

$$u = \frac{r^2 S}{4Tt} \tag{5.8}$$

$$s = \frac{Q}{4\pi T} W(u) \tag{5.9}$$

where s = drawdown in the observation well at radius r; Q = pumping rate; T = transmissibility coefficient; t = time after pumping started; r = distance of the observation well from the pumping well; S = storage coefficient, dimensionless (5×10^{-5} to 5×10^{-3} for confined aquifers, 0.05 to

Well Hydraulics

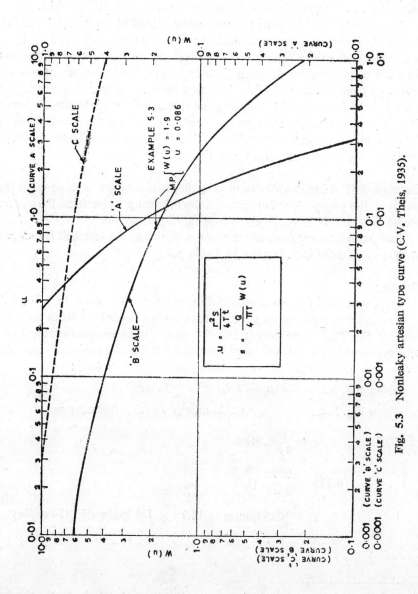

Fig. 5.3 Nonleaky artesian type curve (C.V. Theis, 1935).

0.30 for water table aquifers); u = argument and $W(u)$ = well function, dimensionless.

The plot of u verses $W(u)$ on standard log paper is called the type curve, Fig. 5.3. Tables are also available for $W(u)$ and u, Table 5.1.

The aquifer constant S and T can be determined by observing the drawdowns at several observation wells at a particular instant or the drawdown in a single well over a long period of time, for a constant pumping rate from the main well, by drawing time-drawdown and distance-drawdown curves or from the match point of the type curve. The pumping rate during the test is held constant by changing the setting of a valve in the discharge line and not by changing the pump speed. Fairly long time predictions of drawdown can be made from S and T by the methods of Theis, modified Theis (Jacob, Chow) and other investigators.

Example 5.3: Time-drawdown data in three observation wells towards the north of the pumping well are given in the following. Determine the aquifer constants T and S by the Theis non-equilibrium method.

Pump test data on three observation wells when the test well is pumped at the rate of 2270 lpm. is given in Table 5.2.

Solution

Data processed for Theis non-equilibrium method is given in Table 5.3.

s versus r^2/t is plotted on a transparent log-log paper (which has the same scale as the Theis type curve), Fig. 5.4, and is superimposed on the Theis type curve with the coordinate axes parallel.* From the position where the two curves match, the following match point values are read:

From Fig. 5.3, $W(u) = 1.9$, $u = 0.086$
From Fig. 5.4, $s = 0.335$ m, $r^2/t = 1.5 \times 10^3$ m²/min

$$s = \frac{Q}{4\pi T} W(u)$$

$$0.335 = \frac{2270}{4\pi T} (1.9)$$

$$T = 1025 \text{ lpm/m} = 14.75 \times 10^5 \text{ lpd/m or } \mathbf{1474 \text{ m}^2/day}$$

$$u = \frac{r^2 S}{4 T t}$$

---

*It is convenient to use a 'reversed' type curve '$W(u)$ vs $\frac{1}{u}$' in which case the observed data should be plotted as 's vs $\frac{t}{r^2}$'

Well Hydraulics

Table 5.1 Values of $W(u)$ for values of u (abridged, after Wenzel, 1942)

u	1.0	2.0	3.0	4.0	5.0	6.0	7.0	8.0	9.0
$\times 1$	0.219	0.049	0.013	0.0038	0.0011	0.00036	0.00012	0.000038	0.000012
$\times 10^{-1}$	1.82	1.22	0.91	0.70	0.56	0.45	0.37	0.31	0.26
$\times 10^{-2}$	4.04	3.35	2.96	2.68	2.47	2.30	2.15	2.03	1.92
$\times 10^{-3}$	6.33	5.64	5.22	4.95	4.73	4.54	4.39	4.26	4.14
$\times 10^{-4}$	8.63	7.94	7.53	7.25	7.02	6.84	6.69	6.55	6.44
$\times 10^{-5}$	10.94	10.24	9.84	9.55	9.33	9.14	8.99	8.86	8.74
$\times 10^{-6}$	13.24	12.55	12.14	11.85	11.63	11.45	11.29	11.16	11.04
$\times 10^{-7}$	15.54	14.85	14.44	14.15	13.93	13.75	13.60	13.46	13.34
$\times 10^{-8}$	17.84	17.15	16.74	16.46	16.23	16.05	15.90	15.76	15.65
$\times 10^{-9}$	20.15	19.45	19.05	18.76	18.54	18.35	18.20	18.07	17.95
$\times 10^{-10}$	22.45	21.76	21.35	21.06	20.84	20.66	20.50	20.37	20.25
$\times 10^{-11}$	24.75	24.06	23.65	23.36	23.14	22.96	22.81	22.67	22.55
$\times 10^{-12}$	27.05	26.36	25.96	25.67	25.44	25.26	25.11	24.97	24.86
$\times 10^{-13}$	29.36	28.66	28.26	27.97	27.75	27.56	27.41	27.28	27.16
$\times 10^{-14}$	31.66	30.97	30.56	30.27	30.05	29.87	29.71	29.58	29.46
$\times 10^{-15}$	33.96	33.27	32.86	32.58	32.35	32.17	32.02	31.88	31.76

Example: For $u = 3.0 \times 10^{-4}$, $W(u) = 7.53$

Table 5.2   Pumping test data on three observation wells (Ex. 5.3)

Time since pumping started t (min)	Observed drawdowns (cm)		
	Well $N-1$ $r = 60$ m	Well $N-2$ $r = 120$ m	Well $N-3$ $r = 240$ m
1	20.1	4.9	0.12
2	30.2	11.6	1.22
3	36.9	16.1	2.74
4	41.5	20.4	4.88
6	48.5	26.5	8.23
8	53.4	30.2	11.30
10	56.7	34.2	14.00
14	63.5	39.6	18.00
18	67.0	43.5	21.97
24	72.0	48.2	26.50
30	76.0	51.8	28.96
40	80.8	57.4	34.20
50	84.8	61.0	37.50
60	87.9	64.3	40.25
80	92.7	68.3	45.50
100	96.5	72.5	49.40
120	100.0	76.0	51.80
150	104.2	79.9	55.80
180	107.0	83.0	59.20
210	110.0	85.7	62.00
240	111.2	87.8	65.25

Table 5.3 Data for Theis non-equilibrium method

Well $N-1$ $r = 60$ m		Well $N-2$ $r = 120$ m		Well $N-3$ $r = 240$	
s (m)	$\dfrac{r^2}{t}$ m²/min	s (m)	$\dfrac{r^2}{t}$ m²/min	s (m)	$\dfrac{r^2}{t}$ m²/min
0.201	3.6×10^3	0.049	1.44×10^4	0.0012	5.75×10^4
0.302	1.8×10^3	0.116	7.2×10^3	0.0122	2.87×10^4
0.369	1.2×10^3	0.161	4.8×10^3	0.0274	1.91×10^4
0.415	9.0×10^2	0.204	3.6×10^3	0.0488	1.49×10^4
0.485	6.0×10^2	0.265	2.4×10^3	0.0823	9.6×10^3
0.534	4.5×10^2	0.302	1.8×10^3	0.1130	7.2×10^3
0.567	3.6×10^2	0.342	1.44×10^3	0.140	5.75×10^3
0.635	2.57×10^2	0.396	1.03×10^3	0.180	4.1×10^3
0.670	2.0×10^2	0.435	8.0×10^2	0.22	3.2×10^3
0.720	1.5×10^2	0.482	6.0×10^2	0.265	2.4×10^3
0.760	1.2×10^2	0.518	4.8×10^2	0.29	1.92×10^3
0.808	9.0×10	0.574	3.8×10^2	0.342	1.44×10^3
0.848	7.2×10	0.610	2.88×10^2	0.375	1.15×10^3
0.879	6.0×10	0.643	2.4×10^2	0.403	9.6×10^2
0.927	4.2×10	0.683	1.8×10^2	0.455	7.2×10^2
0.965	3.6×10	0.726	1.44×10^2	0.494	5.75×10^2
1.000	3.0×10	0.760	1.2×10^2	0.518	4.8×10^2
1.042	2.4×10	0.799	9.6×10	0.558	3.83×10^2
1.070	2.0×10	0.830	8.0×10	0.592	3.2×15^2
1.100	1.7×10	0.857	6.85×10	0.62	2.74×10^2
1.112	1.5×10	0.878	6.0×10	0.653	2.40×10^2

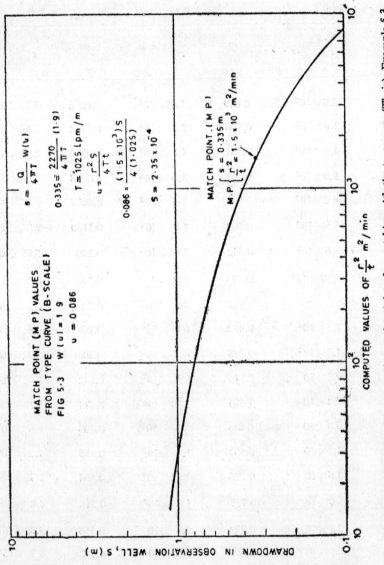

Fig. 5.4 Non-equilibrium curve of observational data for matching with the type-curve (Theis). Example 5.3.

$$0.086 = \frac{(1.5 \times 10^3)\, S}{4\,(1.025)}$$

$$S = 2.35 \times 10^{-4}$$

The assumptions in the Theis equation are:

(*i*) Radial flow into the well with no entrance losses (horizontal stream lines).

(*ii*) Instantaneous release of water from storage upon a lowering of the drawdown curve (no lag) when pumping, i.e. proportional to the rate of change of head.

(*iii*) All water comes from storage (no influent sleepage or aquifer leakage).

(*iv*) Constant coefficient of storage (independent of time and pressure).

(*v*) Non-pumping piezometric surface horizontal.

(*vi*) Complete penetration of the aquifer.

(*vii*) Homogeneous, isotropic aquifer of uniform saturated thickness and infinite aerial extent.

(*viii*) Discharge from well is constant and radius of well is infinitesimal.

Water table aquifer: In confined aquifers, water comes mainly by gravity drainage which is not instantaneous. The storage coefficient reduces with time and ultimately becomes equivalent to the specific yield. Non-equilibrium equation can be applied to unconfined aquifers with the following limitations:

(*i*) The drawdown should be small in relation to the saturated thickness.

(*ii*) the observation well should be at a distance of 0.2 to 0.6 times the saturated thickness of the aquifer.

(*iii*) the minimum time interval after start of pumping for the non-equilibrium equation to be applicable is

$$t > \frac{5\, S_y\, H}{K} \qquad (5.10)$$

For example, if $S_y = 0.15$, $H = 30$m and $K = 10^5$ lpd/m² or 100m³/day/m² or 100 m/day,

$$t > \frac{5\,(0.15 \times 30)}{100}$$

$$> 0.23 \text{ day, or } 5.5 \text{ hr}$$

Limitations (ii) and (iii) are suggested by Boulton.

(*iv*) *Jacob's correction for very thin aquifers under water table conditions.* When water is withdrawn from unconfined aquifer the transmissibility decreases as the aquifer is dewatered and the drawdown is more than that in confined aquifers of equivalent initial transmissibility. Jacob (1944) has shown that adjustment can be made for the effect of the dewatering with the following equation

$$s' = s - \frac{s^2}{2H} \qquad (5.11)$$

where s' = drawdown in an equivalent confined aquifer; s = drawdown observed in water table aquifer and H = saturated thickness of the water table aquifer prior to start of pumping.

If $\frac{s^2}{2H} < 0.003$ m, the correction need not be applied.

Example 5.4: A 30 cm well penetrating a water table aquifer is pumped at the rate of 3800 lpm. Initial saturated thickness of the acquifer is 8.2 m. 18-day pumping tests were conducted and the drawdown values at the end of 18 days in 6 observation wells (3 on a line extending north from the pumped well and 3 to the south) are given in Table 5.4. Determine the aquifer constants T and S. If the observed drawdown in the pumped well is 4.5 m after 18 days determine the well efficiency.

Table 5.4 Drawdown values in 6 observation wells (Ex. 5.4)

Line	Well No.	Distance from the pumped well, r (m)	Observed drawdown after 18 days, s (m)
N	1	15.0	1.80
	2	30.7	1 40
	3	57.7	1.04
S	1	14.9	1.67
	2	30.6	1.31
	3	57.9	1.19

Solution

For very thin aquifers under water table conditions Jacob's correction, for the effect of dewatering of the aquifer and corresponding decrease in transmissibility, has to be applied as shown in Table 5.5.

s' versus r is plotted on a semi-log paper, Fig. 5.5. $\Delta s'$ per log cycle of r is measured and T and S are determined by the Jacob's method, Eqs. (4.82) and (4.83).

Well Hydraulics

Table 5.5 Jacob's correction (Ex. 5.4)

Well No.	Distance from the pumped well, r (m)	Observed drawdown after 18 days, s (m)	Correction for dewatering of the aquifer $\dfrac{s^2}{2H}$ (m)*	Corrected drawdown $s'(m) = s - \dfrac{s^2}{2H}$
N—1	15.0	1.80	0.197	1.603
N—2	30.7	1.40	0.119	1.281
N—3	57.7	1.04	0.066	0.974
S—1	14.9	1.67	0.170	1.500
S—2	30.6	1.31	0.105	1.205
S—3	57.9	1.19	0.086	1.104

*$\dfrac{s^2}{2H} > 0.003$ m, hence the correction has been applied.

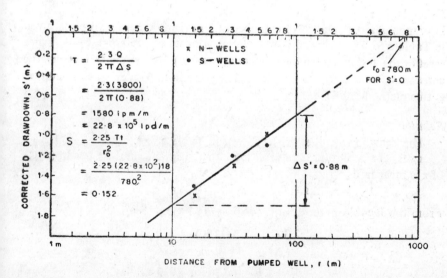

Fig. 5.5 Distance-drawdown graph for water table aquifer (Jacob's correction), Example 5.4.

$$T = \frac{2.3 \, Q}{2\pi \, \Delta s} = \frac{2.3 \, (3800)}{2\pi \, (0.88)} = 1580 \text{ lpm/m} = 22.8 \times 10^5 \text{ lpd/m}$$

$$S = \frac{2.25\,Tt}{r_0^2}$$

$r_0 =$ intercept on the zero drawdown axis $(s' = 0) = 780$ m

$$S = \frac{2.25\,(22.8 \times 10^2)\,18}{780^2} = 0.152$$

Well efficiency: For the pumped well $r = 0.15$ m. From the graph for $r = 0.15 \times 10^2 = 15$ m, $s' = 1.53$ m. For $r = 0.15$ m, $s' = 1.53 +$ rise for 2 cycles, i.e. $2(0.88) = 3.29$ m against an observed drawdown of 4.5 m.

$$\text{Well efficiency} = \frac{3.29}{4.50} \times 100 = 73\%$$

Also, specific capacity of the well $= \dfrac{T}{1.4} = \dfrac{1.58}{1.40} = 1.13$ m^3/min/m

Probable drawdown for a pumpage of 3800 lpm $= \dfrac{3.8}{1.13} = 3.36$ m as against an observed drawdown of 4.5 m

$$\text{Well efficiency} = \frac{3.36}{4.50} \times 100 = 74.6\%$$

which is nearly the same as obtained by the extrapolation of the graph.

Example 5.5: The time-drawdown data from an observation well at Pattamangalam, (Tamil Nadu, India) 12.3 m from a pumped well is given in Table 5.6. The test well is pumped at the rate of 1150 lpm. Static water level in the test well (before pumping started) is 2.18 m. Determine the constants T and S by the modified Theis method. Under what condition is this method valid?

Solution

The drawdown values are as shown in Table 5.7.

s versus t is plotted on a semi-log paper, Fig. 5.6, and Δs per log cycle of t is measured

$$\Delta s = 0.73 - 0.43 = 0.3 \text{ m}$$

From the Jacob's method, Eqs. (4.80) and (4.81),

$$T = \frac{2.3\,Q}{4\pi\,\Delta s} = \frac{2.3\,(1150)}{4\pi\,(0.3)} = 700 \text{ lpm/m}$$

$$= 10.1 \times 10^5 \text{ lpd/m or } \mathbf{1010 \text{ m}^2\text{/day}}$$

$$S = \frac{2.25\,Tt_0}{r^2}$$

$t_0 =$ time intercept on the zero-drawdown axis $(s = 0)$; time for $s = \Delta s$ or a multiple of Δs is read on the graph; say for $s = \Delta s = 0.3$ m, $t = 3.55$ min; then for $s = 0$, $t = 0.355$ min, i.e. one log cycle less.

Table 5.6 Time-drawdown data (Ex. 5.5)

Time t(min)	Depth of measuring point (m)
0	2.18
1	2.42
2	2.42
3	2.46
4	2.50
6	2.55
8	2.59
10	2.63
14	2.67
18	2.69
22	2.71
28	2.72
35	2.75
45	2.82
55	2.83
65	2.86
80	2.87
100	2.92
120	2.94

Table 5.7 Drawdown values (Ex. 5.5)

Time t (min)	Drawdown s (m)	Time t (min)	Drawdown s (m)
1	0.24	22	0.53
2	0.24	28	0.54
3	0.28	35	0.57
4	0.32	45	0.64
6	0.37	55	0.65
8	0.41	65	0.68
10	0.45	80	0.69
14	0.49	100	0.74
18	0.51	120	0.76

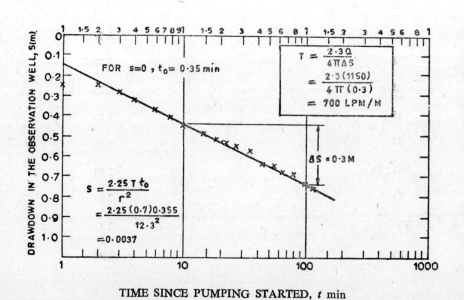

Fig. 5.6 Time-drawdown graph (Jacob's), Example 5.5.

Well Hydraulics

$$S = \frac{2.25\,(0.7)\,(0.355)}{(12.3)^2} = 0.0037$$

This method is valid for $u < 0.01$

$$u = \frac{r^2 S}{4\,T\,t} < 0.01$$

or $\quad t > \dfrac{r^2 S}{0.04\,T} > \dfrac{(12.3)^2\,(3.7 \times 10^{-3})}{0.04\,(0.7)} > 20$ min

The early pump test data (observed drawdown within 20 min) should be ignored while drawing the straight line graph.

Example 5.6: For the data given in the Example 5.3 draw a composite-drawdown graph s versus t/r^2 and determine the aquifer constants T and S by the modified Theis (Jacob's) method by ignoring early pump test data.

Solution

Pump test data on three observation wells processed for drawing the composite drawdown graph is given in Table 5.8.

The composite-drawdown graph is drawn by plotting s versus t/r^2 on a semi-log paper, Fig. 5.7. Δs per log cycle of t/r^2 is measured and aquifer contsants T and S are determined from Eqs. (4.84) and (4.85).

$$T = \frac{2.3\,Q}{4\pi\,\Delta s} = \frac{2.3\,(2270)}{4\pi\,(0.4)} = 1040 \text{ lpm/m}$$

$$= 15 \times 10^5 \text{ lpd/m or } \mathbf{1500 \text{ m}^2/\text{day}}$$

$$S = 2.25\,T\left(\frac{t}{r^2}\right)_0$$

From the graph when $s = 0$, $\quad \left(\dfrac{t}{r^2}\right)_0 = 10^{-4}$ min/m²

$$S = 2.25\,(1.04)\,10^{-4} = 2.34 \times 10^{-4}$$

Chow's Method

Chow (1952) developed a method of solution of Theis equation by avoiding the curve matching technique and not being restricted to 'large values of t and small values of r' as in the Jacob's method.

Chow introduced the function

$$F(u) = \frac{W(u)\,e^u}{2.3} \qquad (5.12)$$

The relation between (Fu), (Wu) and u is shown in Fig. 5.8 and in Table 5.9. $F(u)$ is calculated from the time-drawdown data on an observation well. In Chow's method the time-drawdown data on an observation

Table 5.8 Pump test data on three observation wells (Example 5.6)

Well $N=1$ $r = 60$ m		Well $N-2$ $r = 120$ m		Well $N-3$ $r = 240$ m	
s (m)	$\dfrac{t}{r^2}$ min/m^2	s (m)	$\dfrac{t}{r^2}$ min/m^2	s (m)	$\dfrac{t}{r^2}$ min/m^2
0.201	2.78×10^{-4}	0.049	6.95×10^{-5}	0.001	1.74×10^{-5}
3.302	5.55×10^{-4}	0.116	1.39×10^{-4}	0.012	3.48×10^{-5}
0.369	8.35×10^{-4}	0.161	2.08×10^{-4}	0.027	5.24×10^{-5}
0.415	1.11×10^{-3}	0.204	2.78×10^{-4}	0.049	6.70×10^{-5}
0.485	1.67×10^{-3}	0.265	4.17×10^{-4}	0.082	1.04×10^{-4}
0.534	2.22×10^{-3}	0.302	5.56×10^{-4}	0.113	1.33×10^{-4}
0.567	2.78×10^{-3}	0.342	6.95×10^{-4}	0.140	1.74×10^{-4}
0.635	3.89×10^{-3}	0.396	9.71×10^{-4}	0.180	2.44×10^{-4}
0.670	5.00×10^{-3}	0.435	1.25×10^{-3}	0.220	3.13×10^{-4}
0.720	6.67×10^{-3}	0.482	1.66×10^{-3}	0.265	4.17×10^{-4}
0.760	8.33×10^{-3}	0.518	2.08×10^{-3}	0.290	5.21×10^{-4}
0.808	1.11×10^{-3}	0.574	2.63×10^{-3}	0.342	6.95×10^{-4}
0.848	1.39×10^{-2}	0.610	3.47×10^{-3}	0.375	8.70×10^{-4}
0.879	1.67×10^{-2}	0.643	4.15×10^{-3}	0.403	1.04×10^{-3}
0.927	2.38×10^{-2}	0.683	5.55×10^{-3}	0.455	$1 30 \times 10^{-3}$
0.965	2.78×10^{-2}	0.726	6.93×10^{-3}	0.494	1.74×10^{-3}
1.000	3.34×10^{-2}	0.760	8.32×10^{-3}	0.518	2.08×10^{-3}
1.042	4.17×10^{-2}	0.799	1.04×10^{-2}	0.558	2.61×10^{-3}
1.070	5.00×10^{-2}	0.830	1.25×10^{-2}	0.592	3.13×10^{-3}
1.100	5.88×10^{-2}	0.857	1.46×10^{-2}	0.620	3.65×10^{-3}
1.112	6.67×10^{-2}	0.878	1.67×10^{-2}	0.653	4.17×10^{-3}

well is plotted on a semi-logarithmic paper, an arbitrary point P is chosen on the plotted curve and a tangent to the curve is drawn at P. The drawdown (s_P) and the slope of the tangent to the curve at P (Δs_P), i.e. the drawdown difference per log cycle of time is read on the graph. The value of $F(u)$ is then calculated as

$$F(u) = \frac{s_P}{\Delta s_P} \qquad (5.13)$$

Well Hydraulics

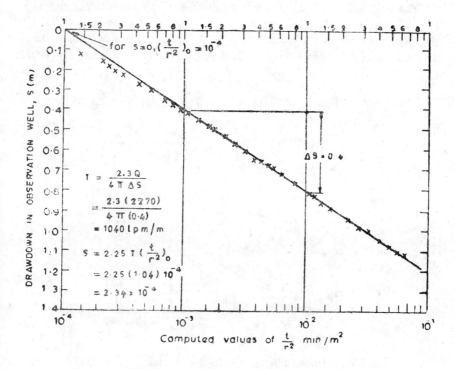

Fig. 5.7 Composite-drawdown graph (Jacob's), Example 5.6.

Table 5.9 Value of $F(u)$ for different values of u (after Chow, 1952)

u	1	2	3	4	5	6	7	8	9
$u \times 1$	0.250	0.157	0.117	0.0898	0.0734				
$\times 10^{-1}$	0.874	0.647	0.532	0.455	0.401	3.360	0.327	0.301	0.276
$\times 10^{-2}$	1.77	1.49	1.33	1.21	1.13	1.06	1.00	0.956	0.913
$\times 10^{-3}$	2.75	2.46	2.28	2.16	2.07	1.99	1.92	1.870	1.82
$\times 10^{-4}$			$F(u)=W(u)/2.30$						

Example: for $u = 3 \times 10^{-2}$, $F(u) = 1.33$

The values of u and $W(u)$ corresponding to $F(u)$ are determined from Fig. 5.8, and then T and S are calculated from Eqs. (5.8) and (5.9). For the data given in Example 5.5, from Fig. 5.9, $s_P = 0.49$ m and $\Delta s_P = 0.3$ m. Then $F(u) = 0.49/0.3 = 1.63$, and from Fig. 5.8, $W(u) = 3.8$ and $u = 0.011$. Hence,

$$T = \frac{Q}{4\pi s_P} W(u) = \frac{1150}{4\pi (0.49)} \times 3.8 = 710 \text{ lpm/m or } \mathbf{1022 \text{ m}^2/\text{day}}$$

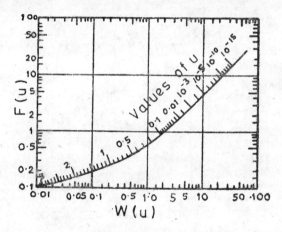

Fig. 5.8 Relation between $F(u)$, $W(u)$ and u (after Chow, 1952).

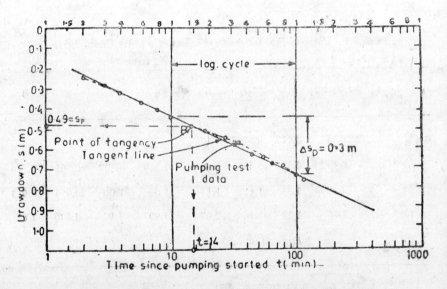

Fig. 5.9 Aquifer constants by Chow's method, Example 5.5.

Well Hydraulics

$$S = \frac{4Ttu}{r^2} = \frac{4 \times 0.71 \times 14 \times 0.011}{(12.3)^2} = 0.0029$$

as compared to $T = 700$ lpm/m and $S = 0.0037$ obtained by Jacob's method.

Example 5.7: The following particulars are obtained from a tubewell site:

Total depth	75 m
Screen diameter	30 cm
Screen setting	50 to 60 m b.g.l.
Static water level	10 m b.g.l.
Transmissibility of the aquifer	8×10^{-4} lpd/m
Storage coefficient of the aquifer	5×10^{-4}
Proposed yield	1300 lpm

It has to be ascertained whether the tubewell can be pumped at this rate continuously for a year, assuming no recharge.

Solution

The resulting drawdown after 1 year of continuous pumping can be estimated by the Jacob's method, Eq. (4.78)

$$s = \frac{2.3\,Q}{4\pi T} \log_{10} \frac{2.25\,T\,t}{r^2\,S} = \frac{2.3\,(1.3)}{4\pi \left(\frac{8 \times 10}{24 \times 60}\right)} \log_{10} \frac{2.25\,(8 \times 10)\,365}{(0.15)^2\,5 \times 10^{-4}} = 41.8 \text{ m}$$

Static water level	= 10.0 m b.g.l.
Drawdown after 1 year of continuous pumping	= 41.8 m
Pumping water level	= 51.8 m b.g.l.

The pumping water level after 1 year of continuous pumping drops below the top of the screen, thus causing partial dewatering. On the other hand if the pumping rate is reduced to 1000 lpm, the resulting drawdown after 1 year of continuous pumping

$$s = 41.8 \times \frac{1000}{1300} = 32.2 \text{ m}$$

and the pumping water level

$$\text{p.w.l.} = 32.2 + 10.0 = 42.2 \text{ m b.g.l.}$$

which is well above the top of the screen. Hence a safe yield of 1000 lpm can be expected from the tubewell. But actually in nature, as more and more water is pumped from storage, the cone of influence spreads till it reaches some source of recharge, thus safeguarding against partial dewatering.

Law of Times

From the Eqs. (4.81), (4.83) and (4.85) it follows that

$$\frac{t_1}{r_1^2} = \frac{t_2}{r_2^2} = \frac{t_3}{r_3^2} = \ldots = \frac{t_n}{r_n^2} \qquad (5.14)$$

that is, the time of occurrence of zero-drawdown or equal drawdown vary directly as the square of the distances of the observation wells from the discharge well and are independent of the rate of pumping. It may also be noted from Eqs. (4.80) and (4.82) that Δs per log cycle of distance is twice Δs per log cycle of time. If from a time-drawdown curve on a particular well Δs is 1.5 m, Δs per log cycle of distance is 3 m. If the drawdown at a distance of 30 m from the pumping well is 1 m, the drawdown at 3 m from the pumping well is 4 m and at 0.3 m from the pumping well is 7 m.

Also from the time-drawdown curve, interpretation of aquifer boundaries can be made. Recharge boundaries cause the plot of observed drawdowns to diverge below the Theis-type curve, Fig. 5.3, and above the time-drawdown straight line plot of the Jacob method, Fig. 5.6. Barrier (impermeable) boundaries cause the plot of observed drawdowns to diverge above the Theis-type curve and below the time-drawdown straight line plot of the Jacob's method. If a change in slope (divergence), depending on whether it is a recharge or barrier boundary is indicated at a time t on the plot, Fig. 5.10, the distance to the boundary from the discharging well is taken as nearly equal to $x/2$, where x is given by (from the theory of image wells, Fig. 5.24)

$$\frac{r^2}{t_0} = \frac{x^2}{t} \qquad (5.15)$$

where r = distance of the observation well from the discharging well;

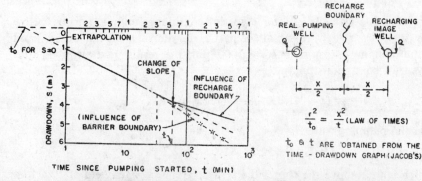

a. TIME-DRAWDOWN GRAPH (JACOB'S)—RECHARGE AND BARRIER BOUNDARIES

b. LOCATION OF RECHARGE (OR BARRIER BOUNDARY)

Fig. 5.10 Illustrative sketches for recharge and barrier boundaries.

Well Hydraulics

t_0 = time for zero drawdown taken from the time-drawdown curve (Jacob) and t = time at which a change of slope is indicated.

Example 5.8: Pump test data on a 60 cm well at Milattur, Tamil Nadu is given in Table 5.10. The well was pumped at the rate of 900 lpm. Determine the aquifer constants T and S and comment on the hydraulic boundary conditions if any.

Solution

The time-drawdown graph s versus t is plotted on a semi-log paper, Fig. 5.11. Δs per log cycle of t is measured and the aquifer constants S and T are determined as follows:

$$T = \frac{2.3\ Q}{4\pi\ \Delta s} = \frac{2.3\ (900)}{4\pi\ (0.59)} = 279 \text{ lpm/m} = 4.02 \times 10^5 \text{ lpd/m or } \mathbf{402\ m^2/day}$$

Table 5.10 Pump test data (Ex. 5.8)

Time since pumping started t (min)	Observed drawdown in the pumped well s (m)
10	3.55
20	3.72
30	3.82
40	3.88
50	3.95
60	4.00
80	4.09
100	4.14
120	4.18
150	4.22
250	4.35
350	4.44
500	4.50
600	4.55
800	4.59
1000	4.62
1300	4.65
1600	4.69
2000	4.72
2500	4.75
3000	4.77
4000	4.80
5000	4.85

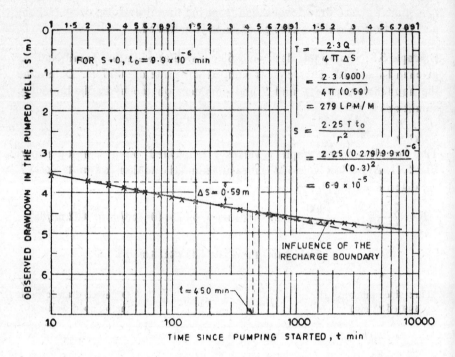

Fig. 5.11 Time-drawdown graph (Jacob's)—Recharge boundary, Example 5.8.

$$S = \frac{2.25 T t_0}{r^2}$$

To find t_0, the time intercept on the zero-drawdown axis ($s = 0$) we can proceed as follows. From Fig. 5.11, for $7\Delta s = 7(0.59) = 4.13$ m, $t = 99$ min for $s = 0$, $t = 99/10^7 = 9.9 \times 10^{-6}$ min.

$$S = \frac{2.25 \left(\frac{4.02 \times 10^2}{24 \times 60}\right) 9.9 \times 10^{-6}}{(0.3)^2} = 6.9 \times 10^{-5}$$

Hydraulic boundary condition: At $t = 450$ min, the plot of observed drawdowns diverges above the Jacob's time-drawdown straight line plot (so that the drawdowns are reduced with continued pumping) indicating the existence of a recharge boundary and the distance to the recharge boundary is given by Eq. (5.15)

$$\frac{r^2}{t_0} = \frac{x^2}{t}$$

$$\frac{(0.3)^2}{9.9 \times 10^{-6}} = \frac{x^2}{450}$$

$$x = 2100 \text{ m}$$

Well Hydraulics 161

The distance to the recharging image well $x = 2100$ m and the distance to the recharging boundary from the pumping well is $x/2 = \mathbf{1050\,m}$ (Fig. 5.10).

Jacob's method gives fair predictions only when steady shape and steady state conditions are developed. This method is specially applicable to artesian conditions since it takes a long period to reach steady shape conditions in water table aquifers. The method is extremely useful in making long term predictions of withdrawals from aquifers whose boundary conditions and storage and transmissibility coefficients are known. When the pumping well penetrates only a part of the aquifer (partial penetration) the transmissibility coefficient obtained from the distance drawdown plot is not reliable, while the time-drawdown plot gives a reliable value of the transmissibility coefficient if boundary effects or leakage do not develop before the data yields a straight line plot. For observation wells affected by partial penetration, the storage coefficient S cannot be determined from the time-drawdown plot.

Example 5.9. A 30 cm well was pumped at the rate of 1080 lpm for one day and the drawdown in the well was 1.85 m. The pumping was stopped and the residual drawdowns during recovery for one hour are given in Table 5.11 below. Determine the aquifer constants S and T.

Table 5.11 Residual drawdown data (Ex. 5.9)

Recovery time t' (min)	Residual drawdown s' (m)
1	0.875
2	0.735
3	0.694
4	9.662
5	0.640
6	0.625
8	0.590
10	0.570
12	0.556
16	0.535
20	0.498
30	0.458
40	0.423
45	0.410
60	0.383

Solution: The data processed for plotting the recovery curve is given in Table 5.12.

Table 5.12 Data processed for plotting the recovery curve (Ex. 5.9)

Time since pumping stopped t' (min)	Time since pumping started $t = 1440 + t'$, (min)	Ratio t/t'	Residual drawdown s' (m)
1	1441	1441	0.875
2	1442	721	0.735
3	1443	481	0.694
4	1444	361	0.662
5	1445	289	0.640
6	1446	241	0.625
8	1448	181	0.590
10	1450	145	0.570
12	1452	121	0.556
16	1456	91	0.535
20	1460	73	0.498
30	1470	49	0.458
40	1480	37	0.423
45	1485	33	0.410
60	1500	25	0.383

The values of s' versus t/t' are plotted on a semi-log paper, Fig. 5.12. The residual drawdown $\Delta s'$ per log cycle of time t/t' is measured and T is determined as

$$T = \frac{2.3\,Q}{4\pi\,\Delta s'} = \frac{2.3\,(1{,}080)}{4\pi\,(0.24)} = 823.6 \text{ lpm/m} = 11.9 \times 10^5 \text{ lpd/m}$$

or **1190 m²/day**.

The storage coefficient can be determined from the observed drawdown in the well (s_{t_1}) when pumping stopped.

Well Hydraulics 163

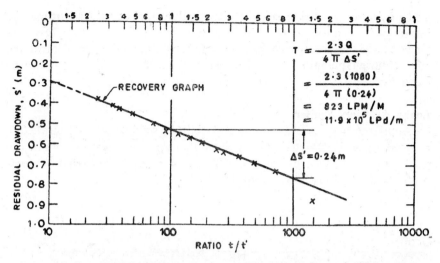

Fig. 5.12 Time-drawdown graph—Theis recovery, Example 5.9.

$$s_{t_1} = \frac{2.3\,Q}{4\pi\,T} \log_{10} \frac{2.25\,Tt_1}{r^2 S}$$

$$1.85 = \frac{2.3\,(1,080)}{4\pi\,(823.6)} \log_{10} \frac{2.25\,(0.823)\,1,440}{(0.15)^2\,S}$$

$$\log_{10} \frac{119000}{S} = 7.708$$

$$S = 0.00233$$

Confirmation of the Boundary from the Recovery Curve

If the recovery curve intercepts the time axis, i.e. $s' = 0$ at some positive values of t/t', Fig. 5.13, it may be interpreted that a recharge boundary is encountered. On the other hand if the recovery curve intercepts the

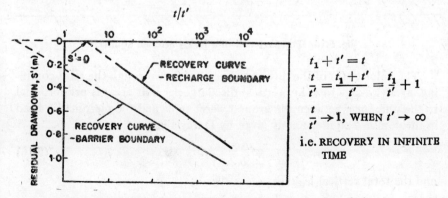

Fig. 5.13 Recovery curves for recharge and barrier boundaries.

axis of drawdown giving some positive value of s' at $t/t' \to 1$, it may be interpreted that an impervious (barrier) boundary is encountered. The existence of the boundary can be confirmed by drawing a time-drawdown curve (during pumping) and also the distance to the boundary can be calculated.

Leaky Artesian Aquifer

Aquifers which are overlain or underlain by semi-permeable strata are referred to as leaky aquifers. In such aquifers a signicant portion of the yield may be derived by vertical leakage or seepage through the semi-confining formations into the aquifer, Fig. 5.14.

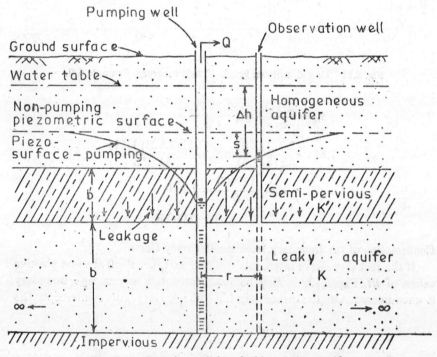

Fig. 5.14 Pumping well in a leakly artesian aquifer.

The velocity (v) of the downward vertical flow through the semi-confining layer (leakage rate, i.e. vertical discharge per unit area) is proportional to the difference between the ground water table and the piezometric head in the confined aquifer and is given by Darcy's law.

$$v = K' \frac{\Delta h}{b'} \qquad (5.16)$$

and the total vertical leakage

$$Q_c = v \, A_c$$

Well Hydraulics

or
$$Q_c = \left(\frac{K'}{b'}\right) \Delta h \, A_c \qquad (5.17)$$

where Q_c = leakage through the confining bed

K' = coefficient of vertical permeability of confining bed

b' = thickness of confining bed through which leakage occurs

A_c = area of confining bed through which leakage occurs

Δh = difference between the piezometric head in the aquifer and in the source bed above confining bed (GWT in Fig. 5.14)

The leakage (recharge) is proportional to the difference in head Δh and is more near the pumping well since the piezometric surface is very low. The ratio K'/b' is called leakance and its reciprocal b'/K' is called the hydraulic resistance c of the confining layer and has the dimension of time. $\sqrt{Tc}$ is called the leakage factor B and has the dimension of length; T is the transmissibility ($= Kb$) of the confined aquifer.

Example 5.10: From the pumping tests of a semiconfined aquifer of thickness 30 m and permeability 20 m/d, it is estimated that the recharge rate from an overlying unconfined aquifer through an aquitard of thickness 2 m, is 50 mm/year. The average piezometric surface in the semiconfined aquifer is 16 m below the water table in the unconfined aquifer. Determine the hydraulic characteristics of the aquitard (semi-confining layer) and the aquifer.

If a well drilled in the aquifer is pumped at the rate of 5000 m³/day, how many square kilometers of recharge area are required to sustain the flow at the estimated recharge of 50 mm/year?

Solution: Recharge through aquitard per unit area

$$Q = K' i A$$

$$\frac{0.050}{365} = K' \frac{16}{2} (1 \times 1) \qquad K' = 1.71 \times 10^{-5} \text{ m/day}$$

Leakance $= \dfrac{K'}{b'} = \dfrac{1.71 \times 10^{-5}}{2} = 8.55 \times 10^{-6} \text{ day}^{-1}$

Hydraulic resistance $c = \dfrac{b'}{K'} = 1.17 \times 10^5$ days

Aquifer

$T = Kb = 20 \times 30 = 600 \text{ m}^2/\text{day}$

Leakage factor $B = \sqrt{Tc} = \sqrt{600 \, (1.17 \times 10^5)} = 8380$ m

To sustain a pumpage of 5000 m³/day the recharge area A in km² required

$$\frac{0.050}{365} A = 5000$$

$$A = 3.65 \times 10^7 \text{ m}^2 \text{ or } \mathbf{36.5 \text{ km}^2}$$

Unsteady Radial Flow—Leaky Artesian Aquifer

Hantush and Jacob (1965) showed that the drawdown of the piezometric head in a leaky confined acquifer is given by the equation

$$s = \frac{Q}{4\pi T} W\left(u, \frac{r}{B}\right) \tag{5.18}$$

and

$$u = \frac{r^2 s}{4 T t} \tag{5.8}$$

$$\frac{r}{B} = \frac{r}{\sqrt{T/(K'/b')}} \tag{5.19}$$

which are very similar to Eqs. (5.8) and (5.9) except that the well function contains the additional term $\frac{r}{B}$. Values of $W\left(u, \frac{r}{B}\right)$ for various values of u and $\frac{r}{B}$ are given by Hantush (1956) in Table 5.13. A family of leaky artesian type curves where values of $W\left(u, \frac{r}{B}\right)$ versus $\frac{1}{u}$ for various values of $\frac{r}{B}$ are plotted on log-log paper by Walton (1962) as shown in

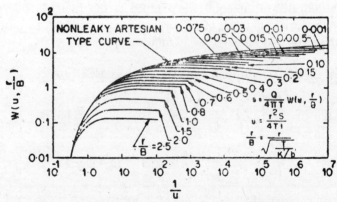

Fig. 5.15 Unsteady state leaky artesian type curves.

Fig. 5.15. The observed time-drawdown data s versus t are plotted on a transparent log-log paper of the same scale. The time-drawdown plot is superimposed on the type curves with the respective axes parallel and a match point is selected after the field-data curve has been matched with one of the curves of $\frac{r}{B}$ in the family of type curves. The coordi-

Well Hydraulics

nates $W\left(u, \dfrac{r}{B}\right), \dfrac{1}{u}$, s and t at the match point and the value of $\dfrac{r}{B}$ for the curve matched (by interpolation if the curve does not exactly coincide) are used to calculate T, S and K' in the Eqs. (5.18), (5.8) and (5.19). It may be noted that as $K' \to 0$, $\dfrac{r}{B} \to 0$ indicating that for an impermeable confining layer Eq. (5.18) reduces to that of non-leaky aquifer (Theis).

Example 5.11: The time-drawdown data for unsteady flow in an observation well at 30 m from the pumped well are given in Table 5.14. The well was pumped at a constant discharge of 800 lpm. The thickness of the aquifer and the top semi-confining layer are 12 m and 4 m, respectively. Determine the aquifer constants.

Table 5.14 Time-drawdown data (Ex. 5.11)

Time since pumping started (min)	Drawdown (m)
5	0.25
10	0.50
30	1.00
60	1.32
100	1.60
200	1.70
400	1.80
600	1.83
800	1.88
1,000	1.91
1,200	1.92

Solution: The time-drawdown data is plotted on a transparent log-log paper of the same scale as the Walton's type curves and superimposed on the family of leaky artesian type curves, Fig. 5.16. The time-drawdown

Table 5.13 Values of $W\left(u, \dfrac{r}{B}\right)$ for different Values of u and $\dfrac{r}{B}$ (abridged; after Hantush, 1956, 1964)

u \ r/B	0.002	0.004	0.006	0.008	0.01	0.02	0.04	0.06	0.08	0.1	0.2	0.4	0.6	0.8	1	2	4	6	8
0	12.7	11.3	10.5	9.89	9.44	8.06	6.67	5.87	5.29	4.85	3.51	2.23	1.55	1.13	0.842	0.228	0.0223	0.0025	0.0003
0.000002	12.1	11.2	10.5	9.89	9.44														
4	11.6	11.1	10.4	9.88	9.44														
6	11.3	10.9	10.4	9.87	9.44														
8	11.0	10.7	10.3	9.84	9.43														
0.00001	10.8	10.6	10.2	9.80	9.42	8.06													
2	10.2	10.1	9.84	9.58	9.30	8.06													
4	9.52	9.45	9.34	9.19	9.01	8.03	6.67												
6	9.13	9.08	9.00	8.89	8.77	7.98	6.67												
8	8.84	8.81	8.75	8.67	8.57	7.91	6.67												
0.0001	8.62	8.59	8.55	8.48	8.40	7.84	6.67	5.87											
2	7.94	7.92	7.90	7.86	7.82	7.50	6.62	5.86	5.29										
4	7.24	7.24	7.22	7.21	7.19	7.01	6.45	5.83	5.29	4.85									
6	6.84	6.84	6.83	6.82	6.80	6.68	6.27	5.77	5.27	4.85									
8	6.55	6.55	6.54	6.53	6.52	6.43	6.11	5.69	5.25	4.84									
0.001	6.33	6.33	6.32	6.32	6.31	6.23	5.97	5.61	5.21	4.83	3.51								
2	5.64	5.64	5.63	5.63	5.63	5.59	5.45	5.24	4.98	4.71	3.50								
4	4.95	4.95	4.95	4.94	4.94	4.92	4.85	4.74	4.59	4.42	3.48	2.23							
6	4.54				4.54	4.53	4.48	4.41	4.30	4.18	3.43	2.23							
8					4.26	4.25	4.21	4.15	4.08	3.98	3.36	2.23							
0.01	4.04			4.04	4.03	4.00	3.95	3.89	3.81	3.29	2.23	1.55	1.13						
2	3.35					3.35	3.34	3.31	3.28	3.24	2.95	2.18	1.55	1.13					
4	2.68					2.68	2.67	2.66	2.65	2.63	2.48	2.02	1.52	1.13	0.842				

Same Values → 0.842

Well Hydraulics

u		2.30	2.29	2.29	2.28	2.27	2.26	2.17	1.85	1.46	1.11	0.839				
6	2.30	Same Values →	2.03	2.02	2.02	2.01	2.00	1.94	1.69	1.39	1.08	0.832				
8	2.03			1.82	1.82	1.82	1.81	1.80	1.75	1.56	1.31	1.05	0.819	0.228			
0.1	1.82				1.22	1.22	1.22	1.22	1.19	1.11	0.996	0.857	0.715	0.227			
2	1.22				0.702	0.702	0.701	0.700	0.693	0.665	0.621	0.565	0.502	0.210			
4	0.702					0.454	0.454	0.453	0.450	0.436	0.415	0.387	0.354	0.177	0.0222		
6	0.454						0.311	0.310	0.310	0.308	0.301	0.289	0.273	0.254	0.144	0.0218	
8	0.311								0.219	0.218	0.213	0.206	0.197	0.185	0.114	0.0207	0.0025
1	0.219									0.049	0.048	0.047	0.046	0.044	0.034	0.011	0.0021	0.0003
2	0.049										0.0038	0.0037	0.0037	0.0036	0.0031	0.0016	0.0006	0.0002
4	0.0038													0.0004	0.0003	0.0002	0.0001	0
6	0.0004																	0
8	0																	

$$\left[\text{Example: for } u = 0.002 \ \frac{r}{B} = 0.2 \ W\left(u, \frac{r}{B}\right) = 3.50\right]$$

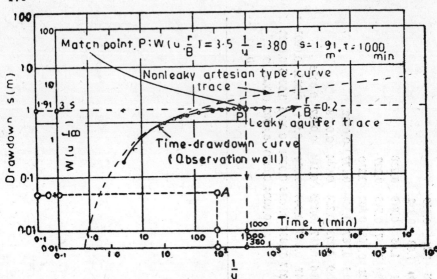

Fig. 5.16 Walton's method for leaky aquifer, Example 5.11.

curve coincides with $r/B = 0.2$ type curve and the coordinates of the match-point P can be read as*

$$W\left(u, \frac{r}{B}\right) = 3.5, \quad \frac{1}{u} = 380. \quad s = 1.91 \text{ m}. \quad t = 1000 \text{ min}$$

$$T = \frac{Q}{4\pi s} W\left(u, \frac{r}{B}\right)$$

$$= \frac{800}{4\pi \times 1.91} \times 3.5 = 117 \text{ lpm/m}$$

$$= 1.68 \times 10^5 \text{ lpd/m}$$

or $168 \text{ m}^2/\text{day}$

$$S = \frac{4Ttu}{r^2} = \frac{4 \times 0.117 \times 1,000}{30^2} \times \frac{1}{380} = 0.0014$$

$r/B = 0.2 \quad \therefore \quad B = \dfrac{20}{0.2} = 150 \text{ m}$

$B = \sqrt{Tc} \quad \therefore \quad c = B^2/T = \dfrac{150^2}{0.117}$

$= 1.92 \times 10^5 \text{ min or } 133.5 \text{ days}$

$K = \dfrac{T}{b} = \dfrac{0.117}{12} = 0.01 \text{ m/min}$

or 14.04 m/day

$c = b'/K' \quad \therefore \quad K' = b'/c = \dfrac{4}{133.5} = 0.03 \text{ m/day}$

---

*Any convenient point like A may be selected. If data on more than one observation well are available, the field plot should be 's vs $\dfrac{t'}{r^2}$.

Well Hydraulics

Steady Radial Flow—Leaky Artesian Aquifer

When the time-drawdown data fall on the flat portions of the family of leaky artesian type curves indicating that the well discharge is balanced by leakage from the semi-confining layer, an equilibrium stage has been reached and the cone of depression is given by the formula (de Glee, 1930, 1951; Hantush and Jacob, 1955)

$$s = \frac{Q}{2\pi T} K_o(r/B) \quad (5.20)$$

and

$$\frac{r}{B} = \frac{r}{\sqrt{T/(K'/b')}} \quad (5.21)$$

where $K_o\left(\frac{r}{B}\right)$ = modified Bessel function of the second kind and zero order

This steady flow is maintained so long as the water level in the formation supplying leakage remains constant. Steady-state leaky artesian type curve is shown in Fig. 5.17 where values of $K_o(r/B)$ versus r/B are plotted on log-log paper. Field data on several observation wells (under steady state conditions), s versus r, are plotted on transparent log-log paper of the same scale as the type curve. The distance-drawdown field plot is superimposed on the type curves keeping the axes parallel and a match point is selected. The match point coordinates $K_o(r/B)$, r/B, s and r are substituted in the Eqs. (5.20) and (5.21) to determine T and K'. The

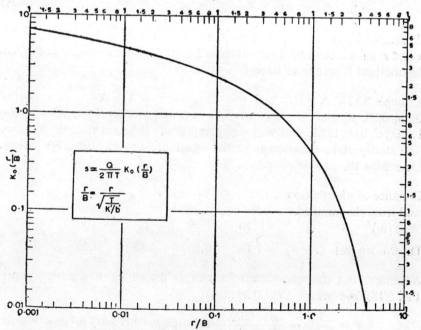

Fig. 5.17 Steady state leaky artesian type curve.

storage coefficient cannot be computed, under such conditions of flow, since the entire yield from the well is derived from leakage sources only. Values of $K_0(r/B)$ for various values of r/B are given in Table 5.15 (Hantush, 1956).

Hantush (1956, 1964) found that if $r/B < 0.05$, Eq. (5.20) can be approximated by

$$s = \frac{2.3\,Q}{2\pi\,T} \log_{10} 1.12 \frac{B}{r} \qquad (5.22)$$

To determine T, a distance-drawdown plot s versus log r, is drawn on semi-log paper, which will be a straight line for $r/B < 0.05$, and T is obtained from Eq. (4.82) as

$$T = \frac{2.3\,Q}{2\pi\,\Delta s}$$

where Δs is the drawdown per log cycle of distance r. If the straight line is extended to give an intercept r_0 on the abscissa where $s = 0$, then

$$1.12 \frac{B}{r_0} = 1, \qquad \text{from Eq. (5.22)}$$

or
$$B = \frac{r_0}{1.12} \qquad (5.23)$$

Since $B = \sqrt{Tc}$, the hydraulic resistance c is given by

$$c = \frac{r_0^2}{1.25\,T} \qquad (5.24)$$

Alternatively, if the coordinates of any point on the straight line plot are s and r, substitution of these values in Eq. (5.22) determines B, and c can be obtained from the relation $B = \sqrt{Tc}$.

Example 5.12: A pump test was conducted on a leaky artesian aquifer 30 m thick situated on an impervious base and overlain by a semi-confining layer 10 m thick. The well was pumped at a constant rate of 1800 lpm. The steady state drawdowns in the observation wells are given below. Determine the aquifer constants.

Distance of observation well from the pumped Well (m)	10	20	60	100	300
Drawdown (m)	0.66	0.55	0.35	0.26	0.07

Solution: The distance-drawdown plot is drawn on a semi-log paper, Fig. 5.18, from which $\Delta s = 0.39$ m,

$$T = \frac{2.3\,Q}{2\pi\,\Delta s} = \frac{2.3 \times 1800}{2\pi \times 0.39} = 1690 \text{ lpm/m or } \mathbf{2433 \text{ m}^2/\text{day}}$$

Well Hydraulics

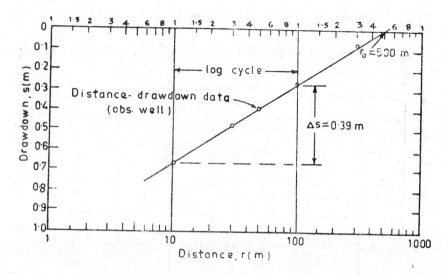

Fig. 5.18 Steady flow in leaky artesian aquifer $\left(\dfrac{r}{B} < 0.05\right)$.

(Example 5.12) (Hantush & Jacob, 1955)

$$K = \frac{T}{b} = \frac{1.69}{30} = 0.0563 \text{ m/min or } 81.2 \text{ m/day}$$

From the plot, the intercept r_0 when $s = 0$ is 500 m. The hydraulic resistance of the semi-confining layer is, therefore

$$c = \frac{r_0^2}{1.25\,T} = \frac{500^2}{1.25 \times 1.69} = 11.83 \times 10^4 \text{ min or } 82.2 \text{ days}$$

The vertical permeability of the semi-confining layer

$$K' = \frac{b'}{c} = \frac{10}{82.2} = 0.123 \text{ m/day}$$

Leakance, $\dfrac{K'}{b'} = \dfrac{0.123}{10} = 0.012$ per day

or 12 lpd/m³

Note: Leakage factor $B = \sqrt{Tc} = \sqrt{1.69\,(11.83 \times 10^4)} = 447$ m.

Since in this method $\dfrac{r}{B} < 0.05$, $r < 0.05 \times 447 = 22.35$ m

and only the data from the observation well at 10 m and 20 m can be used.

Alternatively, from the match point co-ordinates, Fig. 5.19,

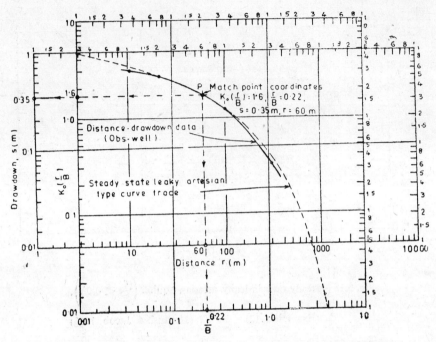

Fig. 5.19 Leaky aquifer constants for steady flow-type curve matching. (Example 5.12)

$K_0\left(\dfrac{r}{B}\right)$, $\left(\dfrac{r}{B}\right)$, s and r are respectively 1.6, 0.22, 0.35 m and 60 m. Substituting in Eq. (5.20)

$$0.35 = \frac{1800}{2\pi T} \times 1.6 = 1310 \text{ lpm/m}$$

$$K = \frac{T}{b} = \frac{1.310}{30} = 0.0437 \text{ m/min or } 63 \text{ m/day}$$

$$\frac{r}{B} = 0.22 \quad \frac{60}{B} = 0.22 \quad \therefore \quad B = 273 \text{ m}$$

$$B = \sqrt{Tc} \quad \therefore \quad c = \frac{B^2}{T} = \frac{273^2}{1.31} = 5.68 \times 10^4 \text{ min or, } 39.5 \text{ days}$$

$$c = \frac{b'}{K'} \quad \therefore \quad K' = \frac{b'}{c} = \frac{10}{39.5} = 0.253 \text{ m/day}$$

Leakance, $\dfrac{K'}{b'} = \dfrac{0.253}{10} = 0.253$ per day or 25.3 lpd/m³

Hantush Inflection Point Method—Unsteady Radial Flow—Leaky Artesian Aquifer

Hantush (1956) developed a method for determining T, S and c from

Table 5.15 Values of $K_o\left(\dfrac{r}{B}\right)$ and $e^{r/B}K_o\left(\dfrac{r}{B}\right)$ for different values of $\left(\dfrac{r}{B}\right)$ (abridged; after Hantush, 1956)

$\dfrac{r}{B}$	$K_o\left(\dfrac{r}{B}\right)$	$e^{r/B}K_o\left(\dfrac{r}{B}\right)$	$\dfrac{r}{B}$	$K_o\left(\dfrac{r}{B}\right)$	$e^{r/B}K_o\left(\dfrac{r}{B}\right)$	$\dfrac{r}{B}$	$K_o\left(\dfrac{r}{B}\right)$	$e^{r/B}K_o\left(\dfrac{r}{B}\right)$
0.010	4.72	4.77	0.10	2.43	2.68	1.0	0.421	1.14
12	4.54	4.59	12	2.25	2.53	1.2	0.318	1.06
14	4.38	4.45	14	2.10	2.41	1.4	0.244	0.988
16	4.25	4.32	16	1.97	2.31	1.6	0.188	0.931
18	4.13	4.21	18	1.85	2.22	1.8	0.146	0.883
0.020	4.03	4.11	0.20	1.75	2.14	2.0	0.114	0.842
22	3.93	4.02	22	1.66	2.07	2.2	0.0893	0.806
24	3.85	3.94	24	1.58	2.01	2.4	0.0702	0.774
26	3.77	3.87	26	1.51	1.95	2.6	0.0554	0.746
28	3.69	3.80	28	1.44	1.90	2.8	0.0438	0.721
0.030	3.62	3.73	0.30	1.37	1.85	3.0	0.0347	0.698
32	3.56	3.68	32	1.31	1.81	3.2	0.0276	0.677
34	3.50	3.62	34	1.26	1.77	3.4	0.0220	0.658
36	3.44	3.57	36	1.21	1.73	3.6	0.0175	0.640
38	3.39	3.52	38	1.16	1.70	3.8	0.0140	0.624
0.040	3.34	3.47	0.40	1.11	1.66	4.0	0.0112	0.609
42	3.29	3.43	42	1.07	1.63	4.2	0.0089	0.595
44	3.24	3.39	44	1.03	1.60	4.4	0.0071	0.582
46	3.20	3.35	46	0.994	1.58	4.6	0.0057	0.570
48	3.16	3.31	48	0.958	1.55	4.8	0.0046	0.559
0.050	3.11	3.27	0.50	0.924	1.52	5.0	0.0037	0.548

Contd.

52	3.08	3.24	52	0.892	1.50
54	3.04	3.21	54	0.861	1.48
56	3.00	3.17	56	0.832	1.46
58	2.97	3.14	58	0.804	1.44
0.060	2.93	3.11	0.60	0.777	1.42
62	2.90	3.09	62	0.752	1.40
64	2.87	3.06	64	0.728	1.38
66	2.84	3.03	66	0.704	1.36
68	2.81	3.01	68	0.682	1.35
0.070	2.78	2.98	0.70	0.660	1.33
72	2.75	2.96	72	0.640	1.32
74	2.72	2.93	74	0.600	1.30
76	2.70	2.91	76	0.601	1.28
78	2.67	2.89	78	0.583	1.27
0.080	2.65	2.87	0.80	0.565	1.26
82	2.62	2.85	82	0.548	1.24
84	2.60	2.83	84	0.532	1.23
86	2.58	2.81	86	0.516	1.22
88	2.55	2.79	88	0.501	1.21
0.090	2.53	2.77	0.90	0.487	1.20
92	2.51	2.75	92	0.473	1.19
94	2.49	2.73	94	0.459	1.18
96	2.47	2.72	96	0.446	1.16
98	2.45	2.70	98	0.443	1.15
0.100	2.43	2.58	1.00	0.421	1.14

Well Hydraulics

the time-drawdown data by reading on the plot (of s versus log t) the values of s_i, t_i and Δs_i, where the subscript i refers to the inflection point i.e. the point where the drawdown (s_i) is one-half of the final or equilibrium drawdown given by the Eq. (5.20), i.e.

$$s_i = \frac{Q}{4\pi T} K_0\left(\frac{r}{B}\right) \tag{5.25}$$

and $\quad u_i = \dfrac{r}{2B} \quad$ or, $\quad \dfrac{r}{2B} = \dfrac{r^2 S}{4Tt_i} \tag{5.26}$

The slope of the (tangent to the) curve at the inflection point (Δs_i) is the drawdown per log cycle time and is given by

$$\Delta s_i = \frac{2.3\,Q}{4\pi T} e^{-r/B} \tag{5.27}$$

Therefore

$$r = 2.3\, B\, (\log \frac{2.3\,Q}{4\pi T} - \log \Delta s_i) \tag{5.28}$$

Finally, it was shown that

$$2.3\frac{s_i}{\Delta s_i} = e^{r/B} K_0\left(\frac{r}{B}\right) \tag{5.29}$$

Values of the function $e^{r/B} K_0\left(\dfrac{r}{B}\right)$ for various values of $\dfrac{r}{B}$ are given in Table 5.4. Hence reading s_i, Δs_i and t_i from time-drawdown plot, the value of the function $e^{r/B} K_0\left(\dfrac{r}{B}\right)$ is determined from the Eq. (5.29) from which the values of $\dfrac{r}{B}$ and $K_0\left(\dfrac{r}{B}\right)$ can be found from Table 5.15. Then T is obtained from Eq. (5.25) and S from Eq. (5.26). B is determined from the value of $\dfrac{r}{B}$, and c is calculated from the relation $B = \sqrt{Tc}$.

Another solution by Hantush (1956) is for the time-drawdown data for several observation wells. The drawdown per log cycle of time (Δs) for the straight line portion of each curve (for each well at a known r) is determined. Then a plot r versus log Δs is drawn on a semi-log paper and a straight line of best-fit is drawn and the slope of this line, i.e. change Δr per log cycle of Δs is determined. The straight line is extended to give the intercept Δs_0 on the abscissa where $r = 0$ and, B and T are determined as

$$B = \frac{\Delta r}{2.3} \tag{5.30}$$

$$T = \frac{2.3\,Q}{4\pi\,\Delta s_0} \tag{5.31}$$

Chart 5.1
Aquifer Performance Test (Radial Flow)

Water table (unconfined) aquifer

- **Steady**
  Equilibrium formula, Dupuit, 1863 and Thiem, 1906
  $$Q = \frac{\pi K(h_2^2 - h_1^2)}{2.303 \log_{10} \frac{r_2}{r_1}}$$

- **Non-steady**
  (a) Theis non-equilibrium formula with certain limitations—Boulton, 1954 & Jacob's correction, 1944
  $$u = \frac{r^2 S}{4Tt}, \quad s' = \frac{Q}{4\pi T} W(u)$$
  $$s' = s - \frac{s^2}{2H}, \text{ if } \frac{s^2}{2H} > 0.003 \text{ m}$$
  (b) Dimensionless type curve method (not included in this book)

Artesian (confined) aquifer

- **Non-leaky**
  - **Steady**
    Equilibrium formula, Dupuit Thiem
    $$Q = \frac{2.72\, T(h_2 - h_1)}{\log_{10} \frac{r_2}{r_1}}$$
    $$T = Kb$$
  - **Non-steady**
    (a) Theis non-equilibrium formula, 1935
    $$u = \frac{r^2 S}{4Tt}, \quad s = \frac{Q}{4\pi T} W(u)$$
    Log-log plot of s vs. r^2/t matched with type curve.

- **Leaky**
  - **Steady** (discharge = leakage)
    (a) Hantush and (Jacob, 1955)
    $$s = \frac{Q}{2\pi T} K_0\left(\frac{r}{B}\right), \quad \frac{r}{B} = \frac{r}{\sqrt{\frac{T}{K'/b'}}}$$
    Log-log plot of s vs r matched with the 'Steady-state leaky artesian type curve',
  - **Non-steady**
    (a) Hantush and Jacob, 1955
    $$s = \frac{Q}{4\pi T} W\left(u, \frac{r}{B}\right)$$
    $$u = \frac{r^2 S}{4Tt},$$
    $$\frac{r}{B} = \frac{r}{\sqrt{\frac{T}{K'/b'}}}$$

Well Hydraulics

(b) Jacob's time-drawdown graph ($u \leqslant 0.01$), 1946
Plot s vs t on semi-log paper (x-axis—t; log scale)
$$T = \frac{2.303\,Q}{4\pi\Delta s}, \quad S = \frac{2.25\,Tt_0}{r^2}$$
and confirm boundaries.

(c) Jacob's distance-drawdown graph ($u \leqslant 0.01$). Plot s vs r on semi-log paper (x-axis—r; log scale)
$$T = \frac{2.303\,Q}{2\pi\Delta s}, \quad S = \frac{2.25\,Tt}{r_0^2}$$

(d) Jacob's composite-drawdown graph ($u \leqslant 0.01$) Plot s vs t/r^2 on semi-log paper (x-axis—t/r^2: semi-log paper
$$T = \frac{2.303\,Q}{4\pi\Delta s}, \quad S = 2.25\,T(t/r^2)_0$$

(e) Chow's method, 1952

(f) Theis' recovery, Jacob 1945 plot s' vs t/t' on semi-log paper (x-axis—t/t': log scale)
$$T = \frac{2.303\,Q}{4\pi\Delta s}, \quad (u \leqslant 0.01)$$

(b) Hantush approximation (1956, 1964),
when $\frac{r}{B} < 0.05$
$$s = \frac{2.303\,Q}{2\pi T}\log_{10} 1.12\frac{B}{r}$$
$$T = \frac{2.303\,Q}{2\pi\Delta s},$$
$$c = \frac{r_0^2}{1.25\,T}$$

(a) Log log plot of s vs t matched with the 'family of leaky artesian type curves' —Walton, 1962

(b) Hantush inflection-point method, 1956

and c is obtained from the relation $B = \sqrt{Tc}$. To determine S, the values of Q, T and $K_0\left(\dfrac{r}{B}\right)$ for the ratio $\dfrac{r}{B}$ are substituted in Eq. (5.25) and s_i is calculated. The corresponding value of t_i is obtained from the time-drawdown plot and S is determined from Eq. (5.26).

A third method by Hantush (1964) is based on a simplified solution of Eq. (5.18) which is valid if $t > 4t_i$ and $\dfrac{Tt}{SB} > 2r$.

Example 5.13: A fully penetrating 20 cm well in a semiconfined aquifer with prompt yield was pumped at a constant discharge of 1200 lpm. The unsteady drawdowns in an observation well at a distance of 40 m from the pumping well were 0.44, 0.77, 1.1, 1.34, 1.53, 1.66, 1.84 & 2.00 m at 30, 60, 100, 200, 300, 400, 600 & 1000 min, respectively. Determine the hydraulic properties of the aquifer, if the thickness of the aquifer is 20 m and that of the top semi-pervious layer 10 m.

Solution—Hantush inflection-point method: A plot of s vs t is drawn on semi-log paper, Fig. 5.20; from the plot

$$s_A = 1.96 \text{ m}, \quad s_i = 1 \text{ m}\left(\approx \dfrac{s_A}{2}\right), \quad t_i = 97 \text{ min}$$

$$\Delta s_i = 1.06 \text{ m}$$

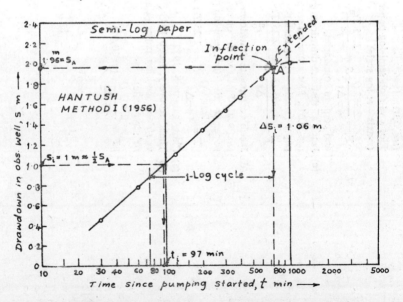

Fig. 5.20 Hantush inflection point method, Example 5.13.

Eq. (5.29):
$$2.3 \frac{s_t}{\Delta s_t} = e^{r/B} K_0\left(\frac{r}{B}\right)$$

$$2.3 \times \frac{1}{1.06} = 2.17 = e^{r/B} K_0\left(\frac{r}{B}\right)$$

From Table 5.15, $\quad K_0\left(\frac{r}{B}\right) = 1.8, \; \frac{r}{B} = 0.19$

Eq. (5.25):
$$s_t = \frac{Q}{4\pi T} K_0\left(\frac{r}{B}\right)$$

$$1 = \frac{1.200}{4\pi T} \times 1.8,$$

$$T = 0.172 \text{ m}^2/\text{min, or } \mathbf{248 \text{ m}^2/\text{day}}$$

$$T = Kb, \; K = \frac{248}{20} = 12.4 \text{ m/day}$$

$$B = \frac{r}{r/B} = \frac{40}{0.19} = 210 \text{ m}$$

$$B = \sqrt{Tc}, \; 210 = \sqrt{248\,c}$$

Hydraulic resistance $c = \mathbf{180 \text{ days}}$

$$c = \frac{b'}{K'}, \; 180 = \frac{10}{K'}, \; K' = \mathbf{0.055 \text{ m/day}}$$

$$\text{Leakance} = \frac{K'}{b'} = \frac{0.055}{10} = 0.0055, \text{ or } \mathbf{5.5 \times 10^{-3} \text{ day}^{-1}}$$

Eq. (5.26):
$$\frac{r}{2B} = \frac{r^2 S}{4T t_i}$$

$$\frac{40}{2 \times 210} = \frac{40^2 S}{4 \times 0.172 \times 97}, \; \mathbf{S = 0.004}$$

Unsteady Radial Flow in an Unconfined Aquifer

The Theis equations are based on the assumption that the water is released from storage in the aquifer in immediate response to the decline in the water table. In unconfined aquifers, gravity drainage may not be immediate, particularly in fine grained formations and with stratifications of slow yielding materials. When a well is pumped in such formations, a typical time-drawdown curve in an observation well is as shown in Fig. 5.21 (log-log plot, USBR, 1977), which can be divided into three segments:

(i) The first segment, of the observed drawdowns in the first few minutes after pumping began, is similar to the Theis type curve for confined aquifer; the pore water is released instantaneously from storage, due to the compression of the aquifer matrix and expansion of the entrapped air (i.e. not by free gravity drainage of the pore space), and the water level

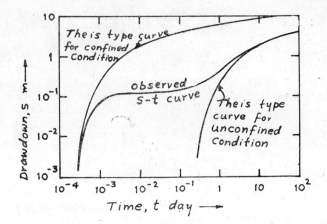

Fig. 5.21 Delayed yield in unconfined aquifer.

in the observation well drops relatively fast. Transmissibility of the aquifer can be determined by the Theis method, and the storage coefficient is equivalent to that of a confined aquifer.

(ii) The second segment of the time-drawdown curves indicates a flattening in slope; the water level drops at a slower rate and almost remains constant for a while (after which it begins to drop faster again) because of the replenishment by gravity drainage from the pore space above the cone of depression which is delayed.

(iii) In the third segment, an equilibrium is reached between the gravity drainage and the rate of fall of the water table. This condition is reached after several minutes to several days after pumping began and can be fitted to the Theis type curve for the unconfined condition, which gives the storage coefficient for the unconfined aquifer i.e. specific yield.

Thus for an unconfined aquifer with delayed yield, a pumping test should be continued sufficiently long to obtain the third segment of the curve and $S(=s_y)$ can be determined by one of the methods of solution of the nonequilibrium equation.

The minimum pumping time required for an accurate estimate of S in an unconfined aquifer depends on the transmissibility (T) of the aquifer. Boulton introduced the term B, defined as

$$B = \sqrt{\frac{T}{\alpha S_y}} \qquad (5.32)$$

where α is an empirical constant characteristic of the aquifer and its reciprocal $1/\alpha$ which has the dimension of time, is called the delay index. The delay index can be obtained for the formation material from Fig. 5.22a. Then knowing r and from an estimate of S_y and T, the minimum pumping time t_{wt} can be obtained from the dimensionless plot of αt_{wt} vs r/B,

Well Hydraulics

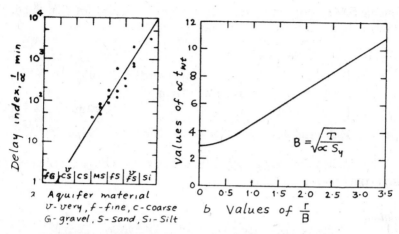

Fig. 5.22 Minimum pumping time in an unconfined aquifer with delayed yield (after Boulton 1963. Pricket 1965).

Fig. 5.22b. Values of the minimum pumping time for different aquifer materials are given by USBR (1977) as a guideline as follows:

Aquifer material	Minimum pumping time, hr
Medium sand and coarser materials	4
Fine sand	30
Silt and clay	170

Delayed yield is not restricted to unconfined aquifers alone. It may occur with leaky aquifers that receive water from the top semi-confining layer with a free water table. The hydraulic characteristics of such an aquifer can be determined by the methods of Boulton (1963) and Kruseman and de Ridder (1970). The r/B value from the best fitting type curve is used to calculate B and then α from Eq. (5.32) with the computed values of T and S_y; For the same r/B value, αt_{wt} is obtained from Fig. 5.22b. Knowing α, t_{wt} is calculated, which indicates the time that yield is no longer delayed and the time that the observed 's vs t curve' should merge with the Theis type curve for the unconfined aquifer. Prickett (1965) has given the following values for the delay index:

Aquifer material	delay index, $1/\alpha$, min
Coarse sand	10
Medium sand	60
Fine sand	200
Very fine sand	1000
Silt	2000

Example 5.14: For an unconfined aquifer consisting of fine sand, the following hydraulic characteristics are estimated.

Transmissibility	100 m²/day
Specific yield	10%
Distance of observation well from pumping well	40 m
Delay index	250 min

What is the minimum pumping time that yield is no longer delayed and Theis equation is applicable.

Solution:

$$B = \sqrt{\frac{T}{S_y} \cdot \frac{1}{\alpha}} = \sqrt{\frac{100}{0.10} \times \frac{250}{60 \times 24}} = 13.15 \text{ m}$$

From Fig. 5.22b, for $\dfrac{r}{B} = \dfrac{40}{13.15} = 3.04$, $\alpha t_{wt} = 9.5$

Therefore

$$t_{wt} = \frac{9.5}{\alpha} = 9.5 \times \frac{250}{1440} = \mathbf{1.65 \text{ day}}$$

Bailer Method

Skibitzke has developed a method for determining the coefficient of transmissibility from the recovery of water level in a well that has been bailed. For large values of t,

$$s' = \frac{V}{4\pi T t} \qquad (5.33)$$

where $V =$ volume of water removed during one bailer cycle $t =$ length of time since bailer was removed

If the residual drawdown is observed at some time after completion of n bailer cycles, removing the same quantity of water in each bailer cycle, then

$$s' = \frac{V}{4\pi T}\left(\frac{1}{t_1} + \frac{1}{t_2} + \ldots + \frac{1}{t_n}\right) \qquad (5.33a)$$

The bailer method is thus appled to a single observation of the residual drawdown after the time since bailing was stopped becomes large.

Slug Method

This is very similar to the bailer method. A quantity of water (V) or slug is injected and the residual heads s at times after injection is completed are noted. An arithmetic plot of residual heads versus reciprocals of times of observation produces a straight line whose slope, appropriately substituted in the equation

$$s' = \frac{V}{4\pi Tt} \tag{5.34}$$

permits computation of the coefficient of transmissibility.

This method has serious limitations since the duration of the test is very short and its use is restricted to artesian aquifers of small to moderate transmissibility ($< 7.5 \times 10^5$ lpd/m).

Application of the Formulae

The actual field conditions may differ from the basic assumptions on which the formulation has been made. If a field data plot on a logarithmic paper does not coincide with a non-leaky artesian type curve, and is displaced in one direction, it indicates possibly recharge or leakage, while displacement in the opposite direction indicates boundary conditions which are affecting the assumption of constant permeability or constant thickness of the aquifer. Various approaches have to be tried and compared such as drawing a time-drawdown curve (when the value of u is very small). A resume of aquifer pumping tests and formulae is given in Chart 5.1. For barrier and recharge boundaries special methods may be adopted such as the 'method of image wells'. Corrections have to be applied for partial penetrations; however, for distance $r > 1.5\ b$, effects of partial penetration are offset.

Multiple Aquifer Performance Tests

When a number of aquifers are encountered in a ground water project, it is necessary to determine the hydraulic characteristics of each aquifer, hydraulic interconnections (vertical leakage), and the quality of water from each aquifer. In advanced countries multiple aquifer performance tests are conducted in a single well using a number of piezometers at different depths of each aquifer with a number of packers.

In the Cauvery delta, each acquifer was tested separately with observation wells in the aquifer tested and also in the aquifers laying above and below it. This method of testing gives an idea of the vertical leakage or hydraulic interconnection of the different aquifers by observing the water level fluctuations in the other observation wells during pumping schedule and the results obtained from the multiple aquifer tests conducted in the Cauvery delta are given in Table 5.16 and Fig. 5.23*.

---

*'Study of Aquifer Systems in Cauvery Delta' by S. Panchanathan, et al. **Paper V-8**, *Int. Symp. on Devpt. of Ground Water Res., Proc. Vol. 3*, **1973, Madras.**

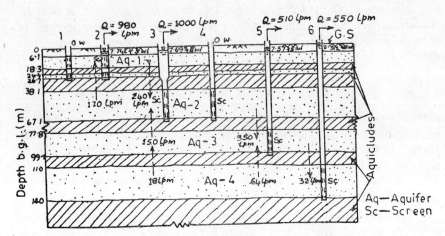

Fig. 5.23 Vertical leakage in Cauvery delta.

Fluctuations in Ground Water Levels

Fluctuations in ground water levels are caused by a pumping well in the vicinity, earthquake, loading and wind. Fluctuations in water levels sometimes occur when railroad trains or trucks pass aquifer test sites. Drawdown data must be adjusted for these changes in loading before they are used to determine the hydraulic properties of aquifers and confining beds. The ground water level rises when there is a decrease in the atmospheric pressure and the water level falls when the atmospheric pressure increases. Drawdown data are adjusted for atmospheric pressure changes occurring during an aquifer test by obtaining a record of atmospheric pressure fluctuations and using the equation

$$BE = \frac{\Delta W}{\Delta B} \times 100 \qquad (5.35)$$

where BE = barometric efficiency (%); ΔW = change in ground water level corresponding to a change in atmospheric pressure (m) and ΔB = change in atmospheric pressure (m of water).

Water levels in wells near surface water bodies are often affected by surface water stage fluctuations either because of a loading effect or a hydraulic connection between the surface water body and the aquifer. As the surface water stage rises, the water level in the well rises and as the surface water stage falls, the water level in the well falls. The ratio of the change in water level in a well to the change due to a loading effect is known as the tidal efficiency and the ratio of the change in the water level in a well to the change in the surface water stage because of a hydraulic connection is known as the river efficiency

$$TE = \frac{\Delta W}{\Delta R} \times 100 \qquad (5.36)$$

Table 5.16 Multiple aquifer performance Tests in the Cauvery delta (at Kiranur Village, Nannilam Taluk, Tanjore District)

Test no.	Well tested no.	SWL RL-m	Aquifer tapped by the test well	Pumping rate Q. lpm	Observation well no.	Aquifers tapped by observation wells	Vertical leakage lpm	Aquifer contributing leakage	Characteristics of the aquifer tapped by the test well				
									b m	T m²/day	K m/day	S	EC micromhos/cm at 25°C
1.	6	+8.312	4	550	5	3	32	3	30.47	135	4.40	—	1,000
2.	5	+7.57	3	510	4	2	350	2	21.62	122	5.87	—	1,055
					6	4	64	4					
3.	3	+7.69	2	1,000	4	2	240	1	28.95	820	28.40	3.3×10^{-4}	778
					5	3	150	3					
					6	4	18	4					
4.	2	+7.742	1	980	1	1	170	2	24.85	522	34.2	6.8×10^{-5}	862
					4	2							

$$RE = \frac{\Delta W}{\Delta R} \times 100 \tag{5.37}$$

where TE = tidal efficiency (%); RE = river efficiency (%); ΔW = change in water level in the well corresponding to a change in the surface water stage (m) and R = change in surface water stage (m).

Jacob (1950) derived expressions relating to barometric and tidal efficiencies and the elasticity of artesian aquifers, and obtained the relations

$$S = n\gamma_w\, b\beta = \frac{1}{BE} \tag{5.38}$$

$$BE + TE = 1 \tag{5.39}$$

[Also see Eqs. (4.70) and (4.72).]

drawdown data are adjusted for surface water stage change during an aquifer test by obtaining a record of surface water stage fluctuations and using Eqs. (5.36) and (5.37).

Well Flow Near Aquifer Boundaries—Image Wells

When a well is located close to a barrier (impermeable) or recharge boundary, marked deviations from a radial flow system occur. Solutions in such cases are readily obtainable by application of the method of images. An image, which may be an imaginary discharging or recharging well, creates a hydraulic system which is equivalent to the effects of a known physical boundary on the flow system. In essence, images enable an aquifer of finite extent to be transformed to one of infinite extent. This enables the radial flow equations to be applied to the modified system.

A common field problem is that of identifying and locating aquifer boundaries. Pumping test data can provide quantitative answers after application of the method of images and non-equilibrium equation. If a pumping well (PW) and an observation well (OW) are located near an unknown impermeable aquifer boundary (B) as shown in Fig. 4.27, an impage pumping well IPW will furnish the equivalent hydraulic flow system. From the law of times,

$$\frac{r^2}{t_p} = \frac{r_i^2}{t_i} \tag{5.40}$$

where r = distance of the observation well from the real well; r_i = distance of the observation well from the image well; t_p = time since pumping began to any selected drawdown before the boundary affects the drawdown, Fig. 5.24 and t_i = time since pumping began where the divergence of the drawdown curve from the type curve equals the selected drawdown, Fig. 5.24.

Half the value of r_i obtained from Eq. (5.40) may be taken as the distance of the impermeable aquifer boundary from the discharging well. Alterna-

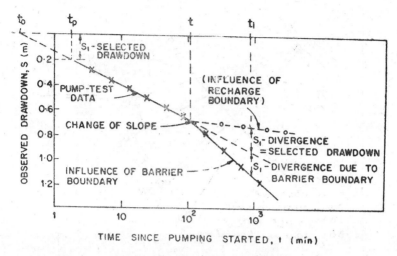

Fig. 5.24. Time-drawdown graph (Jacob's)—location of barrier and recharge boundaries.

tively, if the time (t_0) for zero drawdown is selected from the Jacob's time-drawdown curve and t is the time at which a change of slope (divergence) is indicated, the distance of the aquifer boundary from the discharging well is obtained from the equation

$$\frac{r^2}{t_0} = \frac{r_i^2}{t} \qquad (5.41)$$

which is the same as the Eq. 5.15 used earlier.

The distance r_i obtained from Eq. (5.40) or (5.41) gives an arc on which the image well lies. Data from two or more observation wells are required to locate the image well from the intersection of the three arcs, Fig. 5.25. The aquifer boundary is midway and perpendicular to the line joining the pumped well and the image well.

Example 5.15: The coefficients of transmissibility and storage of a non-leaky artesian aquifer are 5×10^5 lpd/m and 2×10^{-4}, respectively. The aquifer is bounded on one side by a barrier boundary. A fully penetrating production well has been discharging at a constant rate of 1200 lpm. The drawdown in an observation well 50 m from the pumped well due to the effects of the barrier boundary for a pumping period of 2 hours is 2m. Compute the distance from the observation well to the discharging image well associated with the barrier boundary (Fig. 5.22).

Solution: The time since pumping began to cause a drawdown of 2 m in the observation well before the boundary affects the drawdown is given by

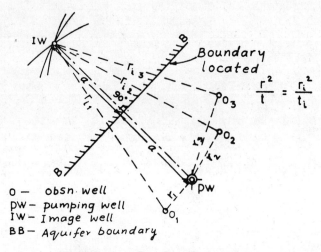

o — obsn well
pw — pumping well
IW — Image well
BB — Aquifer boundary

Fig. 5.25 Location of aquifer boundary.

$$s = \frac{2.3}{4\pi} \frac{Q}{T} \log_{10} \frac{2.25\, T\, t}{r^2 S}$$

$$2 = \frac{2.3\,(1200)}{4\pi \left(\dfrac{5 \times 10^5}{24 \times 60}\right)} \log_{10} \frac{2.25 \left(\dfrac{5 \times 10^5}{24 \times 60}\right) t}{50^2\,(2 \times 10^{-4})}$$

$$\log_{10} 1562\, t = 3.16$$

$$t = 0.925 \text{ min}$$

The distance from the observation well to the discharging image well is given by

$$\frac{r^2}{t_p} = \frac{r_i^2}{t_i}$$

$$\frac{50^2}{0.925} = \frac{r_i^2}{120}$$

$$r_i = 569 \text{ m}$$

Example 5.16: The drawdown is 3 m in an observation well 10 m away from the pumping well (drilled in an artesian aquifer) after 10 min of pumping. What is the time since pumping started, for the same drawdown in another observation well 20 m away from the pumping well?

Solution: Law of times for equal drawdown

$$\frac{r_1^2}{t_1} = \frac{r_2^2}{t_2}$$

Well Hydraulics

$$\frac{10^2}{10} = \frac{20^2}{t_2}, \quad t_2 = 40 \text{ min}$$

Example 5.17: The drawdowns in an observation well 12 m away from the pumping well, are 2.6 m and 2.9 m, after 8 and 80 min, since pumping started. What are the corresponding drawdowns in an observation well 120 m away from the pumping well.

Solution: Drawdown per log-cycle of distance is twice the drawdown per log-cycle of time. From the observation well data at $r = 12$ m, drawdown per log-cycle of time ($t = 8 - 80$ min)

$$\Delta s = 2.9 - 2.6 = 0.3 \text{ m}$$

Drawdown per log-cycle of distance ($r = 12 - 120$ m)

$$\Delta s' = 2 \Delta s = 2 \times 0.3 = 0.6 \text{ m}$$

∴ Drawdown at $r = 120$ m, at $t = 8$ min, is

$$= 2.6 - 0.6 = \mathbf{2.0 \text{ m}}$$

Drawdown at $r = 120$, at $t = 80$ min, is

$$= 2.0 + 0.3 = 2.3 \text{ m}.$$

which can also be obtained from the drawdown at $r = 12$ m at $t = 120$ min, as $2.9 - 0.6 = \mathbf{2.3 \text{ m}}$

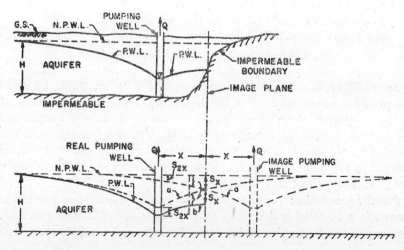

a. Cone of depression if there had been no impermeable boundary
b. Actual cone being depressed by the drawdown components S_x, S_{2x} of the image pumping well which simulates the effect of impermeable boundary

Fig. 5.26 Image well simulating barrier boundary.

Method of Images
Aquifers bounded by an impermeable boundary (aquiclude) and by a recharge boundary (surface water body) are shown in Fig. 5.26 & 5.27.

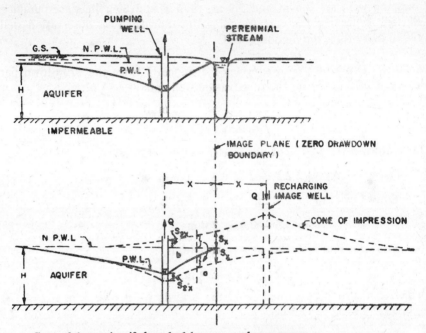

a. Cone of depression if there had been no recharge source
b. Actual cone of depression due to build-up components S_x, S_{2x} of the recharging image well which simulates the effect of recharge boundary

Fig. 5.27 Image well simulating recharge boundary.

These conditions can be modelled mathematically by imagining an equal well placed symmetrically about the image plane which is the plane of the boundary itself. For a barrier boundary the image well is a discharge well and for a recharge boundary the image well is a recharge well. The drawdown at any point is the algebraic sum of the drawdowns resulting from the real (pumping) and image wells. For convenience, the static water level (s.w.l.) in the real well is taken as the datum.

Parallel boundaries: In the event of two parallel boundaries, two image planes are considered which leads to an infinite set of images, Fig. 5.28. However, in practice, pairs of image wells are added until the next pair has negligible influence on the sum of all image well effect out to the point. The image well theory may be applied to such cases by taking into consideration the successive reflections on the boundaries. It is convenient to remember 'ROBS', i.e. 'Recharge boundary—Opposite, Barrier boundary—Same' meaning that the image well is of opposite nature to that of

Well Hydraulics

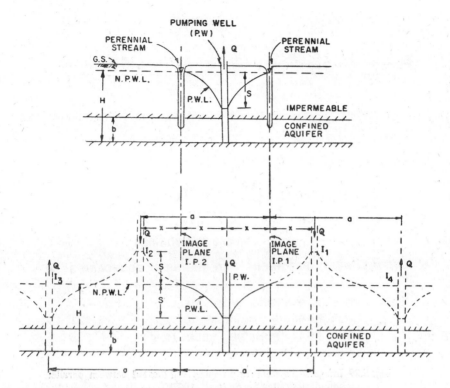

I_1—Recharging image well of P.W. due to I.P. 1 (Recharge boundary)
I_2—Recharging image well of P.W. due to I.P. 2 (Recharge boundary)
I_3—Image pumping well of I_1 due to I.P. 2 (Mask I.P. 1)
I_4—Image pumping well of I_2 due to I.P. 1 (Mask I.P. 2)

Fig. 5.28 Image wells for a pumping well located between two parallel recharge boundaries.

the real (pumping) well in the case of a recharge boundary and of the same nature as the real well in the case of a barrier boundary; also when working the images through one image plane, the other image plane may be masked. For example consider working out the image wells (I) of the real (pumping) well PW due to two parallel boundaries—one barrier B and one recharge R, in Fig. 5.29a. I_1 and I_2 are the images of PW due to the boundaries R and B, respectively; mask B and find I_5 the image of I_2 due to R; mask R and find I_4, I_5 the images of I_1, I_3 respectively due to B; mask B and find I_6, I_7 the images of I_4, I_5 respectively due to R and so on; also I_1, I_3, I_4 and I_5 are recharging image wells and I_2, I_6 and I_7 are discharging image wells. Actually corrections need to be made only for a few of the nearest of the images, the number depending upon the accuracy desired. The image wells (I) of the real (pumping) well (PW) due to two parallel barrier boundaries (B) is shown in Fig. 5.29 b.

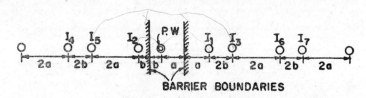

Fig. 5.29 Image well system for pumping well located between parallel boundaries (after Ferris et al., 1962).

Rectangular and wedge shaped boundaries: Image well systems for rectangular and wedge shaped boundaries are shown in Fig. 5.30. Each primary image will produce an unbalanced effect at the opposite boundary. Secondary image wells must be added at appropriate positions until the effects of both real and image wells are balanced at both boundaries.

If θ is the angle included between the boundaries, the exact number of image wells.

$$n_i = \frac{360°}{\theta} - 1 \tag{5.42}$$

For the wedge shaped boundary in Fig. 5.30 b, $\theta = 45°$,

$$n_i = \frac{360}{45} - 1 = 7$$

The image wells are positioned in a circle whose centre is at the apex of the wedge and whose radius is equal to the distance from the real (pumped) well to the wedge apex. The image wells, *I*, are numbered in the sequence in which they are considered and located.

Multiple boundaries: The above procedures can be extended to analyse

Well Hydraulics

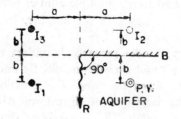

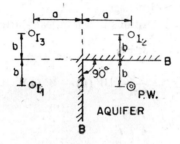

a. Rectangular boundaries

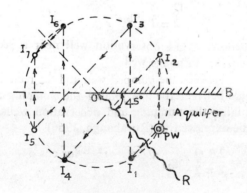

b. Wedge shaped boundary

● — RECHARGING IMAGE WELL
○ — PUMPING IMAGE WELL
◎ — REAL PUMPING WELL (P.W.)
B — BARRIER BOUNDARY
R — RECHARGE BOUNDARY

Fig. 5.30 Image-well systems for pumping well located between rectangular and wedge shaped boundaries (after Ferris et al., 1962).

the pumping effects and locate multiple aquifer boundaries. The complexity of the solution increases with the number, irregularity and different types of boundaries present.

Recharge boundary: Under recharge boundary conditions, water level in the observation well will drawdown initially under the influence of the production well only. When the cone of depression spreads to the recharge boundary, the time-rate of drawdown will change, due to the influence of the image well. The time-rate of drawdown will thereafter continually decrease till eventually equilibrium conditions are reached, when recharge balances discharge.

Under enuilibrium conditions in a non-leaky artesian aquifer

$$s_r = s - s_i \tag{5.43}$$

From Eq. (4.45) $\quad s_r = \dfrac{Q}{4\pi T}[W(u) - W(u_i)] \tag{5.44}$

$$u = \dfrac{r^2 S}{4Tt} \tag{5.45}$$

$$u_i = \dfrac{r_i^2 S}{4Tt} \tag{5.46}$$

From Eq. (4.47) $\quad s_r = \dfrac{2.3\,Q}{2\pi T}\log_{10}\dfrac{r_i}{r} \tag{5.47}$

where s_r = drawdown in the observation well near a recharge boundary and s_i = build-up due to image well

If the distance between the pumped well and the recharge boundary is a and ϕ is the angle between the line joining the pumping and image wells, and the line joining the pumping and observation wells, Fig. 5.31, then Eq. (5.47) can be expressed as (Rorabaugh, 1956)

$$s_r = \dfrac{2.3\,Q}{2\pi T}\log_{10}\dfrac{\sqrt{4a^2 + r^2 - 4ar\cos\phi}}{r} \tag{5.48}$$

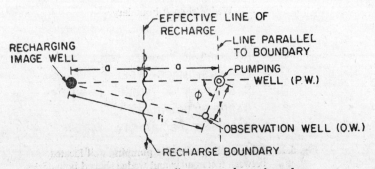

Fig. 5.31 Pumped well near a recharge boundary.

and the drawdown in the pumping well (due to recharge boundary)

$$s_{wr} = \frac{2.3 \, Q}{2 \, \pi \, T} \log_{10} \frac{2a - r_w}{r_w} \qquad (5.49)$$

or
$$s_{wr} \approx \frac{2.3 \, Q}{2 \, \pi \, T} \log_{10} \frac{2a}{r_w} \qquad (5.50)$$

which when compared with Eq. (5.7) shows that in the first approximation this drawdown is equal to that at the face of a well in a circular island aquifer of radius $2a$.

The percentage of pumped water diverted from a source of recharge depends upon the hydraulic properties of the aquifer, the distance of the recharge boundary from the pumping well and the pumping period. If Q_s is the volume of water drawn from the recharge source, the fraction of the well discharge Q_s/Q drawn is given by the dimensionless plot of Q_s/Q versus $a\sqrt{4 \, K \, b \, t/S}$ as shown in Fig. 5.32 (Glover and Balmer, 1954). The same result applies to an unconfined aquifer by replacing the storage coefficient with the specific yield, provided the drawdown is small compared to the saturated thickness of the aquifer.

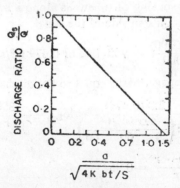

Fig. 5.32 Portion of well discharge furnished by a recharge source (after Glover and Balmer, 1954).

The time required to reach approximate equilibrium conditions may be computed with the equation (Foley et al., 1953)

$$t_e = \frac{a^2 S}{2.25 \, T \, \varepsilon_e \log_{10} \left(\frac{2a}{r}\right)^2} \qquad (5.51)$$

where t_e = time required to reach approximate equilibrium conditions and ε_e = deviation from absolute equilibrium generally assumed as ≈ 0.05.

Example 5.18: From the pumping tests in the area irrigated by the Ganga canal, UP, $S = 0.11$ and $T = 1.5 \times 10^6$ lpd/m. A well is to be sited to intercept the seepage and prevent waterlogging in the reach of the canal where the seepage losses are found to be 105 lps/km.

What should be the distance of the tubewell from the canal so that it does not induce extra seepage from the canal, and the discharge and spacing of tubewells in this area.

If a 30 cm tubewell is located at 250 m from the canal and pumped at the rate of 5000 lpm, determine
  a. the time required to reach approximate equilibrium conditions
  b. the percentage of pumped water furnished by the canal after pumping for one month
  c. the resulting drawdown in the tube well

Solution: The distance of the tubewell from the canal can be determined by trial by assuming $a = 300$ m and estimating r for $s = 0.1$ m along the perpendicular from the tubewell to the canal, i.e. $\phi = 0$. For such a case Eq. (5.48) becomes

$$s = \frac{2.3\,Q}{2\pi T} \log_{10} \frac{2a - r}{r}$$

Assuming a minimum spacing of tubewells along a line parallel to the canal as $5a$, the seepage loss on one side of the canal

$$\frac{105\,(5 \times 300)}{2 \times 1000} = 78.8 \text{ lps. or } 0.079 \text{ m}^3/\text{sec}$$

Assuming that this seepage loss is equal to the discharge from the well

$$0.1 = \frac{2.3\,(0.079)}{2\pi \left(\dfrac{1.5 \times 10^3}{24 \times 60 \times 60}\right)} \log_{10} \frac{2 \times 300 - r}{r}$$

$$\log_{10} \frac{600 - r}{r} = 0.062$$

$$r = 279 \text{ m}$$

For $s = 0.01$ m, $\log_{10} \dfrac{600 - r}{r} = 0.0062$

$$r = 297.6 \text{ m}$$

that is at a distance of 297.6 m from the tubewell towards the canal (or at 2.4 m from the canal towards the tubewell) the drawdown will be just 1 cm. The radius of influence is within the critical distance a, so that the pumping well does not induce extra seepage from the canal. Actually a fraction of the pumpage (as can be determined by constructing a flow net for the discharging well) comes from the landward side and the canal seepage intercepted is less than 78.8 lps. The radius of influence is at a safe margin from the canal and the well does not induce extra seepage from the canal.

Well Hydraulics

The distance r for $s = 0.1$ m along a line parallel to the canal and on either side of the pumping well can be determined by putting $\phi = 90°$ or $270°$ in the Eq. (5.48)

$$s = \frac{2.3\,Q}{2\pi\,T} \log_{10} \frac{\sqrt{4a^2 + r^2}}{r}$$

$$0.1 = \frac{2.3\,(0.079)}{2\pi \left(\dfrac{1.5 \times 10^3}{24 \times 60 \times 60}\right)} \log_{10} \frac{\sqrt{4\,(300)^2 + r^2}}{r}$$

$$\frac{\sqrt{4(3000)^2 + r^2}}{r} = 1.153$$

By trial and error, $r = 1050$.

The distance r for $s = 0.1$ m along a line perpendicular to the canal and on the other side of the pumping well can be determined by putting $\phi = 180°$ in the Eq. (5.48)

$$s = \frac{2.3\,Q}{2\pi\,T} \log_{10} \frac{2a + r}{r}$$

$$0.1 = \frac{2.3\,(0.079)}{2\pi \left(\dfrac{1.5 \times 10^3}{24 \times 60 \times 60}\right)} \log_{10} \frac{2 \times 300 + r}{r}$$

$$\frac{600 + r}{r} = 1.153$$

$$r = 3920 \text{ m}$$

The distances r from the pumping well in the four directions corresponding to a given drawdown can be determined as shown in Table 5.17 and a flownet can be drawn as shown in Fig. 5.33.

As shown in Fig. 5.33, the tubewell is located at 300 m from the canal and the lines of equal drawdown (equipotential lines) are drawn by plotting the distances on the four sides of the well. The flow lines (stream lines) are then drawn by trial and error to form an orthogonal network called flownet. It can be seen from the flownet that the length of the canal which comes under the influence of pumping, with no extra seepage induced, is 1600 m. This distance normally varies from $4a$ to $6a$. The seepage from the canal in this reach at 105 lps/km is $105 \times 1.6 = 168$ lps, and from one side of the canal it is 84 lps. From the flownet, the fraction of the pumpage coming from the landward side $= 79 \times \dfrac{5}{20} \approx 20$ lps. The balance of pumpage $79 - 20 = 59$ lps comes from the canal side against a seepage of 84 lps from the canal. The well is not inducing extra seepage. If the pumpage can be increased to 110 lps, the water coming from the landward side $= 110 \times \dfrac{5}{20} = 27.5$ lps and the balance of $110 - 27.5 = 82.5$ lps is

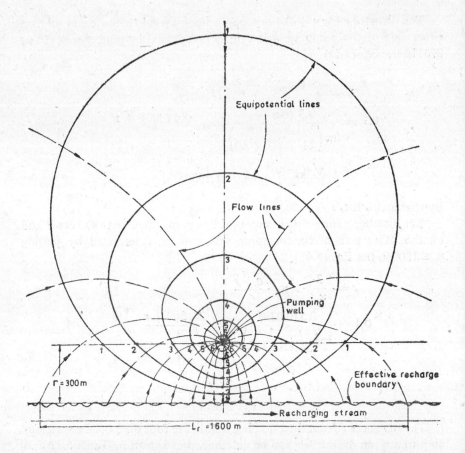

Fig. 5.33 Flownet for the pumping well near Ganga canal, Example 5.18.

drawn from the canal seepage. Still, the well will not milk the canal with a narrow margin of $84 - 82.4 = 1.5$ lps. This can be confirmed by redrawing the flownet with the increased pumpage of 110 lps. Alternatively, the distance of the tubewell from the canal may be slightly reduced, instead of increasing the pumpage.

Spacing of Tubewells

The tubewells should be so spaced that there is no interference of their cones of influence and the spacing between any two tubewells should be more than two times the radius of influence with a safe margin. In the above example tubewells may be located on a line parallel to the canal with a spacing of more than $2r = 2 \times 1050 = 2100$ m, say 3 km apart, with an average pumpage of 100 lps.

(a) If a 30 cm tubewell is located at 250 m from the canal, the time required to reach approximate equilibrium is

Well Hydraulics

Table 5.17 Distance r from the pumping well in the four directions corresponding to a given drawdown

Drawdown $s(m)$	Distance $r(m)$ on the perpendicular from the well towards the canal $s = \dfrac{2.3Q}{2T}\log_{10}\dfrac{2a-r}{r}$	Distance $r(m)$ from the well on a line parallel to the canal and on either side $s = \dfrac{2.3Q}{2\pi T}\log_{10}\dfrac{\sqrt{4a^2+r^2}}{r}$	Distance $r(m)$ from the well on the side away from the canal $s = \dfrac{2.3Q}{2\pi T}\log_{10}\dfrac{2a+r}{r}$
0.1	279	1050	3920
0.2	257	685	1820
0.4	216	410	780
0.6	179	280	443
0.8	145	202	282
1.0	116	148	189
2.0	32.6	34.5	36.6
3.0	8.2	8.3	8.4
4.0	1.98	1.99	1.99

$$t_e = \dfrac{a^2 S}{2.25\, T\varepsilon_e \log_{10}\left(\dfrac{2a}{r}\right)^2} = \dfrac{250^2\,(0.11)}{2.25\,(1.5\times 10^3)\,0.05\,\log_{10}\left(\dfrac{2\times 250}{0.15}\right)^2}$$

$$= 5.78 \text{ days}$$

(b) Percentage of pumpage drawn from the canal after 1 month

$$\dfrac{a}{\sqrt{4Kbt/S}} = \dfrac{250}{\sqrt{4(1.5\times 10^3)\,30/0.11}} = 0.218$$

From the dimensionless plot Fig. 5.28

$$\dfrac{Q_s}{Q} = 0.78, \text{ or } 78\%$$

(c) The resulting drawdown in the tubewell

$$s_{wr} = \dfrac{2.3\,Q}{2\pi T}\log_{10}\dfrac{2a-r_w}{r_w} = \dfrac{2.3\times 5}{2\pi\left(\dfrac{15\times 10^3}{24\times 60}\right)}\log_{10}\dfrac{(2\times 250)-0.15}{0.15}$$

$$= 6.2 \text{ m}$$

Aquifer Test—Recharge Boundary

The distance-drawdown data in several observation wells parallel to the recharge boundary at the end of the test, when water levels are stabilised, are plotted (i.e. s_r vs r) on a semi-log paper and the coefficient of transmissibility is determined from the Jacob's method as

$$T = \frac{2.3 Q}{2\pi \Delta s_r} \tag{5.52}$$

The computed value of T and other known data for the observation well are substituted in Eq. (5.48) and the distance of the pumped well from the recharge boundary a is determined. Since the drawdowns are affected by recharge, the storage coefficient S cannot be computed from the distance-drawdown field data graph. After T and a are computed, the image well may be located and the coefficient of storage can be estimated by a process of trial and error with the Eqs. (5.44) through (5.46). Several values of S are assumed and s_r is determined. The value of S, for which s_r corresponds to the observed value, is taken as the coefficient of storage.

Example 5.19: A production well hydraulically connected to a stream 400 m wide has been discharging continuously at a constant rate of 1800 lpm for a period of three days. The drawdown in three observation wells fully penetrating the water table aquifer, on a line through the production well and parallel to the stream are given in the following. Compute the coefficients of transmissibility and storage of the aquifer and the average infiltration rate of the stream bed.

Distance from production well (m)	18	45	80
Drawdown (m)	0.282	0.177	0.099

Solution: A distance-drawdown graph is drawn on a semi-log plot, Fig. 5.34, and Δs_r per log cycle of r is measured.

$$T = \frac{2.3 Q}{2\pi \Delta S_r} = \frac{2.3 (1.8)}{2\pi (0.26)}$$

$$= 2.54 \text{ m}^3/\text{min/m} = 3.65 \times 10^6 \text{ lpd/m}$$

To determine the distance a, applying Eq. (5.48) to the second observation well

$$s = \frac{2.3 Q}{2\pi T} \log_{10} \frac{\sqrt{4a^2 + r^2}}{r}$$

$$0.177 = \frac{2.3 (1.8)}{2\pi (2.54)} \log_{10} \frac{\sqrt{4a^2 + 45^2}}{45}$$

$$a = 108 \text{ m}$$

Well Hydraulics

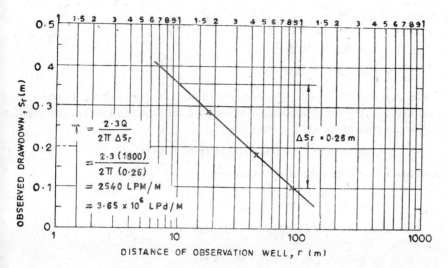

Fig. 5.34 Distance-drawdown graph (Jacob's)—recharge boundary (Example 5.19).

$$r = 45 \text{ m}$$
$$r_i = \sqrt{(2 \times 108)^2 + 45^2} = 221 \text{ m}$$

Assuming $S = 0.003$ (by trial and error process)

$$u = \frac{r^2 S}{4 T t} = \frac{45^2 (0.003)}{4 (2.54 \times 3 \times 24 \times 60)}$$
$$= 13.8 \times 10^{-5}$$

$W(u) = 8.2968$, from Table 5.1

$$u_i = \frac{r_i^2 S}{4 T t} = \frac{221^2 (0.003)}{4 (2.54 \times 3 \times 24 \times 60)}$$
$$= 3.33 \times 10^{-3}$$

$W(u_i) = 5.1399$, from Table 5.1

From Eq. (5.44)

$$s_r = \frac{Q}{4\pi T} [W(u) - W(u_i)] = \frac{1.8}{4\pi (2.54)} [8.2968 - 5.1399]$$
$$= 0.178$$

which is very nearly equal to the observed drawdown in the second observation well. Hence S can be taken as 0.003 for the water table aquifer connecting the production well with the stream.

To find the volume of water drawn from the recharge source $\frac{Q_s}{Q}$:

$$f = \frac{a^2 S}{4 T t} = \frac{108^2 (0.003)}{4 (2.54 \times 3 \times 24 \times 60)}$$
$$= 6.1 \times 10^{-4}$$
$$\sqrt{f} = 2.47 \times 10^{-2}$$

Since $\sqrt{f}$ is very small, the entire pumpage is drawn from the stream (see Fig. 5.32). Taking the length of induced infiltration (L_r) as nearly equal to $5a$, the area of induced infiltration

$$A_r = 400 \, L_r = 400 \, (5 \times 108) = 216{,}000 \text{ m}^2$$

Average infiltration rate of the stream bed

$$I_{ave} = \frac{Qr}{A_r} = \frac{1.8 \, (60 \times 24)}{216000}$$
$$= 0.012 \text{ m/day} = 12 \text{ lpd/m}^2$$

Method of Images for Particular Cases

1. Well adjacent to a stream or line source, Fig. 5.35

For confined nonleaky aquifer with no recharge, applying unsteady state equation, (Jacob's) the drawdown at any time t,

$$s = \frac{2.3 \, Q}{4 \pi \, T} \log \frac{2.25 \, T \, t}{r^2 \, S}$$

or $$s = \frac{2.3 \, Q}{4 \pi \, T} \log \frac{C}{r^2}, \; C = \frac{2.25 \, T \, t}{s}$$

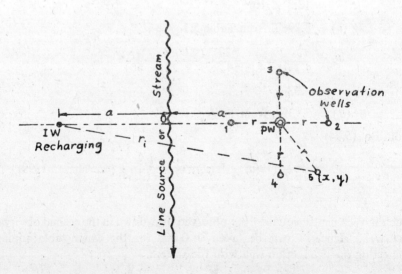

Fig. 5.35 Well in the vicinity of a stream.

Well Hydraulics

a. *The drawdown at the well face due to the pumping and image well,*

$$s_w = \frac{2.3\,Q}{4\pi T}\left[\log\frac{C}{r_w^2} - \log\frac{C}{(2a)^2}\right]$$

$$= \frac{2.3\,Q}{4\pi T}\log\left(\frac{2a}{r_w}\right)^2,$$

or $s_w = \dfrac{2.3\,Q}{2\pi T}\log\left(\dfrac{2a}{r_w}\right),\ R = 2a = r_i$ (5.53)

Applying the steady state equation (Dupuit's)

$$s = \frac{2.3\,Q}{2\pi T}\log\frac{R}{r}$$

The drawdown at the well face due to the pumping and image well,

$$s_w = \frac{2.3\,Q}{2\pi T}\left[\log\frac{R}{r_w} - \log\frac{R}{2a}\right]$$

$$= \frac{2.3\,Q}{2\pi T}\log\left(\frac{2a}{r_w}\right),\ R = 2a = r_i \quad (5.53a)$$

When the unsteady equation is used, the time (t) drops out of the equation, and the drawdown is the same as that given by the Dupuit's equation for the steady state. This shows that the presence of recharge boundary in the aquifer makes the flow steady and the drawdown at the well face or any observation well may be calculated by the Dupuit's equation by putting an appropriate value for the radius of influence, which is $R = 2a$ in the above case.

b. *Drawdown in the observation wells—1, 2, 3 and 4 applying the Dupuit's equation* (neglecting the well radius r_w),

(i) $s_1 = \dfrac{2.3\,Q}{2\pi T}\left[\log\dfrac{R}{r} - \log\dfrac{R}{2a-r}\right]$

or $s_1 = \dfrac{2.3\,Q}{2\pi T}\log\dfrac{2a-r}{r}$, i.e. $R = 2a - r = r_i$ (5.54)

(ii) $s_2 = \dfrac{2.3\,Q}{2\pi T}\left[\log\dfrac{R}{r} - \log\dfrac{R}{2a+r}\right]$

or $s_2 = \dfrac{2.3\,Q}{2\pi T}\log\dfrac{2a+r}{r}$, i.e. $R = 2a + r = r_i$ (5.55)

(iii) $s_{3,4} = \dfrac{2.3\,Q}{2\pi T}\left[\log\dfrac{R}{r} - \log\dfrac{R}{\sqrt{(2a)^2 + r^2}}\right]$

or $s_{3,4} = \dfrac{2.3\,Q}{2\pi T}\log\dfrac{\sqrt{4a^2 + r^2}}{r}$ i.e. $R = \sqrt{4a^2 + r^2} = r_i$ (5.56)

Note : r_i = Distance observation well from the image well.

The above results can be obtained by applying the unsteady state equation (Jacob's), when the time factor drops out.

The steady-state drawdown at any point (x, y) is given by

$$s = \frac{2.3\ Q}{4\pi T} \log \frac{(x+X)^2 + (y-Y)^2}{(x-X)^2 + (y+Y)^2} \tag{5.57}$$

where (X, Y) are the coordinates of the pumping well. When the real well is located at $(a, 0)$, Fig. 5.35,

$$s = \frac{2.3\ Q}{4\pi T} \log \frac{r_i^2}{r^2} = \frac{2.3\ Q}{4\pi T} \log \frac{(x+a)^2 + y^2}{(x-a)^2 + y^2} \tag{5.57a}$$

Example 5.20: A 30 cm well is pumped at the rate of 1000 lpm. The transmissibility of the aquifer is 0.015 m²/s. If the well is located at a distance of 120 m from a stream, what should be the drawdown?

(i) in the pumping well
(ii) in an observation well 80 m away from the pumping well on the side opposite to the stream?
(iii) in an observation well 80 m away from the pumping well, on a line parallel to the stream?

Solution:

(i) Eq. (5.53): $s_w = \dfrac{2.3 \times \frac{1}{60}}{2\pi\ (0.015)} \log \left(\dfrac{2 \times 120}{0.15}\right) = 1.31$ m

Eq. (5.57a): $s_w = \dfrac{2.3\ (1/60)}{4\pi\ (0.015)} \log (120 + 120)^2 = 0.97$ m

(ii) Eq. (5.55): $s_w = \dfrac{2.3 \times \frac{1}{60}}{2\pi\ (0.015)} \log \left(\dfrac{2 \times 120 + 80}{80}\right) = 0.245$ m

Eq. (5.57a): $s_w = \dfrac{2.3 \times 1/60}{4\pi\ (0.015)} \log \dfrac{(200 + 120)^2}{(200 - 120)^2} = 0.245$ m

(iii) Eq. (5.56): $s = \dfrac{2.3\ (1/60)}{2\pi\ (0.015)} \log \dfrac{\sqrt{4 \times 120^2 + 80^2}}{80} = 0.207$ m.

Eq. (5.57a): $s = \dfrac{2.3\ (1/60)}{4\pi\ (0.015)} \log \dfrac{240^2 + 80^2}{80^2} = 0.207$ m.

2. Well in a strip of land bounded by two streams, Fig. 5.36

The Image wells extended to infinity, but for practical purposes it is enough if only a pairs of image wells are considered for estimating drawdown.

Well Hydraulics

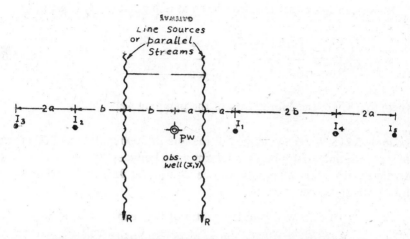

Fig. 5.36 Well in a strip bounded by two parallel streams.

The drawdown at the well face, due to the pumping well and the image wells I_1, I_2, I_3, I_4 and I_5 is

$$s_w = \underset{PW}{\frac{2.3\,Q}{2\pi T}} \underset{I_1}{\log \frac{R}{r_w}} - \underset{I_2}{\log \frac{R}{2a}} - \underset{I_3}{\log \frac{R}{2b}} + \log \frac{R}{2b+2a}$$

$$+ \log \frac{R}{2b+2a} - \log \frac{R}{2b+4a}$$

$$s_w = \frac{2.3\,Q}{2\pi T} \log \frac{2a \cdot 2b\,(2b+4a)}{r_w\,(2b+2a)^2} \tag{5.58}$$

or $\quad s_w = \dfrac{2.3\,Q}{2\pi T} \log \dfrac{2ab\,(2a+b)}{r_w\,(a+b)^2}$, i.e. $R = \dfrac{2ab\,(2a+b)}{(a+b)^2}$ (5.58)

It can be shown that

$$R = \frac{2L}{\pi} \sin \frac{a\pi}{L} \tag{5.59}$$

in the Dupuit's equation, where $L = a + b$ = width of the strip of land.
If the well is located at the centre of the strip of land $a = b$,

$$s_w = \frac{2.3\,Q}{2\pi T} \log \frac{1.5a}{r_w}, \text{ i.e. } R = 1.5a$$

Actually,

$$R = 1.27\,a, \tag{5.60}$$

since another recharging image well (I_6 towards the left, which is at the same distance as I_5 for this case) is not considered.

It can be shown that the drawdown at any point (x, y) is given

$$s = \frac{2.3\,Q}{4\pi T} \log \frac{\cosh \frac{(y-Y)}{L} + \cos \frac{(x+X)}{L}}{\cosh \frac{(y-Y)}{L} - \cos \frac{(x+X)}{L}} \tag{5.61}$$

where (X, Y) are the coordinates of the pumping well.

Example 5.21: Water is pumped at the rate of 3000 lpm from a 30 cm well fully penetrating a strip aquifer 1.6 km wide bounded by two field ditches. Assuming the transmissibility of the aquifer as 0.02 m²/s, what is the drawdown at the well face if

(i) the well is at the centre of the strip

(ii) the well is at 400 m from one ditch

What is the drawdown in an observation well in case (i), located at (300, 400 m) with reference to the pumping well?

Solution:

(i) Dupuit's formula with $R = 1.27\,a$, $a = \frac{L}{2}$

$$R = 1.27 \times \frac{1600}{2} = 1015 \text{ m}$$

$$s_w = \frac{2.3\,Q}{2\pi T} \log \left(\frac{R}{r_w}\right) = \frac{2.3\,(3/60)}{2\pi\,(0.02)} \log \frac{1015}{0.15} = 3.5 \text{ m}$$

(ii) $R = \frac{2L}{\pi} \sin \frac{a\pi}{L} = \frac{2 \times 1600}{\pi} \sin \frac{400\pi}{1600} = 720 \text{ m}$

$$s_w = \frac{2.3\,(3/60)}{2\pi\,(0.02)} \log \frac{720}{0.15} = 3.37 \text{ m}$$

Also $R = \frac{2ab\,(2a+b)}{(a+b)^2} = \frac{2 \times 400 \times 1200\,(800+1200)}{1600^2} = 750 \text{ m}$

which is very near the value obtained previously.

To find the drawdown in the observation well for case (i), with pumping well as the origin of coordinates, from Eq. (5.61)

$$s = \frac{2.3\,(3/60)}{4\pi\,(0.02)} \log \frac{\cosh \frac{\pi\,(400)}{1600} + \cos \frac{\pi\,(300)}{1600}}{\cosh \frac{\pi\,(400)}{1600} - \cos \frac{\pi\,(300)}{1600}} = 1.99 \text{ m}$$

Well Hydraulics

3. Well in an aquifer bounded by two streams at right angles, Fig. 5.37

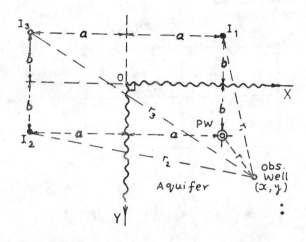

Fig. 5.37 Well in an aquifer bounded by two streams at right angles.

a. *The drawdown at the well face due to the pumping well and three image wells I_1, I_2 and I_3, applying the Dupuit's equation,*

$$s_w = \frac{2.3\,Q}{2\pi T}\left[\log\frac{R}{r_w} - \log\frac{R}{2b} - \log\frac{R}{2a} + \log\frac{R}{\sqrt{a^2+b^2}}\right]$$

or $\quad s_w = \dfrac{2.3\,Q}{2\pi T}\log\left(\dfrac{2\,ab}{r_w\sqrt{a^2+b^2}}\right)$, i.e. $R = \dfrac{2\,ab}{a^2+b^2}$ \hfill (5.62)

If $a = b$, $R = \sqrt{2}a = 1.41\,a$ \hfill (5.62a)

b. *The drawdown in the observation well due to the pumping well and the three image wells, I_1, I_2 and I_3 applying the Dupuit's equation,*

$$s = \frac{2.3\,Q}{2\pi T}\left[\log\frac{R}{r} - \log\frac{R}{r_1} - \log\frac{R}{r_2} + \log\frac{R}{r_3}\right]$$

$$s = \frac{2.3\,Q}{2\pi T}\log\left(\frac{r_1 r_2}{r\, r_3}\right) \hfill (5.63)$$

If the coordinates of the pumping well are (X, Y) and of the observation well (x, y), then

$$s = \frac{2.3\,Q}{4\pi T}\log\frac{[(x-X)^2+(y+Y)^2][(x+X)^2+(y-Y)^2]}{[(x-X)^2+(y-Y)^2][(x+X)^2+(y+Y)^2]} \quad (5.63a)$$

4. **Well in an aquifer bounded by a stream on one side and impermeable barrier at 90°, Fig. 5.38**

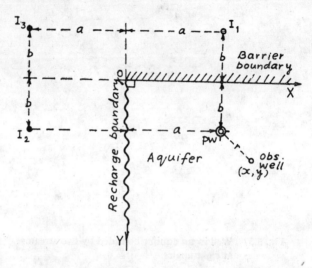

Fig. 5.38 Well in an aquifer bounded by a stream and impervious boundary at right angles.

The drawdown at the well face due to the pumping well and the three image wells I_1, I_2 and I_3, by applying Dupuit's equation,

$$s_w = \frac{2.3 Q}{2\pi T}\left[\log \frac{R}{r_w} + \log \frac{R}{2b} - \log \frac{R}{2a} - \log \frac{R}{2\sqrt{a^2+b^2}}\right] \quad (5.64)$$

or

$$s_w = \frac{2.3 Q}{2\pi T} \log\left(\frac{2a\sqrt{a^2+b^2}}{r_w b}\right),$$

i.e. $\quad R = \dfrac{2a\sqrt{a^2+b^2}}{b}$ \quad (5.64a)

if $\quad a = b,\ R = 2\sqrt{2a} = 2.83a.$

The drawdown at the observation well can be found as in the previous case.

5. **Aquifer bounded by a recharge boundary**

If there is a recharge boundary adjacent to an aquifer, the flow into the well becomes steady, and the Dupuits equation can be used to find the drawdown at the face of the pumping well by substituting the appropriate value for the radius of influence R. The values of R in the Dupuit's equation, are given for a number of boundary conditions, in Fig. 5.39.

Well Hydraulics

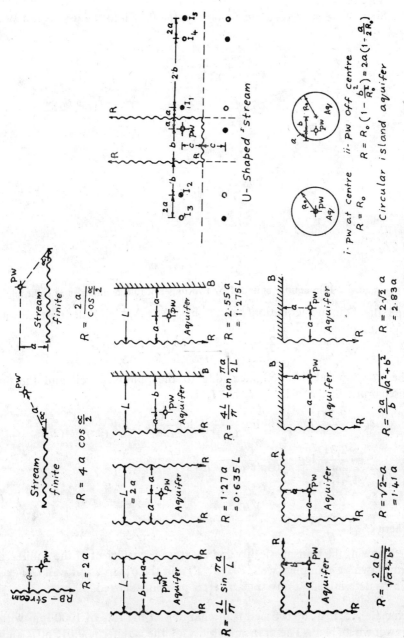

Fig. 5.39 Value of radius of influence in the Dupuit's equation for drawdown in the pumping well.

6. Well in an aquifer bounded by two impermeable barriers at right angles, Fig. 5.40

Since there is no recharge boundary, the flow into the well is not steady

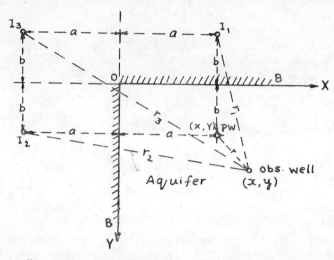

Fig. 5.40 Well in aquifer bounded by two impermeable boundaries at right angles.

and the Dupuit's equation cannot be used. Applying the unsteady-state equation (Jacob's)

$$s = \frac{2.3 Q}{4\pi T} \log \frac{2.25 Tt}{r^2 S}, \quad C = \frac{2.25 Tt}{S}$$

the drawdown at the well face, due to the pumping well and the three discharging image wells I_1, I_2 and I_3, is

$$s_w = \frac{2.3 Q}{4\pi T} \log \frac{C}{r_w^2} + \log \frac{C}{(2b)^2} + \log \frac{C}{(2a)^2} + \log \frac{C}{(2a)^2 + (2b)^2}$$

$$= \frac{2.3 Q}{4\pi T} \log \frac{C^4}{r_w^2\, 64 a^2 b^2 (a^2 + b^2)}$$

or $\quad s_w = \dfrac{2.3 Q}{2\pi T} \log \dfrac{C^2}{8\, r_w ab \sqrt{a^2 + b^2}}$ \hfill (5.65)

where $C = \dfrac{2.25 Tt}{S}$.

Knowing the transmissibility and the storage coefficient of the aquifer, the drawdown at pumping well $PW(X, Y)$ and the observation well $OW(x, y)$ can be determined at any time 't' after pumping started.

Example 5.22: A 30 cm well is discharging at the rate of 1000 lpm with a drawdown of 4 m. The transmissibility of the aquifer is 0.015 m²/s and is bounded by an aquiclude at a distance of 150 m from the well. Determine the drawdown at a distance of 90 m (i) towards the aquiclude, and (ii) on the opposite of the well.

Well Hydraulics

If it were a recharge boundary instead of an aquiclude determine the drawdown in case (i), and in the pumping well.

Solution:

For steady flow conditions, hydraulic head

$$h - h_w = \frac{2.3 Q \log \frac{r}{r_w}}{2\pi T} = \frac{Q \log \frac{r}{r_w}}{2.72\, T}, \text{ Dupuit}$$

(i) *At $r = 90$ m towards the aquiclude*: Hydraulic head due to the pumping well ($r = 90$ m) and the image pumping well ($r_i = 210$ m)

$$= \frac{1/60}{2.72\,(0.015)}\left(\log \frac{90}{0.15} + \log \frac{210}{0.15}\right) = 2.42 \text{ m}$$

Resulting drawdown $= 4 - 2.42 = \mathbf{1.58}$ **m**.

(ii) *At $r = 90$ m on the opposite side of the well*:

$$\text{Hydraulic head} = \frac{1/60}{2.72\,(0.015)}\left(\log \frac{90}{0.15} + \log \frac{390}{0.15}\right) = 2.53 \text{ m}$$

Resulting drawdown $= 4 - 2.53 = \mathbf{1.47}$ **m**.

If it were a recharge boundary (instead of the aquiclude), the resulting drawdown at $r = 90$ m towards the boundary, due to the pumping well and the image recharging well,

$$s = \frac{2.3 Q}{2\pi T}\left(\log \frac{R}{r} - \log \frac{R}{r_i}\right) = \frac{Q}{2.72\, T}\log \frac{r_i}{r}$$

$$= \frac{1/60}{2.72\,(0.015)}\log \frac{210}{90} = 0.15 \text{ m}.$$

Note that this is the same as

$$s = \frac{2.3 Q}{2\pi T}\left(\log \frac{r_i}{r_w} - \log \frac{r}{r_w}\right)$$

$$= \frac{1/60}{2.72\,(0.015)}\left(\log \frac{210}{0.15} - \log \frac{90}{0.15}\right) = \mathbf{0.15}\ \mathbf{m}$$

The drawdown in the pumping well due to recharge boundary

$$s_w = \frac{Q}{2.72\, T}\log \frac{r_i}{r_w} = \frac{1/60}{2.72\,(0.015)}\log \frac{300}{0.15} = \mathbf{1.35}\ \mathbf{m}$$

which is the same as

$$s_w = \frac{1/60}{2.72\,(0.015)}\left(\log \frac{300}{0.15} - \log \frac{0.15}{0.15}\right) = \mathbf{1.35}\ \mathbf{m}.$$

Multiple Well Systems

If there are a number of pumping wells in a given well field, the drawdown at any point is the sum of the drawdowns due to each pumping well, for which the distance of the point from each well and the discharge of each well should be known. The drawdowns depend upon the pumping pattern, i.e. the number of wells, their pumping rate and their array. Solutions may be obtained using equilibrium or non-equilibrium equations, as the case may be. Multiple well systems are used for lowering the ground water table in a given area to facilitate excavation for foundation work, etc. (see Figs. 11.2 and 11.3—multiple well point systems).

Wells may be closely spaced (resulting in mutual interference) and all the wells may be connected to a common supply pipe to meet the large demand for water supply. For an array of a number of equally spaced (a metres apart) fully penetrating wells, all discharging at the same rate (Q), parallel to a line source (at a distance d), Fig. 5.41, the drawdown at any point (x, y) is given by (Forchheimer, 1908),

$$H - h = s = \frac{Q}{2\pi Kb} \log_e \frac{\cosh \frac{2\pi}{a}(x+d) - \cos \frac{2\pi y}{a}}{\cosh \frac{2\pi}{a}(x-d) - \cos \frac{2\pi y}{a}} \tag{5.66}$$

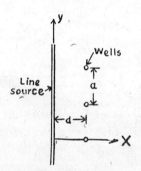

Fig. 5.41 Wells parallel to a line source.

For unconfined aquifers Eq. (5.66) becomes

$$H^2 - h^2 = \frac{Q}{\pi K \pi K} \log_e \frac{\cosh \frac{2\pi}{a}(x+d) - \cos \frac{2\pi y}{a}}{\cosh \frac{2\pi}{a}(x-d) - \cos \frac{2\pi y}{a}} \tag{5.67}$$

Muskat (1973) developed solutions for well discharges for various well patterns localised near the centre of a well field of radius R such that for each well the head at the external boundary can be taken to be H (i.e. R is the radius of influence for each well), Fig. 5.42. In the following solutions it is assumed that all the wells fully penetrate a confined aquifer,

Well Hydraulics

have the same diameter, and drawdown, and discharge for the same period of time. The equations can also be applied to unconfined aquifers by replacing H by $H^2/2b$ and h_w by $\dfrac{h_w^2}{2b}$.

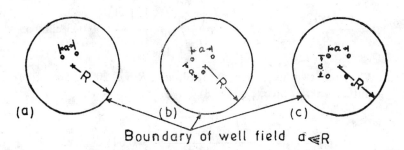

Fig. 5.42 Closely spaced wells in a well field.

(a) Two wells spaced at a distance a $(a \ll R)$, discharging Q_1 and Q_2 for confined and unconfined aquifers, respectively, are given by

$$Q_1 = Q_2 = \frac{2\pi Kb (H - h_w)}{\log_e \dfrac{R^2}{r_w a}} = \frac{\pi K (H^2 - h_w^2)}{\log_e \dfrac{R^2}{r_w a}} \qquad (5.68)$$

where r_w is the radius of each well and $H - h_w = s_w$.

(b) For three wells forming an equilateral triangle of side a $(a \ll R)$,

$$Q_1 = Q_2 = Q_3 = \frac{2\pi Kb (H - h_w)}{\log_e \dfrac{R^3}{r_w a^2}} = \frac{\pi K (H^2 - h_w^2)}{\log_e \dfrac{R^3}{r_w a^2}} \qquad (5.69)$$

(c) For four wells forming a square of side a $(a \ll R)$,

$$Q_1 = Q_2 = Q_3 = Q_4 = \frac{2\pi Kb (H - h_w)}{\log_e \dfrac{R^4}{\sqrt{2}\, r_w a^3}} = \frac{\pi K (H^2 - h_w^2)}{\log_e \dfrac{R^4}{\sqrt{2}\, r_w a^3}} \qquad (5.70)$$

As the number of wells in the group increase, the mutual interference between wells becomes more, with the result the production capacity per well decreases.

Example 5.23: Two 30 cm wells completely penetrate an artesian aquifer of thickness 15 m and are spaced at 200 m. Permeability of the aquifer is 60 m/day. When one well is pumped at the rate 2000 lpm, the drawdown in the well is 5 m. Assume a radius of influence of 300 m and steady state conditions. What will be the discharge when both the wells are pumped keeping the drawdown at 5 m. If a third well is located at 200 m to form an equilateral triangle and all the three wells pumped keeping the drawdown at 5 m, what is the percentage reduction in discharge? What is the percentage reduction in discharge in another well is added to form a square?

Solution: i. When one well is pumped, Eq. (5.7)

$$Q = \frac{2.72\, Ts_w}{\log \frac{R}{r_w}}, \quad T = Kb = \frac{2.72\left(\frac{60}{86400} \times 15\right)5}{\log \frac{300}{0.15}} = 0.043 \text{ cumec.}$$

ii. When the two wells are pumped, $a < R$, i.e. $200 < 300$, there is mutual interference, Eq. (5.68)

$$Q_1 = Q_2 = \frac{2.72\left(\frac{60}{86400} \times 15\right)5}{\log \frac{300^2}{0.15 \times 200}} = 0.041 \text{ cumec.}$$

Decrease in discharge $= \dfrac{0.043 - 0.041}{0.043} \times 100 = 4.65$, say **5%**.

iii. When the three wells are pumped, Eq. (5.69)

$$Q_1 = Q_2 = Q_3 = \frac{2.72\left(\frac{60}{86400} \times 15\right)5}{\log \frac{300^3}{0.15 \times 200^2}} = 0.039 \text{ cumec,}$$

Decrease in discharge $= \dfrac{0.043 - 0.039}{0.043} = 9.3$, say **10%**.

iv. Where the four wells are pumped at the corners of a square of side 200 m keeping the drawdown at 5m, Eq. (5.70)

$$Q_1 = Q_2 = Q_3 = Q_4 = \frac{2.72\left(\frac{60}{86400} \times 15\right)5}{\log \frac{300^4}{\sqrt{2}\,(0.15)\,200^3}} = 0.0386 \text{ cumec.}$$

Decrease in discharge $= \dfrac{0.043 - 0.0386}{0.043} \times 100 - \mathbf{10.22\%}.$

Partial Penetration of the Well

Many artesian wells do not penetrate through the full thickness of the aquifer and many wells are provided with strainers with only a portion of the thickness of the aquifer, Fig. 5.43. In such cases there is a vertical convergence of stream lines near the well, resulting increased entrance velocity with increased drawdown. Kozeny developed a formula for the capacity of a well partially penetrating an artesian aquifer

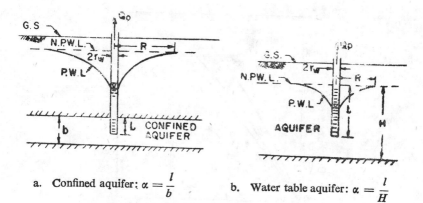

a. Confined aquifer: $\alpha = \dfrac{l}{b}$ b. Water table aquifer: $\alpha = \dfrac{l}{H}$

Fig. 5.43 Partial penetration of wells.

$$Q_p = Q\left[\alpha\left(1 + 7\sqrt{\frac{r_w}{2\alpha b}}\cos\frac{\pi\alpha}{2}\right)\right] \qquad (5.71)$$

where Q_p = discharge in partially penetrated well; Q = discharge in the artesian well for complete penetration; α = penetration, fraction = $\dfrac{l}{b}$, where l = length of strainer and b = thickness of the aquifer and r_w = radius of the well.

Note: If $Q_p = Q$, then $(S_w)_p > S_w$; and if $(S_w)_p = S_w$, then $Q_p < Q$. The radius of influence in calculating Q is assumed to be about 3000 m (for artesian wells). Q_p in unconfined aquifers is obtained from Eq. 5.71 by replacing b by H. The radius of influence in calculating Q in a water table aquifer may be assumed as 300 m. The discharge ratio Q_p/Q for the same drawdown as a function of penetration fraction and well slimness $l/2r_w$ is shown in Fig. 5.44.

Wen Hsiung Li (1954) evolved a simple empirical formula for the ratios of drawdown and discharge in a partially penetrating well based on the data obtained from a series of tests on an electric analog model:

$$\frac{S_{wp}}{S_w} = 1 + \frac{\log\dfrac{b}{r_w}}{\log\dfrac{R}{r_w}}\left[\left(\frac{1}{\alpha}\right)^{0.75} - 1\right] \qquad (5.72)$$

$$\frac{Q_p}{Q} = \frac{1}{1 + \dfrac{\log\dfrac{b}{r_w}}{\log\dfrac{R}{r_w}}\left[\left(\dfrac{1}{\alpha}\right)^{0.75} - 1\right]} \qquad (5.73)$$

applicable for $\alpha > 0.1$.

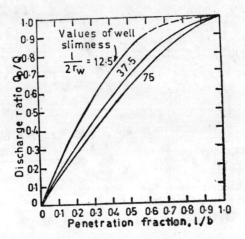

Fig. 5.44 Discharge ratio Q_p/Q vs. penetration.

The equations obtained for the discharge ratio by the various investigators may be expressed in terms of non-dimensional parameters as follows:

Forchheimer : $\dfrac{Q_p}{Q} = \phi(\alpha)$

de Glee : $\dfrac{Q_p}{Q} = \phi\left(\alpha, \dfrac{R}{r_w}, \dfrac{l}{r_w}, \dfrac{R}{b}\right)$

Muskat : $\dfrac{Q_p}{Q} = \phi\left(\alpha, \dfrac{R}{r_w}, \dfrac{b}{r_w}, \dfrac{b}{R}\right)$

Kozeny : $\dfrac{Q_p}{Q} = \phi\left(\alpha, \dfrac{b}{r_w}\right)$

Wen Hsiung Li: $\dfrac{Q_p}{Q} = \phi\left(\alpha, \dfrac{R}{r_w}, \dfrac{b}{r_w}\right)$

In its most general form, the discharge ratio can be expressed as

$$\dfrac{Q_p}{Q} = \phi\left(\alpha, \dfrac{R}{r_w}, \dfrac{b}{r_w}, \dfrac{l}{r_w}, \dfrac{b}{R}\right) \qquad (5.74)$$

The effect of each of the above parameters on the discharge and the drawdown is given in the following.

(i) *Penetration fraction*: An increase in penetration does not give rise to a proportional increase in yield, particularly beyond 50% penetration. For penetration of more than 80%, the effect of partial penetration becomes insignificant. The well depth or screen length has to be chosen carefully from economic considerations.

(ii) *Ratio of radius of influence to well radius*, $\dfrac{R}{r_w}$: The discharge ratio increases with increase in the ratio $\frac{R}{r_w}$. In other words, the relative discharge is more in the case of extensive aquifers, well radius being the same.

(iii) *Ratio of aquifer thickness to radius of the well*, $\frac{b}{r_w}$: As can be seen from Wen Hsiung Li's analysis, a substantial drop in the discharge ratio can be observed for the same penetration when the aquifer thickness is large compared to the well radius.

(iv) *Well slimness*, $\frac{l}{r_w}\left(\text{or }\frac{l}{2r_w}\right)$: For the same penetration fraction, the discharge ratio varies inversely with the well slimness, i.e. the effect of partial penetration is less pronounced in wells of larger size.

(v) *Ratio of aquifer thickness to radius of influence*, $\frac{b}{R}$: The parameter $\frac{b}{R}$ bears an inverse relationship with the discharge ratio, though the effect is relatively less significant. This is due to the fact that the convergence of flow is more intense in case of thick aquifers for the same penetration percentage and hence has a lesser value of discharge ratio.

Example 5.24: A 40 cm well penetrates an aquifer of 30 m thickness and the length of the screened portion is 10 m. The yield is 2000 lpm with a drawdown in the well of 3 m. If the length of the screen is increased to 20 m, what will be the drawdown in the well and the increase in the specific capacity?

Solution: Muskat (1946) developed the equation

$$\left(\frac{Q}{s_w}\right)_p = \frac{Q}{s_w}\left[\alpha\left(1 + 7\sqrt{\frac{r_w}{2\alpha b}}\cos\frac{\pi\alpha}{2}\right)\right] \quad (5.75)$$

$$\frac{2/3}{2/s'_w} = \frac{1/3\left(1 + 7\sqrt{\frac{0.2}{2\times 10}}\cos\frac{\pi}{6}\right)}{2/3\left(1 + 7\sqrt{\frac{0.2}{2\times 20}}\cos\frac{\pi}{3}\right)}$$

Therefore
$s'_w = 1.92$ m (reduced drawdown in the well)

The specific capacity is increased to $Q/s'_w = \frac{2}{1.92} = 1.04$ m²/min

Initial specific capacity, $Q/s_w = 2/3 = 0.67$ m²/min.
Therefore
$$\text{Increase in specific capacity} = \frac{1.04 - 0.67}{0.67} \times 100 = 55\%$$

assuming homogeneous aquifer. This would have little validity in a horizontally stratified aquifer.

Partial Penetration of Wells—Increase in Drawdown

(a) *Confined or artesian aquifer*: The increase in drawdown due to the well not fully penetrating the aquifer or only a portion of the aquifer is screened resulting in higher entrance velocities and convergence of streamlines, Fig. 5.45. The drawdown in the partially penetrating well

$$s_{wp} = s_w + \Delta s_w \tag{5.76}$$

where s_w = drawdown in the pumping well fully penetrating the aquifer and Δs_w = increase in drawdown due to partial penetration of the well.

The disposition of the screen in the aquifer is shown in Fig. 5.45. For steady-state conditions, it can be shown that

(i) when the screen is located at the top or bottom of the aquifer,

$$\alpha = \frac{1}{b}, \; T = Kb,$$

$$\Delta s_w = \frac{Q}{2\pi T} \frac{1-\alpha}{\alpha} \ln \frac{(1-\alpha)l}{r_w}, \; \alpha > 0.2 \tag{5.77}$$

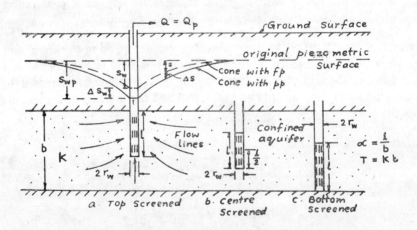

Fig. 5.45 Partially penetrating well in a confined aquifer.

(ii) when the screen is centrally located

$$\Delta s_w = \frac{Q}{2\pi T} \frac{1-\alpha}{\alpha} \ln \frac{(1-\alpha)l}{2r_w} \tag{5.78}$$

(b) *Unconfined aquifer*: For a fully penetrating well, for steady state conditions, from Eq. (5.4),

$$H^2 - h_w^2 = \frac{Q}{\pi K} \ln \frac{R}{r_w} \tag{5.79}$$

For a partially penetrating well (usually bottom half to one-third of the aquifer), $\alpha = l/h_w$, $T = KH$, $(h_w = H - s_w)_{fp}$. Eq. (5.77) modified gives

Well Hydraulics

$$\Delta s_w 2H = \frac{Q}{\pi K}\frac{1-\alpha}{\alpha}\ln\frac{(1-\alpha)l}{r_w} \qquad (5.80)$$

Then,

$$(H^2 - h_w^2)_{pp} = (H^2 - h_w^2)_{fp} + \Delta s_w 2H \qquad (5.81)$$

from which,

$$(s_w)_{pp} = H - (h_w)_{pp} \qquad (5.82)$$

where the subscripts pp = partial penetration and fp = full penetration.

Detailed methods of analysis of the effects of partial penetration on well flow for steady, and unsteady conditions in confined, unconfined, leaky, and anisotropic aquifers have been given by Hantush (1966) and others.

Example 5.25: A 30 cm well penetrates fully a homogeneous aquifer having a permeability of 40 m/day. The saturated thickness of the aquifer is 15 m. What is the drawdown in the well, when pumped at the rate of 1000 lpm? What will be the drawdown if only the bottom half of the aquifer is screened? Assume a radium of influence of 300 m.

Solution: (i) For a fully panetrating well,

$$H^2 - h_w^2 = \frac{2.3\,Q}{\pi K}\log\frac{R}{r_w}$$

$$15^2 - h_w^2 = \frac{2.3\,(1/60)}{\pi\,(40/86400)}\log\frac{300}{0.15}$$

$$15^2 - h_w^2 = 86.5$$

Therefore

$$h_w^2 = 138.5,\ h_w = 11.76\ \text{m}$$

$$s_w = 15 - 11.76 = \mathbf{3.24\ m}$$

(ii) length of well screen $= \frac{1}{2}\times 15 = 7.5$ m from the bottom of the aquifer

$$\alpha = \frac{l}{h_w} = \frac{7.5}{11.76} = 0.638,\ 1-\alpha = 0.362$$

From Eq. (5.80)

$$\Delta s_w 2H = \frac{2.3\,(1/60)}{\pi\,(40/86400)}\frac{0.362}{0.638}\log\frac{(0.362)7.5}{0.15} = 33$$

From Eq. (5.81),

$$(H^2 - h_w^2)_{pp} = 86.5 + 33$$

$$(15^2 - h_w^2)_{pp} = 119.5$$

$$(h_w)_{pp} = 10.26\ \text{m}$$

$$s_{wp} = 15 - 10.26 = 4.74\ \text{m}$$

$$\Delta s_w = 4.74 - 3.24 = \mathbf{1.50\ m}$$

Finite number of wells in a leaky artesian aquifer in a line: This is illustrated in the following example.

Example 5.26: A leaky artesian aquifer 30 m thick having a permeability of 20 m/day is situated above an impervious base and overlain by a semi-confining layer with a resistance of 2000 days against vertical leakage. On the top of this semi-confining layer, a hamogeneous aquifer with constant water table is present. From the leaky artesian aquifer, ground water is abstracted by a series of 7 wells at intervals of 110 m in a straight line. Each well has a diameter of 40 cm, a screened length of 21 m in the middle of the aquifer, and is pumped at a constant rate of 600 m³/day. What is the maximum drawdown in any well?

x:	0.1	0.2	0.3
$K_0(x)$:	2.43	1.75	1.37

Solution: Fig. 5.46

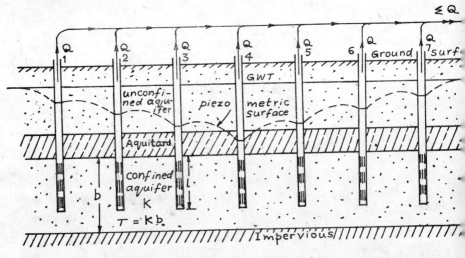

Fig. 5.46 Line of seven wells in a leaky artesian aquifer.

The interference of neighbouring wells is maximum at the central well (No. 4) and the drawdown is maximum in the central well. The drawdown in the central well is due to three components.

(i) its own drawdown, assuming full penetration, s_w
(ii) additional drawdown due to its partial penetration alone,* Δs_w

*The influence of partial penetration is limited to the near vicinity of the well and the mutual influence will consequently be negligible for distances $r > 2b$ or $2H$.

Well Hydraulics

(iii) drawdown (Σs) caused by the neighbouring wells at distances

$$r = 110, 220, 330 \text{ m}$$

Thus the total drawdown in the central well

$$s_{w4} = s_w + \Delta s_w + \Sigma s \quad \text{(due to } 3 \times 2 = 6 \text{ wells)}$$

$$T = Kb = 20 \times 30 = 600 \text{ m}^2/\text{day}$$

$$B = \sqrt{Tc} = \sqrt{600 \times 2000} = 1095.4 \text{ m}$$

$$\alpha = \frac{l}{b} = \frac{21}{30} = 0.7, \; 1 - \alpha = 0.3 \text{ for all wells}$$

For the central well, $\dfrac{r_w}{B} = \dfrac{0.20}{1095.4} < 0.05$

For well No. 3 & 5: $\dfrac{r_3}{B} = \dfrac{110}{1095.4} = 0.1004, \; K_0\left(\dfrac{r_3}{B}\right) = 2.43$

2 & 6: $\dfrac{r_2}{B} = \dfrac{220}{1095.4} = 0.2008, \; K_0\left(\dfrac{r_2}{B}\right) = 1.75$

1 & 7: $\dfrac{r_1}{B} = \dfrac{330}{1095.4} = 0.3012, \; K_0\left(\dfrac{r_1}{B}\right) \; 1.37$

$$s_{w4} = \frac{Q}{2\pi T}\left[\ln \frac{1.12 B}{r_w} + \frac{1-\alpha}{\alpha} \ln \frac{(1-\alpha)l}{2r_w} + 2\left\{K_0\left(\frac{r_3}{B}\right)\right.\right.$$
$$\left.\left. + K_0\left(\frac{r_2}{B}\right) + K_0\left(\frac{r_1}{B}\right)\right\}\right]$$

$$= \frac{600}{2\pi \, 600}\left[2.3\left\{\log \frac{1.12 \times 1095.4}{0.2} + \frac{0.3}{0.7}\log \frac{0.3 \times 21}{2(0.2)}\right\}\right.$$
$$\left. + 2\{2.43 + 1.75 + 1.37\}\right]$$

$$= \frac{1}{2\pi}[2.3\{3.787 + 0.513\} + 2 \times 5.55] = 3.35 \text{ m}$$

Note: $s_w = 1.39$ m, $\Delta s_w = 0.19$ m, $\Sigma s = 1.77$ m.

Circular Battery of Wells in a Leaky Artesian Aquifer

If 'n' wells of equal capacity, spaced at 'a' in a circular array of radius R give a total discharge Q, Fig. 5.47, the drawdown at any point P may be obtained by the method of superposition as

$$s = \sum_{i=1}^{n} \frac{1}{n} \frac{Q}{2\pi Kb} K_0\left(\frac{r_i}{B}\right) \qquad (5.83)$$

where r_i = distance from P to the centre of the various wells. When $R \ll B$, the drawdown inside the circle is given by

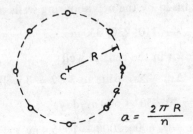

Circular battery of 'n' wells with a combined capacity Q

Fig. 5.47

$$s = \sum_{i=1}^{n} \frac{1}{n} \frac{Q}{2\pi Kb} \ln \frac{1.12 B}{r_i} \qquad (5.84)$$

Neglecting the influence of point abstraction, the drawdown inside the circle is thus constant, equal to the drawdown in the centre C i.e.

$$s_c = \frac{Q}{2\pi Kb} \ln \frac{1.12 B}{R} \qquad (5.85)$$

Due to point abstraction, the drawdown at the well face is larger by

$$\Delta s_w = \frac{1}{n} \frac{Q}{2\pi Kb} \ln \frac{a}{2\pi r_w} \qquad (5.86)$$

and halfway between two adjacent wells smaller by

$$\Delta s_{a/2} = \frac{1}{n} \frac{Q}{2\pi Kb} (0.693) \qquad (5.87)$$

$$a = \frac{2\pi R}{n} \qquad (5.87a)$$

When the wells partially penetrate the aquifer, the drawdown at the well face will be augmented $(\Delta s_w)_{pp}$.

Outside the battery, at a distance 'r' from its centre, the flow conditions are determined by the equation for a single well, i.e.

$$s = \frac{Q}{2\pi Kb} K_0 \left(\frac{r}{B}\right) \qquad (5.88)$$

Well Losses
The ultimate drawdown in a well measured at the designed rate Q of continuous pumping for a given period is different from the theoretical drawdown due to well losses, the total drawdown s_w is made up of

Well Hydraulics

(*i*) head loss resulting from laminar flow in the formation (formation loss),
(*ii*) head loss resulting from the turbulent flow in the zone close to the well face,
(*iii*) head loss through the well screen,
(*iv*) head loss in the well casing

and

$$s_w = BQ + CQ^2 \tag{5.89}$$

$$\frac{s_w}{Q} = B + CQ \tag{5.90}$$

where B = coefficient of formation loss due to (*i*); C = coefficient of well losses due to (*ii*), (*iii*) and (*iv*) (Walton, 1962) and has the dimensions of T^2/L^5; $\frac{s_w}{Q}$ = specific drawdown (reciprocal of specific capacity Q/s_w); BQ = formation loss and CQ^2 = well loss

A plot of s_w/Q versus Q gives a straight line, Fig. 5.48 whose slope gives the well loss coefficient C and the intercept on the s_w/Q axis gives the formation loss coefficient B.

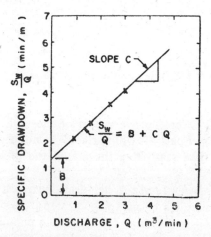

Fig. 5.48 Specific drawdown graph—well and formation loss coefficients.

Rorabaugh (1953) has modified this graphical method when the drawdown is given by $s_w = BQ + CQ^n$, 'n' deviating significantly from 2. See Example 5.30.

Demonstration of Well Losses

1. *From the recovery immediately after pumping is stopped*: If well losses occur in disproportionately high gradients in the immediate vicinity of the

well, the resulting picture will be that of a small, very steep cone of depression, superimposed on the main cone during pumping, Fig. 5.49. When the pumping stops, this small cone will fill up rapidly as the volume of storage is very small and subsequently the main cone will fill up. It can be observed that in the first 1-5 minutes after pumping is stopped, there is rapid recovery after which the recovery pattern simulates the inverse of drawdown.

2. *From the distance-drawdown curve if an observation well is available*: If the aquifer transmissibility is known from a pump test on the well, the drawdown per log cycle of distance Δs can be calculated from Eq. (4.82)

$$T = \frac{2.3\ Q}{2\pi \Delta s}$$

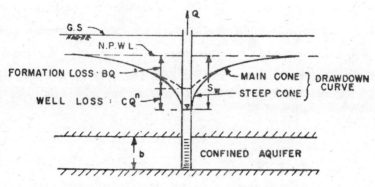

Fig. 5.49 Demonstration of well losses.

The distance-drawdown graph may be plotted and extrapolated towards the well at a distance r_e the effective radius of the well, i.e. the radius of the actual well + gravel pack. It will normally be noticed that the observed drawdown is more than read from the graph, the difference giving the actual well losses.

3. *From a step-drawdown test*: A step-drawdown test is carried out at three to five different rates of pumping or steps, each step being of the same duration, say 60 or 90 min. The maximum pumping rate preferably exceeds the designed pumping rate, though not too much, as too high pumping rates may adversely affect the aquifer framework permanently by excessive turbulance in the immediate vicinity of the well; for instance the soil grains around the well may resettle, but tighter than before. The well loss coefficient is given by

$$C = \frac{\Delta s_2/\Delta Q_2 - \Delta s_1/\Delta Q_1}{\Delta Q_1 + \Delta Q_2} \tag{5.91}$$

where ΔQ = increase in pumping rate at each step; $\Delta Q_1 = Q_2 - Q_1$, $\Delta Q_2 = Q_3 - Q_2$; Δs = difference in drawdown after completion of each step and $\Delta s_1 = s - s_1$, $\Delta s_2 = s_3 - s_2$.

Well Hydraulics

A step-drawdown test is normally carried out on a well designed as a production well to determine the specific capacity versus yield relationship. Evaluation of C gives the well efficiency. It is included in production well drilling contracts where it is generally prescribed to be less than 2000 sec^2/m^5.

Example 5.27: A variable rate well production test was conducted at Sagar, Madhya Pradesh with the results given in Table 5.18. Determine the coefficient of well loss and formation loss and the corresponding percentages of total drawdown in the last case (a) by drawing the specific drawdown curve, (b) by using Eqs. (5.90) and (5.91).

Table 5.18 Results of variable rate well production tests (Ex. 5.16)

Step	1	2	3	4	5
Pumping rate (lpm)	1590	1980	2440	2960	3270
Resulting drawdown	3.69	5.14	7.08	9.63	11.29

Solution: (a) Data processed to draw the specific drawdown curve is given in Table 5.19.

Table 5.19 Data for specific drawdown curve (Ex. 5.16)

Step	Pumping rate Q (m³/min)	Resulting drawdown s_w (m)	Specific drawdown s_w/Q (min/m²)
1	1.59	3.69	2.32
2	1.98	5.14	2.59
3	2.44	7.08	2.90
4	2.96	9.63	3.25
5	3.27	11.29	3.45

A plot of s_w/Q versus Q gives a straight line graph, Fig 5.50. The vertical intercept on the s_w/Q axis gives $B = 1.3$ min/m² and the slope of the

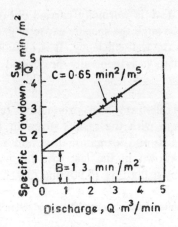

Fig. 5.50 Specific drawdown graph, Example 5.27.

straight line graph gives $C = 0.65$ min^2/m^5. For the last case, step 5, for a pumping rate of 3270 lpm, formation loss $= BQ = 1.3 (3.27) = 4.25$ m well loss $= CQ^2 = 0.65 (3.27)^2 = 6.95$ m sum of formation and well losses $= 11.20$ m against a measured drawdown of 11.29 m percentage of formation loss $= \dfrac{4.25}{11.20} \times 100 = 37.9\%$ of total drawdown percentage of well loss $= \dfrac{6.95}{11.20} 100 = 62.1\%$ of total drawdown.

The average value of $C = 0.65$ min^2/m^5 = 2340 sec^2/m^5 which is higher than the prescribed limit of 2,000 sec^2/m^5.

(b) Data processed to determine the well loss coefficient C by Eq. (5.91) is given in Table 5.20.

Table 5.20 Data to determine well loss coefficient

Step	Q (m^3/min)	ΔQ (m^3/min)	s_w (m)	Δs_w (m)
1	1.59		3.69	
		0.39		1.45
2	1.98		5.14	
		0.46		1.94
3	2.44		7.08	
		0.52		2.55
4	2.96		9.63	
		0.31		1.66
5	3.27		11.29	

Well Hydraulics

for the last step 4-5,

$$C = \frac{\frac{1.66}{0.31} - \frac{2.55}{0.52}}{0.52 + 0.31} = 0.542 \text{ min}^2/\text{m}^5 = 1{,}950 \text{ sec}^2/\text{m}^5$$

which is less than the prescribed limit of 2000 sec^2/m^5.

$$\text{well loss} = CQ^2$$
$$= 0.542 \, (3.27)^2 = 5.8 \text{ m}$$

$$\text{Percentage well loss} = \frac{5.8}{11.29} \times 100$$

$$= 51.4\% \text{ of the total drawdown}$$

$$\text{Formation loss } BQ = s_w - CQ^2 = 11.99 - 5.80 = 5.49$$

$$\text{Percentage formation loss} = \frac{5.49}{11.29} \times 100$$

$$= 48.6\% \text{ of the total drawdown}$$

$$\text{Formation loss coefficient } B = \frac{5.49}{3.27} = 1.67 \text{ min/m}^2$$

The average values of well loss and formation loss are about 42 and 58% respectively. The well should be properly designed and developed to keep the well loss to a minimum. The step-drawdown test enables the determination of the well loss coefficients and also the well efficiency.

Example 5.28: The step-drawdown test data on a well is given on p. 230. Determine the well loss coefficient and the well loss when pumped at the rate of 1260 lpm.

Solution: The test data are plotted on a semi-log paper as shown in Fig. 5.51. The slopes of the time-drawdown curves are extrapolated beyond periods of pumping at each rate and the increments of drawdown are determined for each increase in the rate of pumping for a pumping period of 60 minutes and are given below:

Step	Q (m³/min)	ΔQ (m³/min)	Δs (m)
1	0.450	0.450	0.26
2	1.260	0.810	1.01
3	1.730	0.470	0.98

Time (min)	Depth to water level below M.P. (m)	Pumping rate (lpm)
0	6.87	0
		pumping started
2	7.09	450
5	7.10	450
10	7.11	450
15	7.115	450
20	7.125	450
25	7.13	450
30	7.13	450
40	7.135	450
50	7.135	450
65	7.135	1260
67	7.98	1260
70	8.04	1260
75	8.07	1260
80	8.09	1260
85	8.105	1260
90	8.11	1260
95	8.115	1260
105	8.13	1260
115	8.145	1260
125	8.16	1260
140	8.17	1730
145	9.03	1730
150	9.08	1730
155	9.11	1730
160	9.135	1730
165	9.145	1730
170	9.15	1730
180	9.16	1730
190	9.17	1730
200	9.185	1730
210	9.195	1730

Well Hydraulics

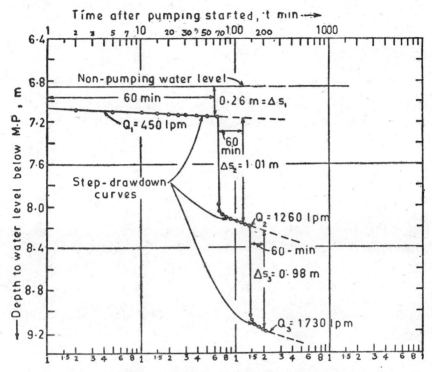

Fig. 5.51 Step-drawdown test, Example 5.28.

The well loss coefficient C for the steps 1 and 2 can be obtained by substituting in the Eq. (5.91) as

$$C = \frac{(1.01/0.81) - (0.26/0.45)}{0.45 + 0.81} = 0.531 \text{ min}^2/\text{m}^5$$

$$\text{or,} = 1912 \text{ sec}^2/\text{m}^5$$

and for steps 2 and 3,

$$C = \frac{(0.98/0.47) - (1.01/0.81)}{0.81 + 0.47} = 0.655 \text{ min}^2/\text{m}^5$$

$$\text{or,} \quad 2357 \text{ sec}^2/\text{m}^5$$

The average value of the well loss coefficient is 2134.

Hence the well loss when pumped at the rate of 1260 lpm is given by

$$\text{Well loss} = CQ^2 = 1912 \left(\frac{1.260}{60}\right)^2 = 0.843 \text{ m}$$

and the total loss (drawdown in the well s_w) when pumped at the rate 1260 lpm is $8.17 - 6.87 = 1.30$ m. Hence the well loss $= \frac{0.843}{1.30} + 100 = 65\%$ of the total loss (the remaining 35% being the formation loss). This percentage of well loss will be around 45% if the pumping rate is limited to around 1000 lpm.

Specific Capacity and Safe Yield

The specific capacity $\left(\dfrac{Q}{s_w}\right)$ of a well is the discharge per unit drawdown in the well and is usually expressed in lpm/m. This is a measure of the effectiveness of a well. From Eqs. (4.78), (5.89) and (5.90)

$$\frac{Q}{s_w} = \frac{1}{\dfrac{2.3}{4\pi T}\log_{10}\dfrac{2.25\,Tt}{r_w^2\,S} + CQ} \tag{5.92}$$

which shows that the specific capacity of a well is not constant but decreases with the increase in pumping rate (Q) and prolonged pumping (t) (Fig. 5.39). The probable drawdown for an unconfined aquifer may be obtained from [by putting $R \approx 300$ m and $r_w \approx 15$ cm in Eq. (5.5)]

$$\frac{Q}{s_w} = \frac{T}{1.2}$$

and for a confined aquifer [by putting $R \approx 3000$ m and $r_w \approx 15$ cm in Eq. (5.7)]

$$\frac{Q}{s_w} \approx \frac{T}{1.6}$$

On an average

$$\frac{Q}{s_w} = \frac{T}{1.4} \tag{5.93}$$

If the well is small in diameter and is pumped at a higher rate, the friction losses will increase. The maximum safe yield of a well or well field is the capacity of the aquifer to supply water without causing a continuous lowering of the water table or piezometric surface and is, therefore, limited by the rate at which the ground water is replenished by rainfall. Excessive lowering of the ground water table by pumping may result in crop failures, depletion of a nearby stream by increased percolation losses, salt water intrusion in coastal aquifers, and occasionally serious settlement of the ground surface.

Also from Eqs. (5.5) and (5.7), Q is inversely proportional to $\log_{10}\dfrac{R}{r_w}$. The percentage increase in discharge for the increase in diameter is given in Table 5.21 and illustrates the fallacy of the idea that big wells necessarily mean proportionately large yields.

Well Efficiency: The efficiency of a well, for a specified duration of pumping, is given by

Well Hydraulics

$$\eta_{well} = \frac{\text{Specific capacity for the observed drawdown}}{\text{Theoretical specific capacity, } Q/BQ}, \frac{Q/s_w}{Q/BQ} \times 100$$

$$= \frac{BQ}{s_w} \times 100\%$$

Table 5.21 Approximate increase in yield corresponding to an increase in well diameter (all other conditions remaining same)

Well diameter D (cm)	5	10	15	20	30	45	60	90
Increase in yield $\Delta Q\%$	0	10	15	20	25	33	38	48

Example 5.29: A 30 cm well penetrates fully an infinite nonleaky artesian aquifer having $T = 6.5 \times 10^5$ lpd/m, $S = 0.0004$. Construct specific capacity curves for the well for $t = 1$ hr, 1 day, 1 month, 1 year and 2 years of draught, when the well is pumped at the rate of (a) 600 lpm, (b) 1200 lpm, and (c) 2000 lpm. Assume a well loss coefficient of 1950 sec²/m⁵. What is the drawdown in the well after pumping for one year at the rate of 1200 lpm?

Solution: From Eq. (5.92), the specific capacity of the well,

$$\frac{Q}{s_w} = \frac{1}{\frac{2.303}{4\pi T} \log_{10} \frac{2.25\, Tt}{r_w^2\, S} + CQ}$$

for (a) $Q = 600$ lpm $= 0.01$ m³/sec.

$$\frac{Q}{s_w} = \frac{1}{\frac{2.303}{4\pi \frac{650}{86400}} \log_{10} \frac{2.25\, Tt}{(0.15)^2 \times 0.0004} + 1950 \times 0.01}$$

Putting $t = 1$ hr, $Tt = \frac{650}{24} \times 1 = 27.1$ m³ and $\frac{Q}{s_w} = \frac{1}{186.2}$ m³/s per m

or $\frac{1}{186.2}(1000 \times 60) = 322$ lpm/m.

Similarly putting $t = 1$ day, 1 month (30 days), 1 year (365 days), and 2 years (730 days), the specific capacities are computed as 273, 234.5, 212, and 207 lpm/m, respectively. Similarly for (b) $Q = 1200$ lpm $= 0.02$ m³/sec, and $t = 1$ hr, 1 day, 1 month, 1 year and 2 years, the specific capacities are computed as 292, 251, 218, 199 and 194 lpm/m, respectively, and for

(c) $Q = 2000$ lpm $= 0.0333$ m³/sec, and $t = 1$ hr, 1 day, 1 month, 1 year and 2 years, the specific capacities are computed as 259, 226, 199, 183, and 179 lpm/m respectively.

The specific capacity curves are thus constructed as shown in Fig. 5.52.

For $Q = 1200$ lpm, $t = 1$ year, $\dfrac{Q}{s_w} = 199$ lpm/m.

Therefore drawdown in the well at the end of 1 year,

$$s_w = \frac{1200}{199} = 6.03 \text{ m}$$

Thus the specific capacity is a useful concept and indicates the productivity of both aquifer and well in a single parameter. A high specific capacity indicates an efficient (well constructed and developed) good yielding well. A decline in specific capacity may indicate failure of well screens (by clogging or so) or declining S or T values due to declining water table or piezometric surfaces. Specific capacity decreases faster in unconfined aquifers since the decline of the water table reduces T.

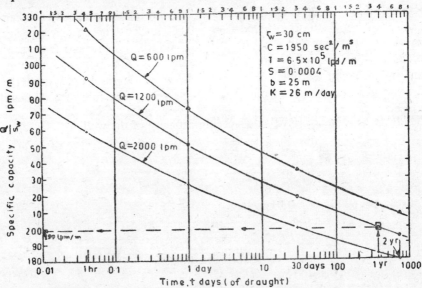

Fig. 5.52 Specific capacity of well versus time, Example 5.29.

Graphical trial-and-error procedure of Rorabaugh (1953) to determine well losses from the step-drawdown test data

Example 5.30: The following discharge drawdown data are obtained from a production well in a step-drawdown test

Q (m³/min):	0.45	1.26	1.73	2.25
s_w (m) :	0.265	1.29	2.325	3.78

Well Hydraulics

Determine the coefficients of formation loss and well loss. Comment on the specific capacity values you obtain.

Solution:

$$s_w = BQ + CQ^n \qquad (5.94)$$

$$\frac{s_w}{Q} = B + CQ^{n-1}$$

$$\frac{s_w}{Q} - B = CQ^{n-1}$$

$$\log\left(\frac{s_w}{Q} - B\right) = \log C + (n-1)\log Q$$

which shows that a plot of $\left(\frac{s_w}{Q} - B\right)$ vs Q on log-log paper should yield a straight line whose slope $m = n - 1$, and when $Q = 1$, $C = \frac{s_w}{Q} - B$. Values of B are tried till a straight line relation plot is obtained (Rorabaugh, 1953), Fig. 5.53.

Q (m³/min):	0.45	1.26	1.73	2.25
$\frac{s_w}{Q}$ (min/m²):	0.59	1.022	1.344	1.68

$\frac{s_w}{Q} - B$:

Try $B = 0$:	0.59	1.022	1.344	1.68
$B = 0.2$:	0.39	0.822	1.144	1.48
$B = 0.3$:	0.29	0.722	1.044	1.38
$B = 0.5$:	0.09	0.522	0.844	1.18
$B = 0.43$:	0.16	0.592	0.914	1.25

$B = 0.43$ min/m² yields a straight line whose slope

$$m = \frac{\Delta y}{\Delta x} = 1.27 = n - 1 \qquad \therefore \ n = 2.27$$

When $Q = 1$, $\frac{s_w}{Q} - B = 0.44 = C$, min²/m⁵

which is < 0.5; hence the production well is properly designed and developed. Further,

Q (m³/min)	:	0.45	1.26	1.73	2.25
BQ, (m)	:	0.194	0.542	3.75	0.97
$CQ^{2.27}$ (m)	:	0.072	0.744	1.53	2.78
$s_w = BQ + CQ^n$ (m)	:	0.266	1.286	2.28	3.75
$\frac{Q}{s_w}$ m³/min per m	:	1.7	0.98	0.75	0.60

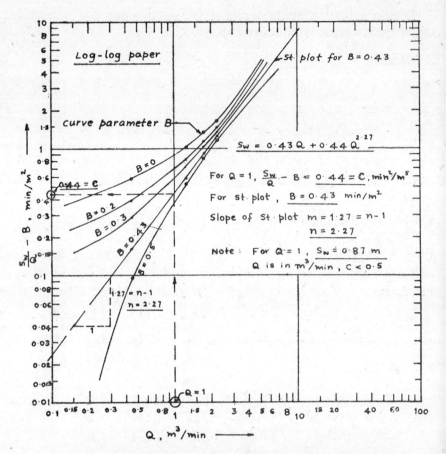

Fig. 5.53 Graphical method of determining well losses, Example 5.30 (Rorabaugh, 1953).

It may be noted that the well loss increases with Q rapidly, and in the last case, it is as high as $\frac{2.78}{3.75} \times 100 = 74\%$, since '$n$' usually lies between 2 and 3.

Since the well loss varies as the nth power of the entrance velocity, the well loss can be kept down by increasing r_w. The step-drawdown test shows how much head is lost in the aquifer and how much in and around the well (screen and gravel pack). Excessive well losses indicate improper design and development of the well or deterioration of the screen. Also the value of B obtained from the test can be used to estimate T of the aquifer, using the appropriate well flow equation relating the formation loss to Q. The specific capacity decreases with increasing Q.

Example 5.31: A 30 cm well penetrates completely an infinite nonleaky

Well Hydraulics 237

artesian aquifer having $T = 600$ m²/day, $S = 0.0004$. The well is continuously pumped at the rate of 60 m³/hr. Assuming a well loss coefficient of 0.5 min²/m⁵, and index $n = 2.3$, determine the specific capacity and the drawdown in the pumping well after 1 year of drought.

Solution: s_w = formation loss + well loss

$$s_w = BQ + CQ^n$$

$$s_w = \frac{2.3\,Q}{4\pi T} \log \frac{2.25\,Tt}{r_w^2\,S} + CQ^n$$

$$\frac{Q}{s_w} = \frac{1}{\frac{2.3}{4\pi T} \log \frac{2.25\,Tt}{r_w^2\,S} + CQ^{n-1}}$$

Substituting the given data, after $t = 1$ year,

$$\frac{Q}{s_w} = \frac{1}{\frac{2.3}{4\pi \frac{600}{86400}} \log \frac{2.25 \times 600 \times 365}{(0.15)^2 \times 0.0004} + (0.5 \times 60)^2 \left(\frac{60}{60 \times 60}\right)^{2.3-1}}$$

$$= \frac{1}{291.82} \text{ m}^3/\text{s per m, or } \frac{1000 \times 60}{291.82} = \mathbf{206\ lpm/m}$$

$$s_w = 291.82 \left(\frac{60}{60 \times 60}\right) = \mathbf{4.864\ m}, \text{ after 1 year}$$

assuming no replenishment; but usually the aquifer will be recharged by natural rainfall or other sources.

Cavity Wells

If a relatively thin impervious formation of a stiff clay layer is encountered at a shallow depth underlain by an extensive thick confined alluvial aquifer, then it is an excellent location for a cavity well. A hole is drilled using the hand boring set, and casing pipe is lowered to rest firmly on the stiff clay layer, Fig. 5.54. A hole of small cross-sectional area is drilled into the sand formation and is developed into a big hollow cavity by pumping at a high rate or by operating a plunger giving a large yield. In the initial stage of pumping, fine sand comes along with water resulting in the formation of a cavity. During development the size of the cavity increases till the velocity of ground water flow at its perimeter becomes small enough to retain the aquifer material in place. With further pumping ultimately equilibrium condition is reached when clean water is discharged. The depth of the cavity at the centre varies from 15–30 cm with 6–8 m radius of the cavity. The flow of water into the cavity is spherical and the yield is low. Cavity wells have shorter life and failure is caused usually due to the collapse of the clay roof. Since the depth is usually small, deep-

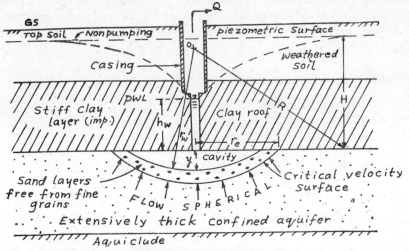

Fig. 5.54 Flow into cavity well.

well pumps are not necessary and thus the capital costs of construction, development and installation of pumpset of a cavity well are low.

Steady flow into cavity well

Assuming the cavity to be a segment of a sphere of radius 'r_w' resting on the top of a confined, homogeneous and isotropic aquifer of infinite areal extent and extensive thickness, the well yield for steady state flow condition is given by

$$Q = \frac{2\pi K y (H - h_w)}{1 - \frac{r_w}{R}} \qquad (5.95)$$

and the width of the cavity

$$r_e = \sqrt{(2r_w - y)\, y} \qquad (5.96)$$

where y = depth of cavity (at the centre); r_w = radius of the cavity; R = radius of influence; K = Permeability of the aquifer; r_e = width of the cavity and Q = yield of the well.

Unsteady flow into cavity well

As the pumping progresses, assuming no recharge into the confined aquifer, water is drawn from storage of the aquifer, from a wider and wider area. The cone of depression expands and the drawdown increases with time for the constant pumping rate, Q, given by

$$s = \frac{Q\sqrt{S_s}}{4K^{3/2}\sqrt{\pi t}} + \frac{Q}{2\pi K r} \qquad (5.97)$$

Well Hydraulics

where s = drawdown in the observation well at a distance 'r' from the cavity well; Q = constant pumping rate and S_s = specific storage coefficient (i.e. for unit aquifer thickness).

Example 5.32: The following data are obtained from a cavity tube well:

Discharge	1800 lpm
Drawdown	4 m
Permeability of the aquifer	50 m/day
Depth of cavity	20 cm
Radius of influence	150 m

Determine the radius and width of cavity.

Solution: Well yield $Q = \dfrac{2\pi K y (H - h_w)}{1 - \dfrac{r_w}{R}}$

$$\dfrac{1800}{100 \times 60} = \dfrac{2\pi (50/86400) \, 0.20 \times 4}{1 - \dfrac{r_w}{150}}$$

Radius of the cavity, $r_w = \mathbf{135.5\,m}$

Width of cavity, $r_e = \sqrt{(2r_w - y) y} = \sqrt{2 \times 135.5 - 0.2) 0.2} = \mathbf{7.36\,m}$

Hydraulics of Open Wells

Theis equation does not apply in the case of shallow dug open wells since there is no instantaneous release of water from the aquifer, most of the water being pumped only from storage inside the well.

In alluvial soil if the water is pumped at a high rate the depression head, i.e. S.W.L.−W.L. inside the well during pumping, well increase, which may cause excess gradients resulting in loosening of sand particles. This limiting head is called critical depression head. The safe working depression head is usually one-third of the critical head and the yield under this head is called the maximum safe yield of the well.

Yield tests; The following tests may be performed to get an idea of the probable yield of the well.

(a) *Pumping test:* The water level is depressed to an amount equal to the safe working head for the subsoil. Then the water level is kept constant by making the pumping rate equal to the percolation into the well. The quantity of water pumped in a known time gives an idea of the probable yield of the well of a given diameter. The test may be carried out on an existing open well. In hard rock areas if

D = diameter of the well; d = depth of water column; Q = pumping rate; and t = time required for emptying the well; then the volume of water stored in the well

$$V = \frac{\pi D^2}{4} \times d$$

Rate of seepage into the well $= \dfrac{Qt - V}{t}$ (5.98)

(b) *Recuperation test*: The water level in the well is depressed by an amount less then the safe working head for the subsoil. The pumping is stopped and the water level is allowed to rise or recuperate. The depth of recuperation in a known time is noted from which the specific yield of the well is calculated as follows.

If the water level inside the rises from s_1 to s_2 (measured below S.W.L.) in time t and if s is the drawdown at any time t, Fig. 5.55, from Darcy's law

$$Q = KAi = KA \frac{s}{L} = CAs$$

where s is the head loss in a length of flow path L and C is a constant $= K/L$ which can be determined after integrating

$$Q\,dt = -A\,ds$$

The negative sign indicates a decrease in drawdown or depression head

$$CAs\,dt = -A\,ds$$

$$C \int_0^t dt = \int_{s_1}^{s_2} -\frac{ds}{s}$$

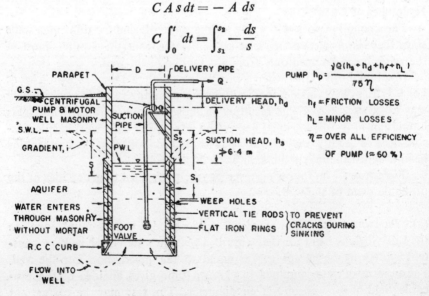

Fig. 5.55 Recuperation test—open well.

Well Hydraulics

which gives

$$C = \frac{2.303}{t} \log_{10} \frac{S_1}{S_2} \qquad (5.99)$$

C is called the specific yield of the well per unit cross-sectional area per unit depression head. If A is the cross-sectional area of the well, the specific capacity of the well (Slitcher)

$$C' = \frac{2.303 A}{t} \log_{10} \frac{S_1}{S_2}$$

If H is the safe working depression head, then the yield of the well

$$Q = C' H \qquad (5.100)$$

or $$Q = C A H \qquad (5.101)$$

This equation assumes vertical impervious well steining and flow into the well only from the pervious bottom. C has a dimension of $1/T$ and has values of 0.25 for clay, 0.50 for fine sand and 1.00 for coarse sand, all per hour.

Example 5.33: A well 3 m in diameter has a normal water level of 3 m b.g.l. By pumping, the water level in the well is depressed to 9 m b.g.l. In a time interval of 4 hours the water level rises by 4.5 m. Determine the specific yield of the well. What is the safe yield of the well if the working depression head is 3.5 m.

Solution: Specific yield of the well, Fig. 5.56

$$C = \frac{2.303}{t} \log_{10} \frac{S_1}{S_2} = \frac{2.303}{4} \log_{10} \frac{9-3}{4.5-3}$$

$$= 0.924 \text{ m}^3/\text{hr}/\text{m}^2 \text{ per metre drawdown}$$

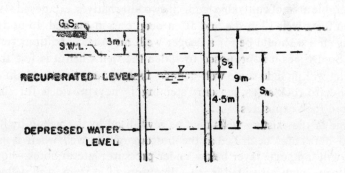

Fig. 5.56 Recuperation in open well, Example 5.33.

Safe yield

$$C = CAH = 0.924 \left(\frac{\pi}{4} \times 3^2 \right) 3.5$$

$$= 22.95 \text{ m}^3/\text{hr} = 382.5 \text{ lpm}.$$

Example 5.34: Determine the diameter of an open well to give a yield of 340 lpm. The working depression head is 3.5 m. The aquifer soil consists of coarse sand.

Solution: For coarse sand, $C = 1$ per hour

$$Q = CAH$$

$$0.34 \times 60 = 1 \left(\frac{\pi}{4} D^2 \right) 3.5$$

$$D = 2.72 \text{ m}$$

From the above worked examples it can be seen that in alluvial soil where an impervious vertical steining is provided to support the soil, the percolation into the well is entirely from the bottom and depends on the area of cross-section of the well. Bigger diameter wells are recommended in such soils to obtain larger yields. In case of wells in rocky substrata with fissures and cracks, the lower portion of steining may be provided with alternate bands of masonry, *pucca* (impervious) and laid dry (without putting cement mortar for joints), and the percolation into the well is mostly from the sides through fissures and cracks. In such wells higher yields are obtained by going deeper as long as the weathering and fractures are evident, rather than making the wells wider or of larger diameter. Larger diameter wells also involve large volume of excavation in rock and the mounds of excavated rock deposited on the ground surface occupy considerable area of cultivable land. If two alternatives are posed—whether to go in for a well 12 m dia and 10 m deep or 6 m dia and 15 m deep, the obvious choice should be for a deeper well, all other conditions remaining same. Sometimes it is proposed to widen the well when it is felt that such widening will include some well defined fissures and fractures. In case of wells in hard rock areas, masonry steining is not provided for the lower fractured rock exposures.

Some of the existing wells may be revitalised by deepening by blasting; vertical bores may be drilled at the bottom of the well when it is found that it will tap some layer below under pressure; lateral bores—horizontal or inclined—may be drilled in the direction of certain well defined fractures yielding water (see section on Well Revitalisation).

Well Hydraulics

QUIZ 5

Part A

1. Select the correct answer(s):
   I. A steady ground water flow condition exists when
      a. the water levels in the wells case to decline
      b. the water levels respond to changes in atmospheric pressure or tides
      c. the water levels drop as the pumping is continued
      d. the Laplace's equation is satisfied
      e. the Darcy's law governs the flow
      f. the pumping rate can be obtained by applying Dupuit's or Thiem's equation
      g. none of these conditions

   2. An unsteady ground water flow condition exists when
      a. the Darcy's law no longer governs the flow.
      b. the Laplace's equation is not applicable
      c. the pumping rate cannot be obtained by applying Dupuit's or Theim's equation
      d. the water levels or piezometric heads drop as the pumping is continued
      e. the rate of flow through the aquifer is changing
      f. water comes mostly from storage of the aquifer
      g. the solutions can be obtained by an RC network analogue
      h. the solutions can be obtained by mathematical modelling
      i. none of the above conditions

   3. The assumptions of the Thiem's equation are
      a. complete penetration of the well with 100% efficiency
      b. the pumping water levels remain constant with time
      c. the pumpage mostly comes from nearby recharge sources
      d. the pumpage mostly comes from storage
      e. the flow is two dimensional
      f. the flow is radial and satisfies Laplace's equation
      g. the flow is entirely laminar
      h. none of the above

   4. The aquifer constants S and T can be determined by pumping at a constant rate and
      a. by observing drawdowns in several observation wells for a very short period immediately after pumping starts
      b. by observing drawdowns in several observation wells at a particular instant immediately after pumping start:
      c. by observing drawdowns in several observation wells at a particular instant after a long duration of pumping
      d. by observing drawdowns in the pumping well for a long period of time
      e. by observing drawdowns in an observation well for a long period of time
      f. from time-drawdown and distance-drawdown data on a few wells for long duration pumping

g. observing the maximum drawdown at the end of a long duration pumping, stop pumping, and observe the drawdown as the ground water table recovers in an hour or so
h. observing the maximum drawdown for a short period of pumping, stopping the pumping and observing the drawdowns over a long period as the ground water table recovers.

5. Pumping tests are conducted to
   a. identify the aquifer boundaries, their nature and distance
   b. make fairly long time predictions of drawdown, assuming no recharge
   c. make fairly long time predictions of drawdown, assuming recharge at a certain percentage of a.a.r.
   d. determine the aquifer constants S and T
   e. determine the efficiency of the well
   f. determine the specific capacity of the well
   g. select a suitable pump for installation
   h. determine hydraulic interconnection between aquifers (vertical leakage), if any
   i. determine the safe sustained yield
   j. all the above purposes

6. The assumptions of the Theis equation are
   a. the well penetrates the entire thickness of the aquifer
   b. the entire pumpage comes from nearby recharge sources
   c. the entire pumpage comes from storage
   d. most of the pumpage comes from overlying or underlying aquifers (by vertical leakage)
   e. the flow is two dimensional
   f. the flow is radial and satisfies Laplace's equation
   g. the flow is entirely laminar

7. The aquifer constants S and T can be determined by
   a. bailing out a known volume (V) of water and observing the drawdown after an appreciable time interval (t) since bailing
   b. bailing out a known volume (V) of water and observing the drawdown immediately after bailing
   c. bailing out the same volume of water (V) n times, and observing the drawdown immediately after the bailing stops
   d. bailing out the same volume of water (V), n times, and observing the drawdown after an appreciable time (t) after the bailing stops
   e. injecting a known volume of water (V) and observing the head after a time interval (t)
   f. all the above methods

8. Jacobs's modifications of the Theis non-equilibrium equation are valid for
   a. small values of u
   b. high values of u
   c. early pump test data
   d. long duration pump test data

e. observation wells close by
   f. distant observation wells
   g. none of of the above conditions
   h. all the above conditions
9. A barrier (impermeable) boundary can be identified when the plot of observed drawdowns diverges
   a. below the Theis type curve
   b. above the Theis type curve
   c. above the time-drawdown straight line plot of Jacob
   d. below the time-drawdown straight line plot of Jacob
   e. coincides with the distance-drawdown plot of Jacob
10. Barrier boundaries can be identified when
    a. the recovery curve (Theis) intercepts the time axis, i.e. $s'=0$ for $\frac{t'}{t} > 1$
    b. the recovery curve (Theis) intercepts the axis of drawdown, i.e. s' is positive at $t/t' \to 1$
11. Recharge boundaries can be identified by
    a, b, c, d, e, of Q. 9, and a, b, of Q. 10 above.
12. When the curve of observed drawdown does not coincide with the Theis type curve
    a. it is influenced by a recharge boundary
    b. it is influenced by a barrier boundary
    c. a significant portion of the pumpage comes from vertical leakage through the confining formations
    d. the vertical leakage is proportional to the drawdown
    e. the observed time-drawdown plot is superimposed on the unsteady leaky-artesian type curves and S, T and K' determined
13. Specific capacity is
    a. a measure of the effectiveness of a well
    b. expressed as discharge per unit drawdown
    c. the same as the well efficiency
    d. determined by step-drawdown test
    e. constant for a given well for any time
    f. decreases with prolonged pumping
    g. decreases with increase in the pumping rate
    h. the same as the specific yield of an aquifer
14. The safe yield of a well can be appreciably increased by
    a. increasing the drawdown
    b. doubling the diameter of the well
    c. providing an artificial gravel pack (shrouding)
    d. increasing the length of strainer
    e. increasing the depth of penetration in the aquifer (if not already completely penetrated)

    f.   optimum spacing of the wells
    g.   pumping for certain hours per day; say 12 to 15 hr/day

15. The well loss coefficient
    a.   is determined from a step-drawdown test
    b.   gives the well efficiency
    c.   has dimensions of L^5/T^2
    d. >2000 for production well contact specifications (in metric units)

16. Slitcher's formula for the yield of open (shallow, dug) wells
    a.   assumes flow into the well only from the bottom
    b.   assumes flow into the well only from the sides
    c.   assumes flow into the well both from the bottom and sides
    d.   is not reliable since the flow into the well sunk in rocky substrata, is mostly through the weathered fractures, fissures and cracks on the sides
    e.   can be used in alluvial areas where the circular impervious well steining is sunk to a shallow depth and the flow is entirely from the bottom
    f.   in (e) above, bigger the diameter, larger the yield
    g.   in (d) above, in the lower portion of the steining, alternate bands of masonry may be laid dry

17. Existing open wells may be revitalised
    a.   by drilling bores at the bottom
    b.   by drilling bores on the sides to intercept the fractures
    c.   by deepening by blasting
    d.   by increasing the area of cross-section of the well
    e.   by all the above methods

II. State whether 'true' or 'false'; if false give the correct statement:
1. The storage coefficient is directly proportional to the porosity and thickness of the confined aquifer.
2. The storage coefficient is directly proportional to the expansibility of the pore water.
3. The storage coefficient is directly proportional to the barometric efficiency.
4. The sum of barometric and tidal efficiencies is equal to unity.
5. For a pumping well located at a distance a from a recharge source, the drawdown is almost the same as that of a well situated at the centre of a circular island aquifer of radius $2a$.
6. Times of occurrence of equal drawdown are proportional to the squares of the distances of the observation wells from the pumping well.
7. The available yield of a tubewell can be doubled by doubling the diameter.
8. The well loss is due to turbulent flow and entrance through the screen while the formation loss is due to laminar flow.
9. The Storage coefficient in a leaky artesian aquifer under steady flow conditions cannot be computed.

Well Hydraulics

PROBLEMS

1. A 30 cm well penetrates 45 m below the static water table. After a long period of pumping at a rate of 1200 lpm, the drawdown in the wells 20 and 45 m from the pumped well is found to be 3.8 and 2.4 m respectively. Determine the transmissibility of the aquifer. What is the drawdown in the pumped well?

(1.71×10^5 lpd/m, 13.5 m)

2. A 20 cm well completely penetrates an artesian aquifer. The length of the strainer is 15 m. What is its yield for a drawdown of 3 m. Assume $K=35$ m/day and $R=300$ m.

  If the diameter of the well is doubled find the percentage increase in the yield, the other conditions remaining the same. (846 lpm, 9.45%)

3. If an artesian well produces 250 lpm with a drawdown of 3 m in the pumping well, what is the rate discharge with 4 m drawdown. Assume equilibrium conditions and negligible well losses.

(333.3 lpm)

4. A well penetrates 30 m into a saturated free aquifer. The discharge is 200 lpm at 6 m drawdown in the pumping well. Assuming equilibrium conditions and a homogeneous aquifer, what is the discharge at 9 m drawdown (on the basis of Dupuit's equation)

(300 lpm)

5. A 30 cm well yielding 300 lpm under a drawdown of 2 m penetrates an aquifer 35 m thick. For the same drawdown, what would be the probable yield of (a) 20 cm well, (b) 40 cm well? Assume a radius of influence of 500 m in all cases.

(285 72, 311.03 lpm)

6. An artesian aquifer 30 m thick has a porosity of 25% and bulk modulus of compression of 2000 kg/cm². Estimate the storage coefficient of the aquifer. What fraction of this is attributable to the expansibility of water?

($S=1.54 \times 10^{-3}$, $S_w=3.52 \times^{-5}$ or 2.28%)

7. A 30 cm well penetrates an aquifer of transmissibility of 2×10^5 lpd/m and a storage coefficient of 0.005. What pumping rate could be adopted so that the drawdown will not exceed 10 m within the subsequent two years of drought?

(800.65 lpm)

8. A 25 cm well penetrates an artesian aquifer of 10 m thick. After 10 hours of pumping at the rate of 1100 lpm the drawdown in the well is 2.6 m and after 48 hours the drawdown is 2.85 m. Determine the transmissibility and storage coefficients of the aquifer. What is the permeability of the aquifer material? After what time will the drawdown be 4.1 m?

(7.94×10^5 lpd/m, 0.00038, 79.4 m/day, 1.51 yr)

9. In the centre of a strip of land 1500 m wide, a fully penetrating well of 40 cm yields 2000 lpm. The aquifer is in direct contact with the boundary ditches, receives no recharge from above or below and has a transmissibility of 1200 lpm/m. What is the drawdown in the well?

  (*Hint*: Use Dupuit's equation) (2.19 m)

10. A 30 cm well 75 m deep is proposed in an aquifer having a transmissbility of 1.5×10^5 lpd/m and a coefficient of storage of 0.004. The static water level is expected to be 20 m below ground level. Assuming a pumping rate of 2000 lpm, what will be the drawdown in the well after (a) 1 year, (b) 2 years?

(32.15, 33.21 m)

11. Tabulated below are the data on an observation well 12.3 m from a 20 cm well which is pumped for test at the rate of 1,150 lpm. Find the transmissibility and

storage coefficient of the aquifer. What will be the drawdown at the end of 180 days (a) in the observation well and (b) in the pumped well. Use the modified Theis method; under what conditions is this method valid?

Time (min)	2	3	5	7	9	12	15	20	40	60	90	120
Drawdown (m)	2.42	2.46	2.52	2.58	2.61	2.63	2.67	2.71	2.79	2.85	2.91	2.94

(1080 m²/day, 4.12×10^{-11}, 3.89 m, 5.06 m instantaneously)

12. The following information is obtained on a 30 cm production well:

$T = 500$ m³/day/metre

$S = 0.15$

$Q = 1600$ lpm

$t = 2$ days

Determine the radius of the cone of depression assuming a drawdown of 0.003 m. (122.4 m)

13. A pump test was conducted on a fully penetrating well in a non-leaky artesian aquifer. Determine the drawdown in an observation well at 10, 50 and 200 min from the time of starting the pump, given the following data:

Constant discharge from the well = 2800 lpm

Distance of observation well from the pumping well = 20 m

Storage coefficient of aquifer = 7.24×10^{-4}

Transmissibility of aquifer = 1500 lpm/m

(0.71 m, 0.95 m, 1.15 m)

14. A 40 cm well was pumped at the rate of 2000 lpm for 200 min and the drawdown in an observation well 20 m from the pumping well was 1.51 m. The pumping was stopped and the residual drawdowns during recovery in the observation well for 2 hours are given in the following table. Determine the aquifer constants S and T.

Time since pumping stopped (min)	Residual drawdown (m)	Time since pumping stopped (min)	Residual drawdown (m)
2	0.826	45	0.180
3	0.664	50	0.159
5	0.549	55	0.155
10	0.427	60	0.149
16	0.351	70	0.146
20	0.305	80	0.140
25	0.271	90	0.134
30	0.241	100	0.131
35	0.220	110	0.131
40	0.201	120	0.131

(2×10^{-4}, 1284 m²/day)

Well Hydraulics

15. The coefficients of transmissibility and storage of a non-leaky artesian aquifer are 3×10^5 lpd/m and 2×10^{-4}, respectively. The aquifer is bounded on one side by a barrier boundary. A fully penetrating production well has been discharging at a constant rate of 1,000 lpm. The drawdown in an observation well 40 m from the pumping well due to the effects of the barrier boundary for a pumping period of 2 hr is 2.5 m. What should be the approximate distance of the pumping well from the barrier boundary?

$$\left(\frac{x}{2} \approx \frac{r_i}{2}, r_i = 634 \text{ m}\right)$$

16. A production well tapping a non leaky artesian aquifer has been discharging at a constant rate for five years. A recharge boundary is situated at 10 km from the well. The coefficients of transmissibility and storage of the aquifer are 2.5×10^5 lpd/m and 4×10^{-4}, respectively. Compute the percentage of the well yield drawn from the source of recharge.

(88%)

17. A 30 cm well, located at a distance of 1 km from a recharge boundary, has been discharging at a constant rate from a water table aquifer. The coefficients of transmissibility and storage of the aquifer are 1.8×10^5 lpd/m and 0.2 respectively. Compute the time that must elapse the water level in the well stabilise.

(59.86 days)

18. The barometric efficiency of a well in an artesian aquifer infinite in areal extent is 60%. If the thickness of the aquifer is 45 m and the porosity 30%, estimate the coefficient of storage of the aquifer.

$(S = 1.1 \times 10^{-4})$

19. The tidal efficiency of a well in an artesian aquifer overlain by an extensive body of tide water is 40%. If the thickness of the aquifer is 50 m and the porosity 30% estimate α, the reciprocal of the modulus of elasticity of the aquifer skeleton.

$(9.4 \times 10^{-6} \text{ cm}^2/\text{kg})$

20. An infinite non-leaky artesian aquifer, $T = 25,000$ lpd/m, $S = 0.0004$, $r_w = 15$ cm, $t = 1$ hr, 1 day, 1 month, 1 year. Construct a graph showing the relation between specific capacity and time.

$(2.52, 2.01, 1.65, 1.46) \times 10^4$ lpd/m)

21. A variable-rate well-production test was conducted with the following result:

Step	Pumping rate (m^2/day)	Resulting drawdown (m)
1	2295	3.69
2	2954	5.14
3	3516	7.08
4	4273	9.63
5	4704	11.24

Determine the well loss coefficient and the well loss for a pumping rate of 3000 m^3/day.

(2421 sec^2/m^5, 2.91 m)

22. A variable-rate well-production test was conducted with the following results:

Step	Pumping rate (m^3/day)	Resulting drawdown (m)
1	1308	2.71
2	1723	3.60
3	2192	4.89
4	2726	6.40
5	3358	8.39

Determine the average well loss and formation loss coefficients.

(1756.5 sec^2/m^5, 168.5 sec/m^2)

23. An aquifer is delimited by two converging barrier boundaries, the angle of the wedge being 30°. Compute the number of image wells associated with the wedge-shaped boundary system.

(11)

24. A 30 cm well is discharging at the rate of 1000 lpm with a drawdown of 4.5 m. The tranmissibility of the aquifer is 0.015 m^2/sec and is bounded by an aquiclude at a distance of 120 m from the well. Determine the drawdown (i) at a distance of 80 m from the well towards the aquiclude, and (ii) at a distance of 80 m on the opposite side of the well (Assume steady state conditions).

25. Show that for a pumping well located at a distance a from a recharge source, the drawdown is almost the same as that of a circular island aquifer of radius 2a.

26. A circular island of 800 m radium has an effective rainfall of 6 mm/day. A central well of 30 cm diameter is pumped at a constant rate of 600 lpm. K for the island aquifer is 20 m/day. The depth of sea around the island is 10 m. Determine the drawdown in the well and at the water divide.

(5.17 m, 0)

27. (a) Describe briefly the image well theory.
    (b) An aquifer is bounded by two converging boundaries at an angle of 36°, one being a barrier boundary and the other a recharge boundary. Compute the number of image wells and mark them nearly in a sketch:

(9)

28. A 40 cm well penetrates an aquifer of 30 m thick and the length of the screen is 10 m. The yield is 3000 lpm with a drawdown of 2.5 m. If the length of the screen is increased to 20 m, what will be the drawdown in the well and the increase in the specific capacity?

(1.6 m, 50%)

29. (a) What is the effect of partial penetration on the drawdown in the well?
    (b) A 30 cm well fully penetrating an artesian aquifer of 30 m depth and fully

screened yields 1800 lpm with a drawdown of 3 m in the well. What will be the resulting drawdown when the screen extends from 7.5 m to 22.5 m depth below the top of aquifer?

(4.44 m)

30. A leaky artesian aquifer 40 m thick and having a permeability of 20 m/day is situated above an impervious base and overlain by a semi-confined layer with a resistance of 2000 days against vertical water movement. On top of this semi-confined layer a homogeneous aquifer with constant water table is present. From the leaky artesian aquifer, ground water is abstracted by a series of nine wells at intervals of 110 m in a straight line. Each well has a diameter of 40 cm, a screened length of 20 m in the middle of the aquifer and is pumped at a constant rate of 400 lpm. What is the maximum drawdown in any well?

31. A confined aquifer with a transmissibility of 1,000 lpm/m is situated above an impervious base and overlain by a semi-confining layer with a resistance of 2000 days against vertical leakage and above this layer a homogeneous aquifer with constant water table is present. From the leaky aquifer water is pumped at a rate of 800 lpm by a fully penetrating well of 30 cm diameter. What is the drawdown in the well and at a distance of 800 m from the well, assuming the well yield is balanced by the leakage.

Hint: Use Eq. (5.22) and Table 5.15

(1.2 m, 0.124 m)

32. Determine the yield from a 30 cm well fully penetrating an unconfined aquifer of 30 m thickness and permeability of 12 m/day. The drawdown in the well is 4 m and rainfall penetration is 1 mm/hr.

Hint: Yield=recharge within the radius of influence.

33. Two tubewells of 20 cm diameter are spaced at 120 m distance and penetrate fully a confined aquifer of 12 m thickness. Calculate the discharge when only one well is discharging under a depression head of 3 m. What will be the percentage decrease in the discharge of this well, if both the wells are discharging under the depression head of 3 m. Assume the radius of the circle of influence for each well as 200 m and permeability of the aquifer 40 m/day.

Hint: Use Eq. (5.55)

(1186.5 m³/day, 6.3%)

34. Three tubewells 20 cm diameter are located at the vertices of an equilateral triangle of side 120 m and penetrate fully a confined aquifer of 20 m thickness. Calculate the discharge when only one well is discharging under a depression head of 3 m. What will be the percentage decrease in the discharge of this well, if all the three wells are discharging under a depression head of 3 m. Assume the radius of the circle of influence for each well as 200 m and permeability of the aquifer as 45 m/day.

(2224.8 m³/day, 11.85%)

35. Five tubewells of 20 cm diameter are equally spaced along the boundary of a circular well field of radius 200 m. The wells fully penetrate an artesian aquifer of thickness 20 m, permeability 40 m/day and storage coefficient 0.0002. Calculate the drawdown in the wells after 4 hours of the start of pumping all the wells at the rate of 800 lpm. What is the drawdown at the centre of the well field?

36. During a recuperation test, the water level in an open well was depressed by 2.5 m by pumping. It recouped 1.8 m in 80 min. Find:
   (a) the yield from the well of 4 m/diameter under a depression head of 3 m, and
   (b) the diameter of the well to yield 480 lpm under a depressed head of 2m.

(600 lpm 4.37 m)

37. A non-leaky artesian aquifer is 30 m thick. A production well fully penetrating the aquifer is continuously pumped at a constant rate of 100 m³/hr for a period of 1 day. The observed drawdown in a fully penetrating observation well at a distance of 80 m from the production well are given in Table 5.22. Compute the coefficients of transmissibility, permeability and storage of the aquifer.

Table 5.22 Drawdowns in a fully penetrating observation well

Time after pumping started (min)	Drawdown in observation well (cm)	Time after pumping started (min)	Drawdown in observation well (cm)
1	14	60	70
2	22	80	75
3	27.3	100	80
4	31	300	83
6	34	500	100
7	38	700	103
8	40	900	106
10	43	1000	108
30	60	1440	114

(1375 m²/day, 45.8 m/day, 1.345×10⁻⁴)

38. A Production well fully penetrating a water table aquifer is pumped at a constant rate of 100 m³/hr for a period of 200 days. The observed drawdowns in fully penetrating observation wells at the end of pumping period are given below. Compute the coefficients of storage and transmissibility of the aquifer.

Distance of observation well from production well, m	2	20	200
Drawdown, m	2.75	1.80	0.90

(977 m²/day, 0.11)

39. A leaky artesian aquifer has a transmissibility of 6.5×10^5 lpd/m. The semi-confining layer at the top has a resistance of 2000 days against vertical water movement. A water table aquifer with constant water table exists over the semi-confining layer. A 20 cm well completely penetrates the leaky aquifer and the drawdown in the well due to pumping is 3.5 m. Determine the rate of pumping and also the drawdowns in the two observation wells at 60 m and 180 m from the pumping well. Assume that the entire pumpage comes from leakage.

Hint: Use Eq. (5.22) and Table 5.15.

(1050 lpm, 1.13 m, 0.73 m)

40. (a) Explain the theory of images as applied to ground water hydraulics

(b) Using the image well theory, determine for the following cases, the number of

Well Hydraulics

image wells with their locations for a given pumping well in the real aquifer. Also derive the equations for the drawdown:
(i) Semi-infinite aquifer with recharge boundary on one side.
(ii) Two recharge straight boundaries meeting at 90°.
(iii) Two unlike straight boundaries meeting at 90°.

41. The drawdown is 2 m in an observation well 20 m away from the pumping well after 15 min of pumping. At what time the same drawdown will occur in another well 40 m away?

(1 hr)

42. The drawdowns in an observation well 15 m away from the pumping well are 3 and 4 m after 10 and 100 min of starting pumping. What are the corresponding drawdowns in an observation well 150 m away from the pumping well?

(1, 2 m)

43. In the centre of a strip of land 1400 m wide a fully penetrating well of 0.5 m is pumped at the rate of 2500 lpm from a confined aquifer having a transmissibility of 0.015 m²/s. The aquifer is in direct contact with two boundary ditches. What is the drawdown at the well face?

(3.6 m)

44. An aquifer is bounded by a recharge boundary and a barrier boundary as shown in Fig. 5.57, the angle of the widge being 45°. Compute the number of image wells associated with this multiple boundary system. Locate all the image wells in a neat sketch. Indicate how you would determine the drawdown in the 40 cm well when pumped at rate of 1200 lpm. Assume a transmissibility of 300 m²/day.

45. Locate the image wells for a pumping well in an alluvial valley between a stream and mountain range in Fig. 5.58. Indicate how you would determine the drawdown in the well at any time 't' (after pumping started) when pumped at the rate Q, the values of transmissibility and storage coefficients for the aquifer being known.

46. The coefficients of transmissibility and storage of a nonleaky artesian aquifer are 520 m²/day and 1.21×10^{-3}, respectively. The aquifer is bounded on one side by a barrier boundary. A fully penetrating production well has been discharging at a constant rate of 2 m³/min. The additional drawdown in an observation well 50 m away from the pumping well due to the effects of barrier boundary is found to be 0.5 m (i.e. the divergence in the Jacob's time-drawdown plot) for a pumping of 90 min. What should be the approximate distance of the pumping well from the barrier boundary.

$$\left(\approx \frac{r_i}{2} = 69 \text{ m} \right)$$

47. (a) How do you ensure that the 'tubewell construction' has been satisfactory? Describe any test you would conduct for this purpose.

(b) An infinite nonleaky artesian aquifer has a transmissibility of 600 m²/day and storage coefficient 4×10^{-4}. Calculate the specific capacity of 30 cm fully penetrating well after 1 year of pumping at the rate of 1600 m³/day. Assume a well loss coefficient of 1960 sec²/m⁵. What is the drawdown in the well after 1 year, assuming no replenishment? What is the well efficiency?

(271 m³/d.m., 5.9 m, 88.6%)

48. Say True or False; if false give the correct statement:
   1. The aquifer constants S and T can be determined by observing the drawdowns at several wells at a particular instant or drawdown in a single well at time intervals, for a constant pumping rate.
   2. Theis equation assumes that the entire passage is instantaneously released from storage inside the tubewell.

3. The drawdown per log-cycle of time is twice the drawdown per log-cycle of distance.
4. Presence of a recharge boundary in the aquifer tends to offset the unsteady character of the flow.
5. Data on three observation wells are required when a well is pumped, to delineat the aquifer boundary.
6. Image wells transform a finite aquifer into an infinite aquifer so that the drawdown at any point can be determined by applying Theis equations.
7. The specific capacity for a pumping well is constant for any time.
8. Partial screening is to decrease the entrance velocity.
9. As the length of the well screen decreases, the specific capacity of the well increases.

(√, ×, ×, √, √, √, ×, ×, ×)

6

Ground Water Analog Models

As the ground water development increases, problems of well field management will become dangerously critical in many places and studies on optimum well spacing will be required in order to minimise mutual interference between pumped wells. Where the hydrogeological conditions are complicated by non-isotropic materials, variable infiltration, boundary limits and by all manner of natural and artificial effects ranging from influent or effluent rivers to well abstraction and irrigation seepage, the solutions are so difficult and time consuming as to be impracticable. Under such circumstances the simulation of an aquifer by a 'model' affords an indirect method of varying the inputs so that by measurement of subsequent response of the inter-related system one can provide answers to the various kinds of questions posed such as

(i) How extensive will the cone of influence be if the abstraction rate is increased by a factor, say 3.

(ii) If a new well is sunk at a point A and pumped at x lpd, what extra drawdown will be produced in an adjacent well B situated y metres away.

(iii) How far apart should two wells be spaced so as to produce optimum yields.

Analog Models

Analog models physically simulate the field prototype condition either by using actual aquifer materials for direct representation or by indirectly simulating the aquifer and flow by various physical analogies, such that the partial differential equations that describe the response of the two systems, i.e. prototype and analog model, are similar. Analog models applicable to ground water problems are of two types: (a) mechanical analogs and (b) electric analogs.

MECHANICAL ANALOG MODELS

(i) *Sand models in glass flumes*: There have long been used for the solution of seepage problems and will provide useful information. They involve the application of Darcy's law and the equation for simulation of the model is

$$Q_r = K_r\, i_r\, L_r^2 \tag{6.1}$$

where r = subscript for the model-prototype ratio; L_r = length ratio and L_r^2 = area ratio.

(ii) *Viscous flow model (parallel plate or Hale-Shaw model)*: The viscous flow of liquid between closely spaced parallel plates simulates two-dimensional ground water movement. The viscous flow model consists of two closely spaced parallel glass plates or transparent perspex sheets with a capillary space of constant width of 1 to 1.5 mm and viscosity of the fluid, such as glycerine, appropriate to the permeability of the aquifer.

The loss of head between two closely spaced parallel plates is given by Poiseuille's equation

$$h = \frac{12\mu v L}{\gamma B^2} \tag{6.2}$$

where h = loss of head; μ = dynamic viscosity of the liquid; γ = specific weight of the model fluid ($=\rho g$); B = space between parallel plates; L = length of the plates and v = velocity of flow of the model fluid.

Putting $B = 2a$ in Eq. (6.2)

$$v = \frac{a^2 g}{3\nu}\frac{h}{L} \tag{6.3}$$

$$v = K_m \frac{h}{L} \tag{6.4}$$

where $\nu \left(=\frac{\mu}{\rho}\right)$ = kinematic viscosity of the model fluid; ρ = density of the model fluid and $K_m \left(=\frac{a^2 g}{3\nu}\right)$ = model permeability.

Eq. (6.3) is analogous to Darcy's law

$$v = Ki$$

The velocity ratio between model and prototype

$$v_r = \frac{v_m}{v_p} = \frac{a^2 g}{3\nu K} \tag{6.5}$$

field permeability

$$K = k\frac{\nu}{\mu}$$

where $k\,(= cd^2)$ = coefficient of intrinsic or physical permeability depending on soil characteristics alone [see Eq. (4.12)] and μ, γ = viscosity and specific weight of water in the field, respectively.

Time scale

$$v_r = \frac{a^2 \rho_r}{3k\mu_r} \quad (6.6)$$

$$T_r = \frac{L_r}{v_r} \quad (6.7)$$

where r = subscript for model-prototype ratio and L_r = linear scale ratio.

The plate spacing and fluid can be selected to correspond to a desired permeability. The model may be of horizontal or vertical type, but the latter has greater application. It can represent two phase flow by using liquids of different density. This makes the viscous flow model ideal for simulation of conditions involving fresh and saline waters and prediction of the interface movement under changing conditions of abstraction and replenishment. The water table or piezometric levels in wells can be observed in the vertical model and the flow lines can be seen and photographed if dye is added to the viscous liquid. Storage is simulated by the addition of small storage reservoirs connected to the interspace. The parallel plate model is isotropic with respect to permeability; but local variations in width of the interspace may be made by placing obstructions between the plates to simulate non-homogeneity. The liquid used may be water with some substance added to increase its viscosity, glycerine or any petroleum oil.

ELECTRIC ANALOG MODELS

Various models such as conductive paper (teledeltos) and electrolytic tanks make use of the analogy between the subsoil flow (Darcy's law) and flow of electricity (Ohm's law) but to a large extent they have been superceded by networks consisting of a mesh of resistors linked together at nodal points to give a square, rhombic or polygonal pattern. Currents may be fed into the mesh in accordance with the predetermined inputs of infiltration and leakage or may be led out of the mesh as simulation of well abstraction or effluent flow to rivers.

RC Network Analog

Further refinement is made by adding to the resistor mesh, at nodal points, grounded capacitors which represent the coefficient of storage in the aquifer. Recharge boundaries are simulated by common connections and impermeable boundaries by open ends. Each node in RC (resistor-capacitor) network analog represents a location for a well. Pumping at a constant rate at any well is simulated by injecting a negative current pulse of scaled duration at the corresponding node in the analog. The transient response of all other nodes is studied by observing their varying potentials with the help of a cathode ray oscilloscope. These transients represent the time-drawdown characteristics of the aquifer at the corresponding locations. The electrical analog is then modified step by step in its values of resistors and capacitors, such that all observed drawdown histories in the wells for the corresponding causative pumpages are correctly simulated by the electrical transients at the corresponding nodes. If this step by step modification is done completely and satisfactorily, then the analog can be relied upon to compute any drawdown at defined locations and times, for given excitations of the aquifer. However, this modification process is invariably very arduous and frustrating and perfect simulation of the aquifer is very difficult to achieve.

The partial differential equation governing the unsteady flow in an aquifer is

$$\nabla^2 h = \frac{S}{T}\frac{\partial h}{\partial t} \tag{6.8}$$

The equation for transient flow of electric current is

$$\nabla^2 V = \frac{C}{\sigma}\frac{\partial V}{\partial t} \tag{6.9}$$

where h = drawdown in the well; V = voltage drop; S = coefficient of storage of the aquifer; T = transmissibility of the aquifer; C = electrical capacitance; σ = electrical conductance and ∇^2 = operator $\left(\frac{\partial^2}{\partial x^2} + \frac{\partial^2}{\partial y^2}\right)$ in two-dimensional flow.

The aquifer is subdivided into squares of equal area $\Delta x\, \Delta y$. A resistor-capacitor network with a square pattern as shown in Fig. 6.1(a) and network junctions at nodes as defined in Fig. 6.1(b) is set up. The intersections of grid lines are called nodes. The junctions consists of 4 resistors and 1 capacitor connected to a common terminal; the capacitor is also connected to the ground. The infinitesimal $\Delta x\, \Delta y = a_g^2$, in which a_g is the width of the grid interval. The area a_g^2 is small compared to the size of the model (representing the entire aquifer) and the behaviour of the discontinuous network closely simulates the continuous aquifer. Because the model consists of discrete elements, a finite difference form of the partial differential equation for the unsteady ground water flow is used for comparison with the equation for the flow of electric current. The finite difference form of the partial differential equation [Eq. (6.8)] is

$$\frac{h_2 + h_3 + h_4 + h_5 - 4h_1}{a_g^2} = \frac{S}{T}\frac{\partial h}{\partial t}$$

which can be written as

$$T\left(\sum_{2}^{5} h_i - 4h_1\right) = a_g^2 S \frac{\partial h}{\partial t} \tag{6.10}$$

where h_1 is the head at node 1 and h_i the heads at nodes 2-5.

Ground Water Analog Models

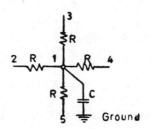

b. NETWORK JUNCTION AT NODE

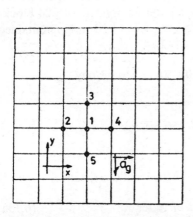

a. R C. NETWORK-SQUARE PATTERN

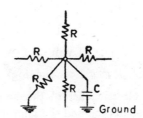

c. SIMULATION OF LEAKAGE BY ADDITIONAL RESISTOR GROUND

Fig. 6.1 RC Network analog.

The relation of electrical potentials in the vicinity of the junction, according to Kirchoff's current law, can be expressed as

$$\frac{1}{R}\left(\sum_{2}^{5} V_i - 4V_1\right) = C \frac{\partial V}{\partial t} \tag{6.11}$$

where V_i, V_1 = the electrical potentials at the ends of resistors 2—5 and 1, respectively; R = resistance and C = capacitance.

From Eqs. (6.10) and (6.11), the analogy between electrical system and aquifer system is apparent and by inspection of the two equations the analogous parameters in the two systems are

$$T = \frac{1}{R} \tag{6.12}$$

$$h = V \tag{6.12a}$$

$$a_g^2 S = C \tag{6.12b}$$

$$Q = I \tag{6.12c}$$

Scaling factors are necessary to provide for proportionality of the analogous parameters in the two systems. To illustrate this, consider an aquifer of 6×4 km extent and of uniform thickness of 50 m, with a permeability of 3 m/day and $S = 10^{-3}$. The aquifer may be assumed under uniform pressure at every point. The aquifer can be divided into a number of equal blocks, the size of each block being determined on a compromise between economy and accuracy required. Let the aquifer be divided into 24 blocks each of 1 km $\times$ 1 km $\times$ 50 m size. If a differential head of 1 m is maintained between the faces of any block, the flow will be

$$Q = KiA = 3\,\frac{1}{1000}(1000 \times 50) = 150\ \text{m}^3/\text{day}$$

The transmissibility of each block,

$$T = Ky = 3 \times 50 = 150\ \text{m}^2/\text{day}$$

Assuming the following scales

R: 1 megohm $= \dfrac{1}{150}$ day/m² (hydraulic resistance) (6.13)

V: 1 volt $= 1$ m head of water (6.13a)

t: 1 sec (analog) $= 1000$ days (aquifer) (6.13b)

The electric current through the resistor of 1 megohm for 1 volt drop is 1 microampere and the charge in 1 sec is 1 microcoulomb. The volume of flow in the aquifer at 150 m³/day for 1000 days is $150 \times 1000 = 1.5 \times 10^5$ m³. Hence the scale factors are

$$1\ \text{microampere} = 150\ \text{m}^3/\text{day} \quad (6.13c)$$

$$1\ \text{microcoulomb} = 1.5 \times 10^5\ \text{m}^3 \quad (6.13d)$$

Each aquifer block can store $S \times 1000 \times 1000 = 10^6\,S = 10^6(10^{-3}) = 10^3$ m³ of water per metre increase of head.

$$\text{Capacitance of 1 microfarad} = \frac{1\ \text{microcoulomb}}{1\ \text{volt}} = \frac{1.5 \times 10^5\ \text{m}^3}{1\ \text{m head}}$$

$$= 1.5 \times 13^5\ \text{m}^3/\text{m head of water} \quad (6.13e)$$

Hence each aquifer block will be represented by $10^3/(1.5 \times 10^5) = 0.0067$ microfarad, i.e.

$$a_g^2 S = 0.0067\ \mu F \quad (6.13f)$$

The electrical equivalent to each block will be a network of four resistors each of 1/4 megohm and one capacitor of 0.0067 microfarad, meeting a one common node, with the other end usually earthed. Each node is connected to its neighbour through the reisistors.

The analogue is constructed on a peg board on which is placed a map of the area. The map usually shows transmissibility in addition to physical features such as roads and towns. A square grid matching the holes in the peg board is overlaid on the map by attaching four resistors to the peg. The capacitor for that node is usually placed on the reverse side of the board. A typical R.C. network is shown in Fig. 6.1. Leakage can be simulated in the analogue model by the addition of resistors connected to ground and to each node of the network, Fig. 6.1c. Barrier boundaries can be simulated by an open circuit. Resistors connected to the nodes along the edge of the analogue model and to the ground, simulate horizontal leakage through boundaries of the aquifer. Leakage of water through a stream bed can be simulated with a resistor connected to appropriate nodes and to ground.

To the resultant RC mesh are applied transient currents of 'square form' type from a pulse generator and the resultant voltage response is observed on an oscilloscope, Fig. 6.2, or traced by a x-y plotter, depending on whether 'fast time' or 'slow-time' equipment is available. Once the model is verified, i.e. faithfully produces historically observed trends after adjustments, it is possible to vary the duration and rate of pumping, or

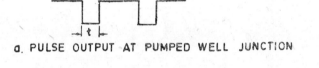

a. PULSE OUTPUT AT PUMPED WELL JUNCTION

b VOLTAGE RESPONSE AT OBSERVATION JUNCTION

c VOLTAGES OBSERVED ON OSCILLOSCOPE

Fig. 6.2 Excitation response of R.C. analog.

of influen seepage from natural or artificial sources, and observe the resultant change of voltage (water level in the aquifer) with time at any point within the mesh. The excitation-response apparatus is comprised of four major parts: a power supply, a wave form generator, a pulse generator, and an oscilloscope.

The RC model is the most versatile and is finding great utility in long term prediction of effects of pumpage on ground water levels and on stream flow. The USGS uses many thousands of nodes to represent a ground water basin. A typical resistor-capacitor network polygonal (non-square) grid used by USGS is shown in Fig. 6.3.

A case history of an R.C. network analog model constructed to determine the effects of continued pumping of ground water from the unconfined aquifer of Pochampad Ayacut in Andhra Pradesh is given below:*

---

*The author gratefully acknowledges the work of M.Y.A. Beig and S. John Prabhakar who have kindly permitted to reproduce here the case history of their paper 'Analog Simulation of An Unconfined Aquifer in Pochampad Ayacut, Andhra Pradesh', in the *Journal of the Inst. of Engrs.* (India), Civil Engg. Divn., Vol. 61, Sept. 1980.

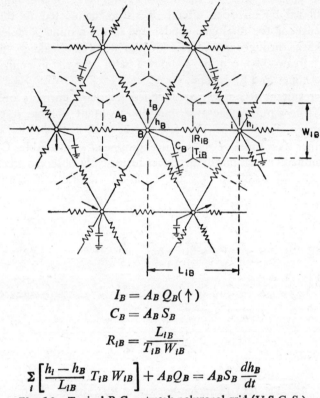

$$I_B = A_B Q_B (\uparrow)$$
$$C_B = A_B S_B$$
$$R_{iB} = \frac{L_{iB}}{T_{iB} W_{iB}}$$

$$\sum_i \left[\frac{h_i - h_B}{L_{iB}} T_{iB} W_{iB} \right] + A_B Q_B = A_B S_B \frac{dh_B}{dt}$$

Fig. 6.3 Typical R.C. network polygonal grid (U.S.G.S.).

(i) Area of Ayacut modelled = 67 km². The wedge shaped area (quite fertile and productive) is bounded on the two sides by Peddavagu and Koratlavagu and on the third side by the Pochampad main canal.

(ii) Ground water in this basin is mostly under unconfined conditions in the weathered zone of the formations. Pump tests in a few representative open wells were conducted and the data was analysed by Papadopulos and Cooper method and the average values of the formation constants were obtained as $T = 79.5$ m²/day and $S = 4.55 \times 10^{-3}$.

(iii) The area was divided into a network of square grids of size 1 km × 1 km as shown in Fig. 6.4 which resulted in 91 nodal points.

(iv) Scale factors were arrived as follows:
(a) Internal resistor of the analog model
$$R = 10 \text{ k}\Omega$$

Therefore
$$R = \frac{1}{T}, \quad 10^4 \ \Omega = \frac{1}{79.5 \text{ m}^2/\text{day}}$$

Ground Water Analog Models

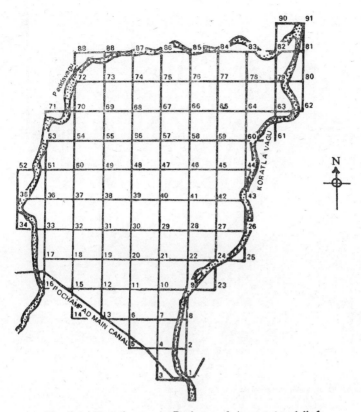

Fig. 6.4 Koratla area in Pochampad Ayacut (modelled).

$$\text{or } 1\Omega = \frac{1}{79.5 \times 10^4} \text{ day/m}^2$$

(b) The water table contours of Jan 1971 as shown in Fig. 6.5 indicated that the maximum and minimum water table elevations were at 304 and 270 respectively giving a head difference of 34 m. Hence a scale of hydraulic head (h) was adopted as

$$1 \text{ m} = 1 \text{ V}$$

(c) Therefore $1 \text{ amp} = \dfrac{1 V}{1 \Omega} = \dfrac{1 \text{ m head}}{1/(79.5 \times 10^4) \text{ day/m}^2}$

Therefore 1 amp current = 7.95×10^5 m³/day (pumping rate)

(d) Therefore 1 coulomb = 1 amp × 1 sec

Assuming, 1 sec = 2.86×10^5 days

1 coulomb = 7.95×10^5 m³/day × 2.86×10^5 days

1 coulomb = 2.27×10^{11} m³

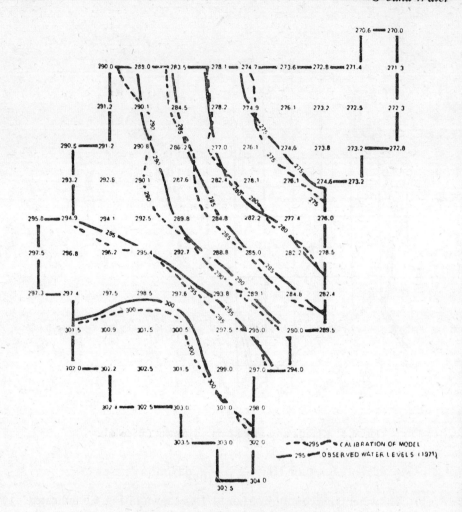

Fig. 6.5 Initial water level elevations.

(e) A capacitance of 1 farad = $\dfrac{1 \text{ coulomb}}{1 \text{ volt}} = \dfrac{2.27 \times 10^{11} \text{ m}^3}{1 \text{ m}}$

Therefore 1 farad = 2.27×10^{11} m²

Capacitance required, $C = a_g^2 S = 1000 \times 1000 \text{ m}^2 \times 4.55 \times 10^{-3}$

$= 4.55 \times 10^3$ m² (i.e. m³ per m head of water)

Therefore

$$C = \dfrac{4.55 \times 10^3 \text{ m}^2}{2.27 \times 10^{11}} = 2 \times 10^{-8} \text{ farads or } 2 \times 10^{-2} \text{ micro-farads}$$

Therefore

$$C = 0.02\, \mu F$$

For preliminary studies the values of the boundary resistors and capacitors were kept same as those of internal ones. A calibrating resistor $R_l = 4.7 \text{ k}\Omega$ was used for simulating pumping rate. For example for a pumping rate of $Q = 1.69 \times 10^2$ m³/day, $I = \dfrac{1.69 \times 10^2}{7.95 \times 10^5} = 0.213 \times 10^{-3}$ amp and the excitation voltage across the calibrating resistor $V_R = V_1 - V_2 = I \times R_l = (0.213 \times 10^{-3}) \times (4.7 \times 1000) = 1$ volt, when a pulse was sent to the pumping well junction or node P_1; the response (i.e. trace of the voltage drop against time) at an observation node O_1 due to the excitation at the pumping node P_1 was observed on the oscilloscope screen from which the corresponding time-drawdown relationship was obtained.

Calibration of the analog model: At each node, a solder lug was bolted which served as a common terminal for four resistors and one capacitor. All the capacitors were grounded to the common earth grid. The two streams were simulated as recharge boundaries by terminating the portions of the network along the boundaries in short circuits. Since the Pochampad main canal was excavated beyond the average depth of weathering in the region and as the canal is also lined, it was assumed that there was no under flow. The boundary was simulated as a barrier boundary with no flow across it by an open circuit.

Taking the confluence of the two at node 91 as the datum, a potential difference of 34V between the nodes 91 and 1, and a potential difference of 32 V between nodes 91 and 32 were maintained and the resulting pattern of declines in potential across the various nodes were measured and transformed into hydraulic potentials. Fig. 6.5 shows the initial water level elevations as observed on the analog model and the actual water table levels collected from field data.

(vi) *Experimental set-up and operations of the analog model*: The excitation response apparatus consisted of a pulse generator, power supplies and an oscilloscope (Fig. 6.6 and 6.7). The pulse generator generated pulses at regular intervals of time and by adjusting the amplitude and width of the pulse, the required pumping rate could be simulated. To simulate the boundary conditions, a transistorised regulated twin power supply was used. An oscilloscope was used to measure the response at the observation well to the different pumping rates at the pumped wells. Four arbitrary nodes were chosen as pumping centres with pumping rates of 1.69×10^2 m³/day (at nodes 7 and 9) and 3.38×10^2 m³/day (at nodes 20 and 76). If one observation well was influenced by more than one pumping centre, the values were algebraically added.

One of the pumping nodes was first excited with an assigned value of discharge and the effect (i.e. the response as voltage vs. time which gave the time-drawdown curve to a different scale) at any other node was observed on the oscilloscope as shown in Fig. 6.8. Time-drawdown curves

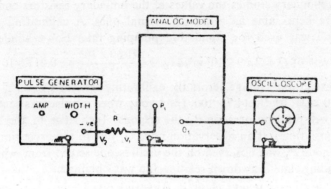

Fig. 6.6 Experimental set up.
P_1—Pumped well, O_1—Observation well
R_1—Calibrating resistor
V_1, V_2—Floating voltages across R_1

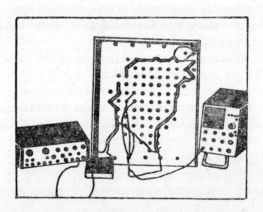

Fig. 6.7 Experimental set-up for analog model.

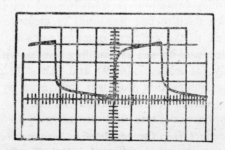

Fig. 6.8 Time drawdown curve trace.

Ground Water Analog Models

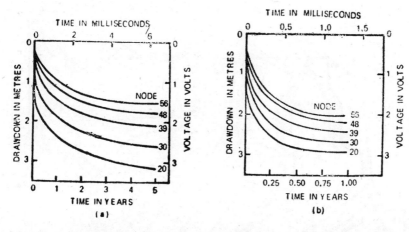

Fig. 6.9 Drawdown curves for various node points due to pumping at node 20; Pumping rate: (a) 1.69×10^2 m^3/day, (b) 3.38×10^2 m^3/day.

at various observation nodes due to the two different rates of pumping at node 20 are shown in Fig. 6.9. The predicted pattern of water table contours at the end of 1, 5 and 10 years when the four pumping centres were pumped simultaneously with different pumping rates are shown in Fig. 6.10. There was a maximum decline of 15 m head along the Pochampad main canal and of about 5 m at the confluence of the two streams. A marked shift in the contours was observed for a pumping period of 1 year and very little shift for 5 and 10 years.

(vii) *Conclusion*: Thus analog model can be used for examining the various choices of ground water development and to plan spacing of wells.

Scale Factors for R.C. Network Analog

The following scale factors were defined for R.C. network analog models (Bermes, 1960 and Pricket, 1975).

Resistance scale $S_R = \dfrac{1}{TR}$ day/m^2 per ohm (6.14a)

Head scale $S_H = \dfrac{H}{V}$ m per volt (6.14b)

Time scale $S_t = \dfrac{t_p}{t_m}$ days per sec (6.14c)

Discharge scale $S_Q = \dfrac{Q}{I}$ m^3/day per amp (6.14d)

Capacitance in farad $C = a_g^2 S \dfrac{S_H}{S_Q S_t}$ (6.15)

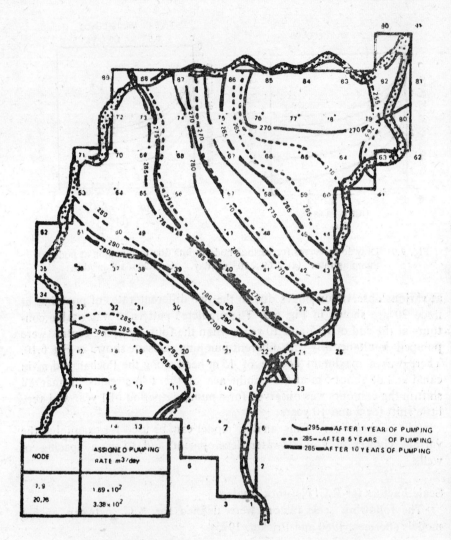

Fig. 6.10 Analog model solution showing water level declines.

$$\frac{S_H}{S_Q} = S_R = \frac{1}{TR}$$

Therefore

$$C = a_g^2 S \frac{S_R}{S_t} \quad (6.15a)$$

Resistance in ohm $\quad R = \dfrac{S_Q}{S_H T} \quad (6.16)$

a_g = nodal spacing in aquifer, m

S = storage coefficient of aquifer

Ground Water Analog Models 269

Vertical flow between aquifer layers can be simulated through resistor R_v connecting two or more resistance networks, Fig. 6.6. Values of R_v can be obtained from

$$R_v = \frac{S_Q}{S_H \left(\frac{K'}{b'}\right) a_g^2} \qquad (6.17)$$

where b' = vertical nodal spacing or the distance between layers and K' = coefficient of vertical permeability.

Irregular shapes of boundary are duplicated by 'vector volume technique'. The values of the resistors and capacitors adjacent to boundaries can be computed from the 'vector volume' technique described by Karplus (1958) as

$$R_x = R_b \frac{\Delta x}{\Delta y}, \quad R_y = R_b \frac{\Delta y}{\Delta x} \qquad (6.18)$$

where R_x = Value of the x-direction resistor adjacent to a boundary; R_y = value of the y-direction resistor adjacent to a boundary; R_b = value of resistors in interior portions of the model and Δx, Δy = length of the aquifer represented by the resistors in x and y directions, respectively.

The magnitudes of capacitors are directly proportional to the areas of the portions of the aquifer they represent. The value of the capacitor adjacent to a boundary is given by

$$C_b = A_v S \frac{S_R}{S_t} \qquad (6.19)$$

where A_v = area of the aquifer represented by the capacitor.

Example 6.1: An R.C. network analog has to be constructed to simulate a confined aquifer 6 km × 4 km extent having a uniform saturated thickness of 50 m, permeability of 3 m/day, and storage coefficient of 10^{-3}. The aquifer may be divided into grids of 1 km × 1 km. 1 sec. in the analog should represent 1000 days for the aquifer. Resistors of 1 megohm are available and 1 m head of water should represent 1 volt in the analog. Work out the scale factors for resistance and discharge. What capacitance is required for each block?

Solution: $T = Kb = 3 \times 50 = 150$ m²/day, $S = 10^{-3}$, $R = 1 \times 10^6 \, \Omega$

$$S_R = \frac{1}{TR} = \frac{1}{150 \,(1 \times 10^6)} = 6.67 \times 10^{-9} \text{ day/m}^2 \text{ per } \Omega$$

$$S_H = \frac{H}{V} = \frac{1 \text{ m}}{1 V} = 1 \text{ m per volt}$$

$$\frac{S_H}{S_Q} = \frac{1}{TR}$$

$$S_Q = S_H\, TR = 1 \times 150 \times 10^6 \text{ m}^3/\text{day per amp}$$
$$= 150 \text{ M m}^3/\text{day per amp}$$

$$S_t = \frac{1000 \text{ day}}{1 \text{ sec}} = 1000 \text{ day per sec.}$$

$$C = a_g^2\, S\, \frac{S_R}{S_t} = (10^3 \times 10^3)\, 10^{-3}\, \frac{1}{(150 \times 10^6)\, 1000}$$
$$= \frac{1}{150 \times 10^6} \text{ farad or } 0.0067\ \mu\text{F}$$

No. of nodes $= \dfrac{6 \text{ km} \times 4 \text{ km}}{1 \text{ km} \times 1 \text{ km}} = 24$

This problem has been worked from first principles earlier.

Example 6.2: An R.C. network analog has to be constructed to simulate a confined aquifer of 40 km × 60 km with an average thickness of 30 m, permeability 25 m/day, and storage coefficient 4×10^{-4}. The maximum head is 40 m. The model can be represented by 40×60 nodes. Resistors of 3000 Ω and capacitors of 0.01 μF are available; model voltage $= 8V$. Work out the scale factors. If a calibrating resistor of 2500 Ω is used for simulating pumping rate, determine the current pulse and excitation voltage to simulate a pumping rate of 1000 m³/day at a particular node.

Solution: $T = Kb = 25 \times 30 = 750 \text{ m}^2/\text{day}, S = 10^{-3}$

$R = 3000\ \Omega,\ C = 0.01\ \mu F$

$$S_R = \frac{1}{TR} = \frac{1}{750 \times 3000} = 0.44 \times 10^{-6} \text{ day/m}^2 \text{ per } \Omega$$

$$S_H = \frac{H}{V} = \frac{40}{8} = 5 \text{ m per volt}$$

$$C = a_g^2\, S\, \frac{S_R}{S_t}$$

$$0.01 \times 10^{-6} = (10^3 \times 10^3)\, 4 \times 10^{-4} \times \frac{1}{(2.25 \times 10^6)\, S_t}$$

$S_t = 1.78 \times 10^4$ days per sec.

$$\frac{S_H}{S_Q} = \frac{1}{TR},\ S_Q = TR\, S_H = 750 \times 3000 \times 5$$
$$= 1.125 \times 10^7 \text{ m}^3/\text{day per amp}$$

To Simulate a pumping rate of 1000 m³/day

$$I = \frac{100}{1.125 \times 10^7} = 8.9 \times 10^{-5} \text{ amp}$$

Excitation voltage $V = IR = (8.9 \times 10^{-5})\, 2500 = \mathbf{0.222\ V.}$

R.C. Analog with Continuous Resistance Medium

A method has been developed using continuous resistance medium of suitable liquids to represent the hydraulic resistance of the aquifer, discrete capacitors being used at designed locations. The continuity of the liquid resistor ensures better simulation. Assuming the scales

1 cm in analog $= X$ metres in aquifer horizontally
$\qquad\qquad\qquad\;\; Y$ metres in aquifer vertically

1 volt $\qquad\quad = h$ metres head of water

1 sec in analog $= t$ days in aquifer

then the other scales are

1 ampere $\;\; = K\rho hY$ m³/day (flow rate)

1 mho $\qquad = K\rho Y$ m²/day (transmissibility)

1 coulomb $= K\rho hYt$ m³ (storage)

Capacitance intensity to be provided in the analog $= \dfrac{SX^2}{K\rho Yt}$ farad/cm²

(6.20)

where $\rho =$ resistivity of the medium (analog liquid), ohm-cm; $K =$ permeability of the aquifer, m/day and $S =$ storage coefficient of the aquifer, fraction.

The frequency and duration of current pulses can be adjusted for given pumping durations by time scaling as given above. Simultaneous injection of pulses of different magnitudes and phase relationships to each other is possible with independent pulse generators isolated from each other. The response transients are photographed for further analysis.

New type of direct simulation analog

A circular tank 80 cm in diameter with an insulating boundary contains an electrolyte about 10 cm deep. A number of copper rods of various diameters (6-20 mm) penetrating the electrolyte to various depths at different locations form the nodes of the electrical network. The electrolytic medium is the continuous resistance field. To each probe (called penetrode) is connected a variable capacitor to simulate storage effects. By varying the diameter, position and penetration of each penetrode, it has been found possible to create a network which should be a scaled version of any other given network of discrete resistors and capacitors. Simulation of the aquifer is complete when the excitation-response pairs are reproduced exactly at homologous points. The only data used are the histories of pump excitations and aquifer responses and no data relating to hypothetically derived parameters such as S and T, will be necessary.

Dipole pumping test

Two wells drilled near each other, say within about 10 m, constitute the dipole, with both wells connected at the top by a pressure pipe connection. Water pumped out of the aquifer from one well is used to recharge the aquifer through the other well. The pressure differential which causes flow in the dipole-aquifer circuit (Fig. 6.11) and the corresponding flow rate are measured continuously. Steady state conditions are obtained in comparatively short time.

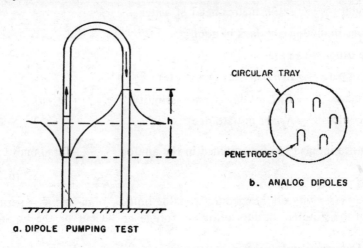

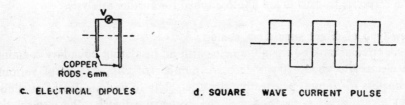

Fig. 6.11 Dipole pumping test.

The same number of dipoles as in the aquifer are employed in the analog model. The pattern of excitation-response pairs are simulated in its transformed and scaled version by manually operating to vary the locations and penetrations of the limbs (penetrodes) of the analog dipoles. All the hydrologic characteristics of the aquifer are reflected in the excitation-response pair at each dipole in the aquifer.

It is also possible to transduce the pressure differential flow pair signals from the aquifer dipoles into voltage difference current pairs, and tele-

Ground Water Analog Models

metrically feed the servos of the analog model to operate the dipole limbs. Prediction of any required response can be done by exciting the model in the required manner and observing the responses at topologically corresponding points in the model tank, and then interpreting the results in terms of the aquifer parameters.

In a more sophisticated version of this technique electrical dipoles are excited electrically by injecting a square wave current pulse across them. The transient responses (in the form of time-varying electrical potentials) of a number of electrodes buried in the aquifer, Fig. 6.11 are telemetered to the model tank, in addition to the transmission of the scaled excitation pulses also. Thus the analog tank receives a number of electrical excitation-response pairs from all the dipoles and electrodes buried in the aquifer. Quick acting servos by automatically operating the analog dipoles, capacitors and electrodes create a homologous R.C. net simulating the aquifer at every instant.

This analog model, after linking with a suitably programmed digital computer can be used to transmit commands to the aquifer pump for regulating their pumping rates and durations, satisfying any criteria (such as limiting drawdowns in a number of wells) that may be stipulated for the economic exploitation of the aquifer. Depending upon the instantaneous aquifer parameters and recharge conditions, and constrained by the policy criteria set up for the aquifer, automatic regulation of pumping in an aquifer can be achieved (see also 'Mathematical Model for a Basin').

Numerical Solution: The analytical solution of the Laplace equation

$$\nabla^2 \phi = 0, \quad \phi = Kh$$
$$\nabla^2 h = 0$$

for an area with complex or irregular boundary conditions, becomes difficult. In that case, the differential equation can be solved numerically to find the values of 'h' at different points in the area, draw the equipotential lines and flowlines, and thus draw a flow net. The area is divided into a network of equal squares $a \times a$, smaller the squares greater is the accuracy), as shown in Fig. 6.12.

For a node at $0 = \left(\dfrac{\partial h}{\partial x}\right)_{1-0} = \dfrac{h_1 h_0}{a}, \quad \left(\dfrac{\partial h}{\partial x}\right)_{0-3} = \dfrac{h_0 - h_3}{a}$

$$\left(\dfrac{\partial^2 h}{\partial x^2}\right)_0 = \dfrac{\partial}{\partial x}\left(\dfrac{\partial h}{\partial x}\right) = \dfrac{\left(\dfrac{\partial h}{\partial x}\right)_{1-0} - \left(\dfrac{\partial h}{\partial x}\right)_{0-3}}{a}$$

$$= \dfrac{h_1 - h_0 - h_0 + h_3}{a^2} = \dfrac{h_1 + h_3 - 2h_0}{a^2}$$

Similarly, $\left(\dfrac{\partial^2 h}{\partial y^2}\right)_0 = \dfrac{h_2 + h_4 - 2h_0}{a^2}$

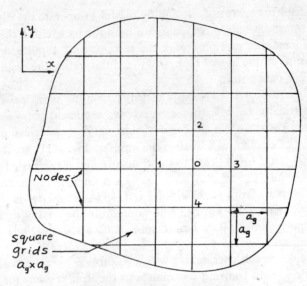

Fig. 6.12 Square network.

Substituting in the Laplace equation

$$\nabla^2 h = \frac{\partial^2 h}{\partial x^2} + \frac{\partial^2 h}{\partial y^2} = 0 \qquad (6.21)$$

at node $\leftarrow 0$, $h_1 + h_2 + h_3 + h_4 - 4h_0 = 0$

or
$$h_0 = \frac{\sum\limits_{1}^{4} h_i}{4}$$

which is in the *finite difference* form. The finite-difference solution of the Laplace equation thus states that 'h' at each node must be equal to the average of h at the four surrounding nodes. The difference

$$h_0 - \frac{\sum\limits_{1}^{4} h_i}{4} = r_e \approx 0 \text{ (very small)} \qquad (6.22)$$

r_e is called residual.

The procedure for finding the correct value of head at each node is to

(i) define the boundary conditions by assigning proper 'h' values and assign tentative values of h for the rest of the nodes of the chosen grid.
(ii) determine the residual at each node as per Eq. (6.22)
(iii) apply corrections to the assumed values to reduce the residuals to some small value.

(iv) after reducing r_e at a particular node to minimum and altering h_0, alter the residuals of adjacent nodes by the change in h_0. For an impervious boundary $\frac{\partial \phi}{\partial n} = 0$, since the velocity normal to the boundary must be zero.

For a vertical impervious boundary

$$\frac{\partial \phi}{\partial x} = 0 \quad \text{or} \quad \frac{\partial h}{\partial x} = 0$$

Therefore $\phi = Kh$

When h_0 is on the boundary,

i.e. $\dfrac{h_1 - h_3}{2a} = 0 \quad \text{or} \quad h_1 = h_3$

$$\nabla^2 h = h_1 + h_2 + h_3 + h_4 - 4h_0 = 0$$

or, $\qquad h_2 + 2h_1 + h_4 - 4h_0 = 0$

$$h_0 = \frac{h_2 + h_4 + 2h_1}{4} \tag{6.23}$$

The residual at the node on the boundary is the difference between the assigned value and that computed as above.

Similarly for a horizontal impervious boundary (h_0 on the boundary),

$$\frac{\partial h}{\partial y} = 0$$

$$h_0 = \frac{h_1 + h_3 + 2h_2}{4} \tag{6.23a}$$

After determining the correct values of h at each node, equipotentials can be drawn (by interpolation) and streamlines drawn as orthogonals, to form a flownet, from which the flow rates can be determined. If the boundaries of the area constitute streamlines, time may be saved by solving the flow system in terms of ψ (instead of h) and drawing the equipotentials as orthogonals to the streamlines.

Special techniques have been developed to most efficiently 'relax' the system from its residuals, hence the name relaxation method (southwell, 1946). Manual operation of the relaxation method is time consuming. The work can be shortened by using 'block relaxation'. The accuracy of the relaxation method depends on the density of the network (typically 50 to 200 nodes). The method can also be applied to heterogeneous medium with irregular boundaries or axisymmetric systems. In areas where the flow is concentrated, subdivided networks may be used to produce greater node density. Automatic relaxation can be obtained through iteration by digital computers. The manual relaxation method has been superseded by R.C. network analog and digital computers.

If the thickness of the aquifer and its permeability is known, the hydraulic gradient, seepage velocity and flow rate across any section of the aquifer can be estimated.

Example 6.3: A line of wells equally spaced at 1.2 km is located between a stream and barrier boundary, which are parallel and 1 km apart. The line of wells is at 0.6 km from the stream. The transmissibility of the aquifer may be assumed as 160 m²/day. When the wells are pumped, the water level in the wells is about 6 m below the stream. Determine the water table elevations and flow into the wells by drawing a flow net.

Solution: Due to symmetry, the flow system will be a repetition of the pattern in the rectangle $ABCD$ Fig. 6.13. $BC, CD,$ and AD are streamlines. AB is an equipotential and is assigned values of 100; the pumping well W is assigned a value 0. The h-values within the area are evaluated for a 6×10 grid by relaxation method performing a few iterations, and the final levels (along with the minimum residuals towards left carrying $\pm$ sign) are given in Fig 6.14. The equipotential lines are drawn by interpolation, streamlines

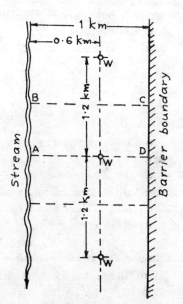

Fig 6.13 Line of pumping wells in an alluvial valley between a stream and mountain range.

Ground Water Analog Models

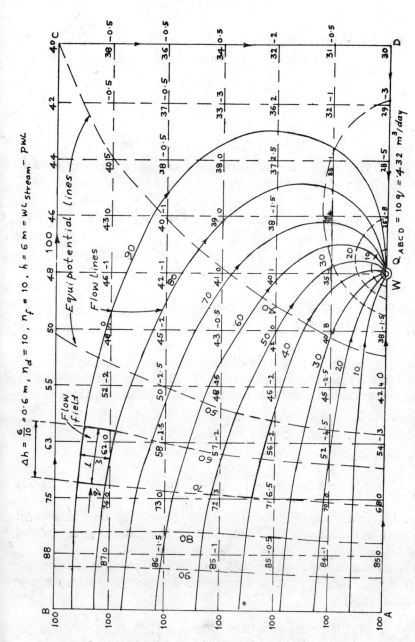

Fig. 6.14 Flownet for flow into the well in the rectangle *ABCD*.

sketched as orthogonals to the equipotentials, and a flownet is thus obtained.

The flow in each flow channel is

$$q = Tiw = T \cdot \frac{\Delta h}{l} \cdot w, \quad \Delta h = \frac{h}{n_d} = \frac{6}{10}$$

For the flow field in Fig. 6.14,

$$l = 4 \text{ units}, \quad w = 1.8 \text{ units}$$

$$q = 160 \frac{0.6}{4} \times 1.8 = 43.2 \text{ m}^3/\text{day}$$

Due to symmetry, the flow into the well

$$Q = 2(10q) = 20 \times 43.2 = \textbf{864 m}^3/\textbf{day}$$

Note: If the aquifer were confined with a uniform thickness of 20 m and permeability of 15 m/d (or T of the confined aquifer known)

$$Q_{ABCD} = Kh \frac{n_f}{n_d} \cdot b = n_f \Delta h T = \left(15 \times 6 \times \frac{10}{10}\right) 20$$

$$= 10 \times \frac{6}{10} (15 \times 20) = 1800 \text{ m}^3/\text{day}$$

Flow into the well

$$Q = 2 \times 1800 = \textbf{3600 m}^3/\textbf{day}$$

Note: The equations cannot be applied in the case of unconfined aquifer since T is not constant due to variable saturated thickness.

For a homogeneous, isotropic aquifer infinite in areal extent, with coefficient of storage S and transmissibility T, with a steady rate of recharge P, the differential equation for two-dimensional flow (Stallman, 1956) is flow

$$\frac{\partial^2 h}{\partial x^2} + \frac{\partial^2 h}{\partial y^2} = \frac{S}{T} \frac{\partial h}{\partial t} - \frac{P}{T} \tag{6.24}$$

and in the finite difference from

$$\sum_{1}^{4} h_i - 4h_0 = \left(\frac{S}{T} \frac{\Delta h_0}{t} - \frac{P}{T}\right) a^2 \tag{6.25}$$

where Δh_0 is the change in head at node '0' during the time interval Δt. The head distribution in time and space in an aquifer of given size and slope can be found by relaxation or interaction method.

The differential equation for two-dimensional flow in a nonhomogeneous aquifer infinite in areal extent with a study rate of recharge P (Stallman, 1956) is

$$T\left(\frac{\partial^2 h}{\partial x^2} + \frac{\partial^2 h}{\partial y^2}\right) + \frac{\partial T}{\partial x} \frac{\partial h}{\partial x} + \frac{\partial T}{\partial y} \frac{\partial h}{\partial y} = S\left(\frac{\partial h}{\partial t}\right) - P \tag{6.26}$$

Ground Water Analog Models

It is assumed that all parameters vary in space and P varies both in space and time.

The entire aquifer can be divided into a network of square, triangular, or hexagonal grids which are so small that the variation of head can be taken to be linear, Fig. 6.15.

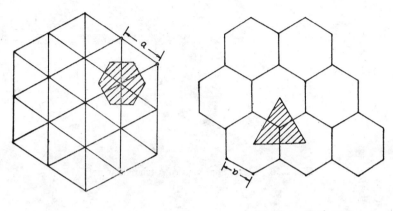

(a) Triangular net (b) Hexagonal net

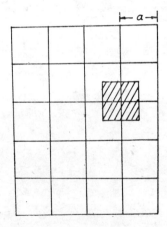

(c) Square net

Fig. 6.15 Aquifer divided into network.

Considering the hexagonal area of a triangular net (shaded area), flow through one side of length $l\left(\dfrac{a}{\sqrt{3}}\right)$, Fig. 6.16.

$$q_1 = Tiw = T\frac{h_0 - h_1}{a} \cdot \frac{a}{\sqrt{3}} = T\frac{h_0 - h_1}{\sqrt{3}}$$

[By adding the flow through all the six sides, the total flow leaving the shaded hexagon]

$$\Sigma q = \frac{T}{\sqrt{3}} \left(6h_0 - \sum_1^6 h_i\right) \qquad (6.27)$$

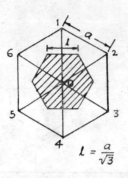

Fig. 6.16 Hexagonal area of a triangular net.

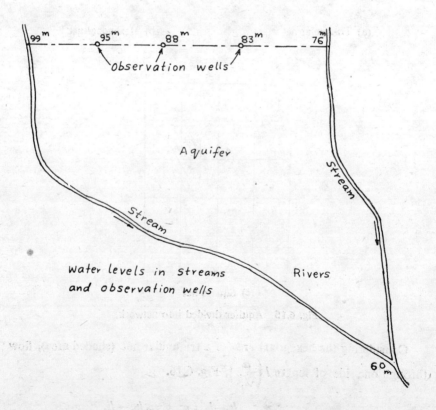

Fig. 6.17 Aquifer bounded by two streams.

If there is no recharge, the total flow leaving the shaded prims would be zero, when

$$6h_0 - \sum_1^6 h_i = 0$$

or
$$h_0 = \frac{\sum_1^6 h_i}{6} \qquad (6.28)$$

which states that the head 'h' at any node must be equal to the average of h at the six surrounding nodes. This can also be obtained by the method of finite differences.

As an example, for an aquifer, bounded by two streams as shown in Fig. 6.17, the water levels in the rivers and in the observation wells at the upstream boundary of the aquifer are given. The distribution of water levels or piezometric heads can be obtained by fitting a triangular network and assuming linear variation between the two given values of heads. The aquifer boundaries are slightly deformed to adjust for the straight sides of the triangles. Tentative values of heads are assumed at the interior nodes and then altered, such that the residuals (i.e the difference between the assigned value and the average of the six surrounding values) are reduced to a minimum. From the final value of heads at the nodes (arrived after two-three iterations), the equipotential lines can be drawn, and the flow lines sketched orthogonal to the equipotential lines to obtain the flownet, Fig. 6.18, from which the flow velocities and flow rates across any section of the aquifer can be determined.

In ground water flow problems, triangular nets are preferred, since the shaded hexagon approaches a circular shape which compares with radial flow problems.

If there is a uniform steady recharge P over the area of the shaded hexagon $\left(\frac{\sqrt{3}}{2} a^2\right)$, then the Eq. (6.27) becomes

$$\frac{T}{\sqrt{3}} (6h_0 - \sum_1^6 h_i) = \frac{\sqrt{3}}{2} a^2 \cdot P$$

or
$$h_0 = \frac{\sum_1^6 h_i + C_1}{6} \qquad (6.29)$$

where $C_1 = \frac{3}{2} \frac{a^2 P}{T}$.

In case of semi-confined aquifers with piezometric head 'h'' the flow rate into the shaded area is

$$\frac{K'}{b'} S(h_0' - h_0)$$

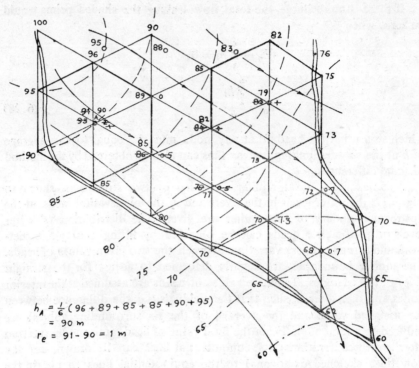

Fig. 6.18 Triangular network and flownet for the aquifer.

Then from the Eq. (6.27),

$$h_0 = \frac{\sum_1^6 h_i + h_0' C_2}{6 + C_2} \qquad (6.30)$$

where $C_2 = \frac{3}{2} \frac{K'/b'}{Kb} a^2$.

Variable thickness in unconfined aquifer

In unconfined aquifers the saturated thickness is not a constant and this variation in thickness has to be taken into account. In Eq.

$$q_1 = Tiw = T \frac{h_0 - h_1}{a} \cdot \frac{a}{\sqrt{3}}$$

$T = K \times$ average thickness of the aquifer between 0 and 1

$$= K \cdot \frac{h_0 + h_1}{2}$$

$$q_1 = \frac{K}{2\sqrt{3}} (h_0^2 - h_1^2)$$

Total discharge leaving the shaded hexagon

$$\Sigma q = \frac{K}{2\sqrt{3}} \left(6h_0^2 - \sum_1^6 h_i^2\right) \qquad (6.31)$$

If there is no recharge,

$$h_0^2 = \frac{\sum_1^6 h_i^2}{6}$$

If there is a uniform recharge P over the shaded area $\frac{\sqrt{3}}{a} a^2$

$$\frac{K}{2\sqrt{3}} \left(6h_0^2 - \sum_1^6 h_i^2\right) = \frac{\sqrt{3}}{2} a^2 P$$

or

$$h_0^2 = \frac{\sum_1^6 h_i^2 + C_3}{6} \qquad (6.32)$$

where $C_3 = \dfrac{3a^2 P}{K}$

Pumping wells in the area

When there are pumping wells in the area, the drawdown due to each well can be determined from the steady state logarithmic formula. This should be subtracted from the results obtained by numerical analysis.

QUIZ 6

I. Select the correct answer(s):
    1. An analogue model of an aquifer system is constructed to study
        a. the effects of influent or effluent seepage, return flow from irrigation, etc.
        b. the effects of increasing the abstraction rate at any time in future.
        c. the rate of ground water extraction to abate sea water intrusion in coastal aquifers.
        d. the effect of pumping new wells on the existing wells
        e. the future pumping patterns
        f. the optimum well spacing
        g. the effects of artificial recharge
        h. the water balance at any time
        i. long predictions of drawdown
        j. all the above items
    2. An analog model is said to be
        a. simulated when the partial differential equations governing the ground water flow system and that describing the analog system are similar
        b. verified when the past records are fed to the analog system, the analog model produces results which have occurred during the past

c. predicting the probable drawdowns for an increased extraction rate in future or new pumping patterns
d. helpful in efficient utilisation and management of ground water resources
e. all the above items

3. An electric analog model utilises the similarity between
   a. the Darcy's law and the Kirchoff's law
   b. the Darcy's law and the Ohm's law
   c. the Darcy's law and the Coulomb's law
   d. the Darcy's law and the Laplace's equation

4. In an RC network analog
   a. the variation in piezometric head corresponds to the voltage drop
   b. the storage function is simulated by the ground capacitor
   c. the permeability is simulated by the electrical conductance
   d. the transmissibility is simulated by the electrical resistance
   e. the ground water flow is simulated by the electric current
   f. the aquifer is divided into square or polygonal grids which correspond to the resistor grids, square or polygonal, respectively and the areas are simply scaled
   g. all the above simulations.

5. In mathematical modelling of ground water flow problems, a finite difference form of the partial differential equations governing the two systems is used because
   a. the grid areas being very small, the discontinuous network simulates a continuous aquifer
   b. the model consists of discrete elements
   c. of the similarity between Ohm's law and Kirchoff's law
   d. the two systems are governed by Laplace's equation

6. The dipole pumping test consists of connecting two drilled wells by a pipe, the discharge from one being the recharge into the other, and
   a. the two wells being located at optimum spacing
   b. the two wells being close by, say 10 m apart
   c. the two wells being 600 m apart
   d. for all spacings

II. State whether 'true' or 'false'; if false give the correct statement.
   1. In a Hale-Shaw model, the model permeability is directly proportional to the viscosity of the model fluid and inversely proportional to the plate spacing.
   2. In analysis of ground water flow problems by an R.C. network polygonal grid, the variation of piezometric head in the node under question (B) can be found out by writing the equation

$$\sum_{\text{all sides of polygon } B} TiW = \left(\frac{\text{change in ground water storage}}{\text{time interval}}\right)_B + \left(\text{Rate of surface inflow or outflow}\right)_B$$

PROBLEMS

1. Work out the scale factor for resistance and capacitors required for an R.C. network analog to simulate a confined aquifer 10×50 km having S and T of 0.001 and 1000 m²/day, respectively. Resistors of 2500Ω are available. The aquifer is represented by 20×100 nodes and 1 Sec in the analog should represent 10000 days for the aquifer.

$$(1\Omega = 4 \times 10^{-7} \text{ day/m}^2, 0.01 \text{ }\mu F)$$

2. Work out the scale factors for an R.C. network analog to represent a confined aquifer 20×60 km of an average thickness of 30 m, having K and S of 30 m/day and 0.001, respectively. The maximum head is 50 m. The model can be represented by 20×60 nodes. Resistors of 4000Ω and capacitors of 0.02 μF are available; model voltage $=10$ volts. If a calibrating resistor of 3000Ω is used for simulating pumping rate, determine the current pulse and excitation voltage to simulate a pumping rate of 1080 m³/day at a particular node.

   Ans. $(1 \Omega = 2.78 \times 10^{-7} \text{ day/m}^2, 1 \text{ amp.} = 1.8 + 10^7 \text{ m}^3/\text{day}, 1 \text{ V} = 5 \text{ m},$
   $1 \text{ sec} = 1.39 \times 10^4 \text{ days}; 6 \times 10^{-5} \text{ amp } 0.18 \text{ V})$

7

Sea Water Intrusion

Sea water intrusion in coastal aquifers occurs when permeable formations outcrop into a body of sea water and when there is a landward hydraulic gradient. Sea water can be prevented from intruding by maintaining a head of fresh water above it. According to Ghyben-Herzberg principle the interface will occur at a depth h_s below mean sea level (m.s.l.), Fig. 7.1, given by

$$(h_s + h_f)\gamma_f = h_s\gamma_s$$

since the pressures at the point of interface should be equal, and $\gamma = \rho g$,

$$h_s = \frac{\gamma_f}{\gamma_s - \gamma_f} h_f = \frac{\rho_f}{\rho_s - \rho_f} h_f = \frac{1}{G_s - 1} h_f \qquad (7.1)$$

where h_s = depth below m.s.l. where interface occurs; h_f = elevation of the ground water table above m.s.l.; ρ_f = density of fresh water and ρ_s = density of sea water

Taking the specific gravity of fresh water as 1 and that of sea water as 1.025 ($= G_s$), from Eq. (7.1),

$$h_s = 40 h_f \qquad (7.2)$$

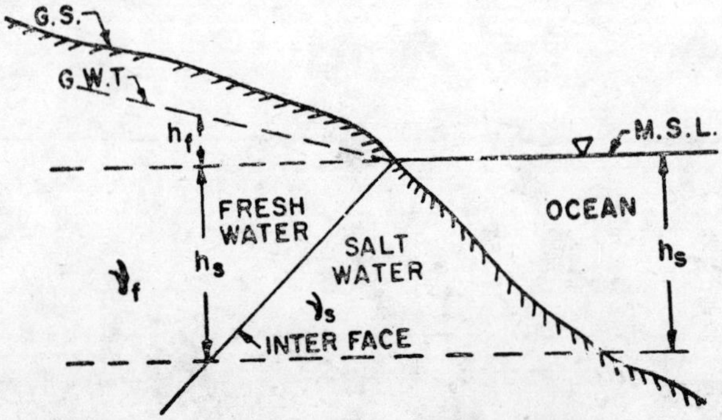

Fig. 7.1 Fresh water-salt water interface.

that is, for a rise or fall of g.w.t. or piezometric head by 1 m will induce a fall or rise respectively of 40 m in the underlying salt water level, even though the response may be greatly delayed.

The net of flow lines and equipotential lines shown in Fig. 7.2 depicts the actual intrusion. Since the total pressure along an equipotential line is constant and the flow lines are sloping upward, the depth to the interface given by the Ghyben-Herzberg relation is less than the actual depth, but the difference is small for flat gradients (Hubbert, 1940).

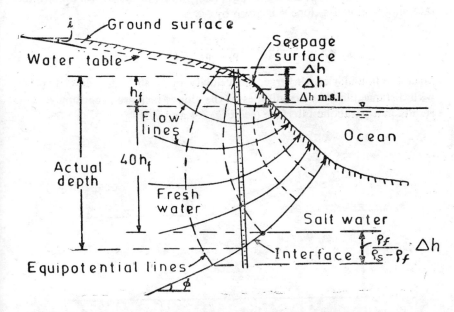

Fig. 7.2 Discrepancy between actual depth to salt water and depth calculated by Ghyben-Herzberg relation. (after Hubbert, 1940)

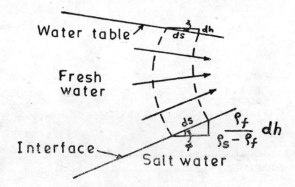

Fig. 7.3 Relation between slopes of the water table and the fresh water-salt water interface.

Slope of Interface

If the water table slope is 'i' Fig. 7.3, then from Darcy's law

$$\sin i = \frac{dh}{ds} = \frac{h}{s} = \frac{v}{K} \qquad (7.3)$$

where v is the velocity of fresh water flow and K is the permeability of the strata. Along this slope, the water table elevation decreases in the direction of flow and hence according to Eq. (7.2), the fresh water-salt water interface should rise. Its slope ϕ is given by

$$\sin \phi = \frac{\rho_f}{\rho_s - \rho_f} = \frac{v}{K} \qquad (7.4)$$

Since the boundaries converge, the velocity of fresh water flow increases with distance, and the magnitudes of the slopes increase accordingly. This results in a parabolic interface (see Fig. 7.4 and 7.5).

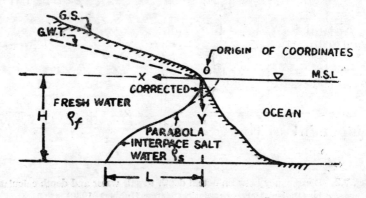

Fig. 7.4 Shape of interface.

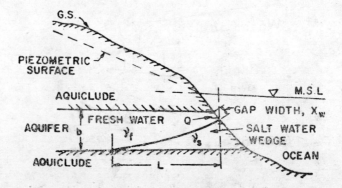

Fig. 7.5 Salt water wedge in a confined aquifer.

Sea Water Intrusion

The interface between fresh water and salt water is theoretically a stream line and no flow across it is possible. However, observations indicate that a narrow mixing zone along the interface exists. This zone results from dispersion occurring in the interface produced by tides, seasonal water table fluctuations, and from molecular diffusion.

Fresh water due to its lower specific gravity will normally float above a layer of salt water without mixing. But when the ground water levels drop, there will be a tendency for sea water to move landward, displacing the fresh water in the aquifer. A landward sloping fresh water-salt water interface will be formed with the depth governed by the Ghyben-Herzberg relationship. As water levels continue to decline, the prism of sea water, designated as the sea water wedge, will continue to advance until conditions of equilibrium are established. When intrusion takes place, the wedge tends to flatten out and the toe moves somewhat faster than the remaining portion of the interface. The intrusion is a very slow process.

Shape of Interface

The interface has a parabolic shape which is very similar to the Dupuit's parabola and is given by (Rumer and Harleman, 1963)

$$y = \sqrt{2\left(\frac{q}{K'}\right)x + 0.55\left(\frac{q}{K'}\right)^2} \qquad (7.5)$$

$$K' = K\frac{\rho_s - \rho_f}{\rho_f} = K(G_s - 1) = 0.025\,K \qquad (7.6)$$

where q = seaward fresh water flow per unit width of ocean front; K = permeability of the aquifer and x, y = coordinates of the interface with the origin at the contact of m.s.l. with the land surface, Fig. 7.4.

When $\qquad x = 0, \quad y_0 = 0.741\,\dfrac{q}{K'}$

When $\qquad y = 0, \quad x_0 = -0.275\,\dfrac{q}{K'}$

The total length of intrusion measured from $x = 0$, is given by

$$L = \frac{K'H^2}{2q}, \text{ when } L > H \qquad (7.7)$$

where H is the thickness of the aquifer.

The time t (in days) required for the toe of the wedge to move the length L is given by

$$t = \frac{\bar{t}nK'H^3}{q_{ul}^2} \qquad (7.8)$$

where $\bar{t}$ = dimensionless factor to be obtained from a master curve; n = porosity of the aquifer material and q_{ul} = ultimate fresh water flow per unit width.

It is seen that the wedge toe velocities are very small.

Salt water wedge in a non-leaky artesian aquifer is shown in Fig. 7.5. Replacing H by b in Eq. (7.7), the seaward fresh water flow is given by

$$q = \frac{K'b^2}{2L} \qquad (7.9)$$

where b is the saturated thickness of the confined aquifer. The length of the intruded wedge is inversely proportional to the fresh water flow. The width of the gap (X_w) through which fresh water escapes to the ocean is approximately given by

$$X_w = \frac{q}{2K'} \qquad (7.10)$$

Glover (1964) developed the following approximate equation for the shape of the fresh water-salt water interface:

$$y^2 - \frac{2qx}{K'} - \frac{q^2}{K'^2} = 0 \qquad (7.11)$$

When $x = 0$, $\qquad y_0 = \frac{q}{K'} \qquad (7.11a)$

When $y = 0$, $\qquad x_0 = \frac{q}{2K'} \qquad (7.11b)$

x_0 = width of gap through which fresh water escapes to the ocean (X_w).

If the coordinate distances (x, y) from the shoreline to any point on the interface are found, the fresh water lost into the sea (q, Q) can be determined.

Example 7.1: By conductivity measurements in a well in a coastal aquifer extending 4 km along the shore, the interface was located at a depth of 20 m below msl at 100 m from the shore, inland. The depth of the homogeneous aquifer is 30 m below m.s.l. and has a permeability of 50 m/day. What is the rate of fresh water flow into the sea and the width of gap at the shore bottom through which it escapes into the sea? What is the position of the toe of the saltwater wedge? Use Glover's method.

If due to ground water exploitation, the fresh water flow into the sea is reduced by 80% how far the toe will eventually move?

Solution:

(i) Glover: $\qquad y^2 - \frac{2qx}{K'} - \frac{q^2}{K'^2} = 0$

$$K' = \frac{\rho_s - \rho_f}{\rho_f} \cdot K = \frac{1025 - 1000}{1000} \times 50 = 0.025 \times 50 = 1.25$$

For a point located on the interface $P(100, 20)$,

$$20^2 - \frac{2q \times 100}{1.25} - \frac{q^2}{(1.25)^2} = 0$$

Solving this quadratic, $q = 2.475$ m³/d per m of shore line

$$q \approx \frac{K'H^2}{2L} = \frac{K'y^2}{2x} = \frac{1.25 \times 20^2}{2 \times 100} = 2.5 \text{ m}^3/\text{day per m of shore line.}$$

Fresh water lost into the sea

$$Q = q \times \text{permeable stretch of shoreline}$$
$$= 2.5(4 \times 1000) = \mathbf{10000 \text{ m}^3/\text{day.}}$$

(ii) Width of gap

$$x_0 = \frac{q}{2K'} = \frac{2.5}{2(1.25)} = \mathbf{1 \text{ m}}$$

(iii) $$L = \frac{K'H^2}{2q} = \frac{1.25 \times 30^2}{2 \times 2.5} = \mathbf{225 \text{ m}}$$

(iv) $$q' = (100 - 80)\% \text{ of } q = 0.2q = \frac{1}{5}q$$

$$L' = \frac{K'H^2}{2q'}, L' = 5L = 5L = 5 \times 225 = \mathbf{1125 \text{ m.}}$$

Dispersion

Dispersion of the interface is the phenomenon that the fresh water-salt water interface is not sharp, but is represented by a narrower or wider transition zone in which the amount of total solids increases more or less rapidly from the landward to the seaward side. Calculations can be made for lines of equal total solids, say 1,500 or 2,000 ppm and this can be assumed as the interface. Data from electrical logging of bore holes can be used to demarcate the fresh water-sea water interface.

A dispersion zone and the dynamics of ground water flow in coastal areas cause a variation of the location of the interface between fresh and salt water from what would be expected by a strict application of Ghyben-Herzberg principle.

Example 7.2: From the collection and interpretation of hydrologic, geologic and geochemical data, the following information was obtained.

Width of aquifer	2.8 km
Thickness of aquifer	30 m
Porosity of aquifer material	10%
Difference in specific gravity	0.03
Permeability of the aquifer	48.9 m/day

From conductivity measurements in two observation wells located at 150 and 225 m from the shore (landward side) the 1,500 ppm line was found to be located at 15 and 22.5 m respectively, below the top of the aquifer. Determine the fresh water-sea water interface.

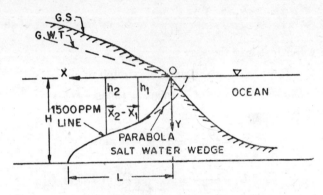

Fig. 7.6 Fresh water-salt water interface, (Ex. 7.1).

Solution: From Darcy's law, Fig. 7.6

$$Q = KAi$$

$$q = K'(1.y)\frac{dy}{dx}$$

$$q\int_{x_1}^{x_2} dx = K'\int_{h_1}^{h_2} y\, dy$$

$$q = \frac{K'(h_2^2 - h_1^2)}{2(x_2 - x_1)} = \frac{48.9(0.03)(22.5^2 - 15^2)}{2(225 - 150)} \tag{7.12}$$

$$= 2.76 \text{ m}^3/\text{day/metre width of coast line}$$

Length of intrusion, from Eq. (7.7),

$$L = \frac{K'H^2}{2q}$$

$$= \frac{48.9(0.03)\, 30^2}{2(2.76)} = \mathbf{239 \text{ m}}$$

At the time of investigation, the toe of the interface will be located at 239 m landward from the shore. The depth of the interface in between the toe and the shore can be computed from Eq. (7.12) and the complete interface can be located.

Since L is inversely proportional to q, if further ground water exploitation is proposed at the rates of $Q/4$, $Q/3$, $Q/2$ and $3Q/4$, ($Q = 2,800\, q$), the seaward fresh water flows will be $3Q/4$, $2Q/3$, $Q/2$ and $Q/4$, and the corresponding lengths of intrusion will be $4L/3$, $3L/2$, $2L$ and $4L$, respectively.

Sea Water Intrusion

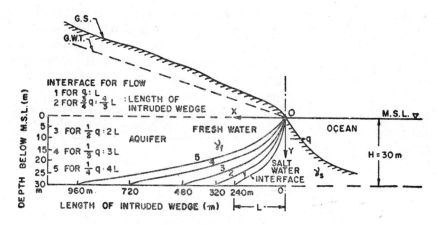

Fig. 7.7 Advancement of salt water wedge with increased ground water exploitation.

Using $3q/4$, $2q/3$, $q/2$ and $q/4$ in Eq. (7.11), the depth of interface can be worked out for any ground water exploitation. The position and shape of the interface for the various cases have been worked out and shown in Fig. 7.7. The time when the toe will advance and reach its final length of intrusion or any other length in between can be calculated from the Eq. (7.8). This will show the rate of landward sea water intrusion and will be very useful for planning ground water exploitation in coastal areas.

Intrusion can be controlled by reducing pumping, by increasing supply, or by forming some type of barrier. A physical barrier may be formed by actually constructing a cut off wall, using sheet piling, concrete, puddle clay, etc. A protective trough may be developed by a line of pumping wells, properly spaced along the coast, Fig. 7.8. These wells produce a mixture of saline and fresh water resulting in a considerable waste of fresh water. The pumping costs involved and the waste of otherwise usable waters are major factors to be considered in evaluating the practicability of protecting a ground water basin by maintaining a pumping trough.

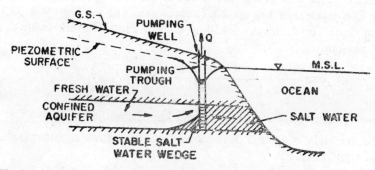

Fig. 7.8 Pumping trough developed by line of pumping wells along the coast.

A ground water mound or ridge may be created by concentrated recharge of the aquifer. A series of spreading grounds or injection wells, or a combination of both, could be utilised, as dictated, by the geologic conditions encountered. The use of fresh water barriers by ground water recharge to prevent sea water intrusion, is practised extensively on the sea coast of southern California. The fresh water barrier should be for enough inland to force all the wedge back seaward. Otherwise, the fresh water will separate the wedge and force the landward wedge still farther inland, creating a saline wave, Fig. 7.9.

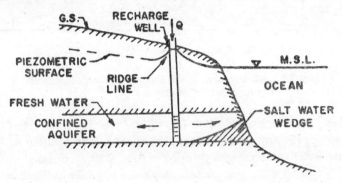

Fig. 7.9 Ground water ridge created by a line of recharge wells along the coast.

A continuous programme for the collection and interpretation of hydrologic, geologic and water quality data has to be initiated to halt and abate sea water intrusion in all basins known to be affected at present. It was found by UNDP investigations that along the Madras coast at the rates of extraction of ground water during 1966, the movement of sea water interface was at the rate of 120 m/year and the limits of extractions were recommended to halt and abate this intrusion. Due to overpumping to meet the ever increasing industrial and agricultural demands, the ground water levels in the Madras coastal aquifers dropped by more than 25 to 30 m during 1970 causing a reversal of the hydraulic gradient and movement of seawater interface towards land at the rate of 120 to 150 m per year. The GWT was falling at a gradient of 1 m per km towards sea. The interface was located at 60 m depth (below m.s.l.) at 3 km inland. From Glover's method, Eq. (7.11), assuming $K = 50$ m/day

$$q \approx \frac{K'y^2}{2x} = \frac{(0.025 \times 50)\, 60^2}{2 \times 3000} = 0.75 \text{ m}^3/\text{day/m}$$

or

$$0.75 \times 1000 = 750 \text{ m}^3/\text{day per km of coastline}$$

The methods suggested for controlling the seawater intrusion into coastal aquifers are:

Sea Water Intrusion   295

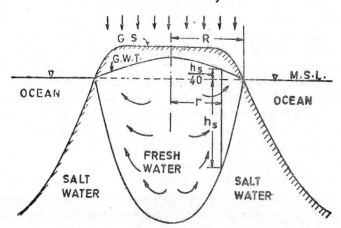

a. FRESH WATER LENS IN AN ISLAND AQUIFER

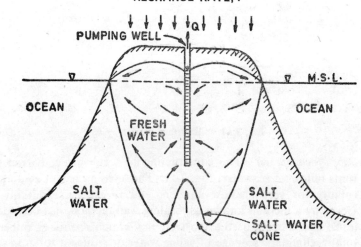

b. SALT WATER CONE DUE TO PUMPING

Fig. 7.10 Oceanic island aquifers.

(i) *Modification of pumping*, i.e. reduction and rearrangement of pattern of pumping draft. By this method the over draft causing intrusion can be eliminated enabling the ground water levels to rise above m.s.l. and to maintain a seaward gradient.

(ii) *Artificial recharge* by spreading areas (for unconfined aquifers) and recharge wells (for confined aquifers).

(iii) *Pumping Trough*—A line of pumping wells constructed parallel to the coast would form a trough and the gradients created would limit sea water intrusion to a stationary wedge inland of the trough, as shown in Fig. 7.8 for a confined aquifer. This reduces the usable storage capacity of the basin and is costly to install and operate.

(iv) *Pressure Ridge*—In an unconfined aquifer recharge from spreading areas could create a fresh water ridge, see Fig. 15.8. In a confined aquifer, a line of recharge wells could form a ridge of the piezometric surface, Fig. 7.9. A small amount of recharge water would run waste into the sea, the remainder moving landward to supply part of the pumping draft.

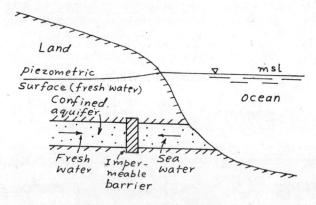

Fig. 7.11 Subsurface barrier.

(v) *Subsurface barrier*—The construction of a subsurface barrier prevents inflow of seawater, Fig. 7.11. This method has limited applicability for very shallow aquifers and is expensive initially. In Germany a method known as 'sealing wall method' has been developed and is being used extensively in German open cast mines to totally eliminate seepage of saline water or reduce it to an insignificant amount. While the initial cost is there, the recurring cost will be nil.

Oceanic Island Aquifers

Since oceanic islands are relatively permeable, sea water is in contact with ground water on all sides. Ground water replenishment is entirely by

Sea Water Intrusion

rainfall and only a limited amount of fresh water is available. A fresh water lens, Fig. 7.10a is formed by the radial movement of ground water towards the coast. An approximate fresh water boundary can be determined, from the boundary condition $h_s = O$ where $r = R$, given by

$$h_s^2 = \frac{P}{2K'\rho_s/\rho_f} (R^2 - r^2) \qquad (7.13)$$

$$K' = K \frac{\rho_s - \rho_f}{\rho_f}$$

where h_s = depth below m.s.l. to the fresh water boundary (interface) at radius r; K = permeability of the aquifer; r = radial distance to the point where the depth is h_s; R = radius of the island; P = recharge rate from rainfall.

If the specific gravities of sea water and fresh water are taken as 1.025 and 1.0 respectively, then the denominator in Eq. (7.13) becomes 0.05125 K. Thus the depth to salt water at any location is a function of the rainfall recharge, size of the island and permeability of the aquifer. Substantial fresh water supplies are more likely to be found on large islands with large recharge and low permeability.

When an island well is pumped at a rate as to lower water table near to m.s.l., the fresh water-salt water equilibrium is disturbed and the salt water will rise as a cone to enter the well, Fig. 7.10 b. To prevent sea water intrusion, the drawdown should be a minimum so as to draw fresh water only from top of the lens. Many islands get water supplies from infiltration galleries consisting of horizontal collecting tunnels.

Pumping Island Aquifer—Steady Unconfined Flow with Rainfall

To derive an equation for the phreatic surface of an island aquifer of radius R with an effective rainfall rate P, when a central well is pumped at a rate Q_w, Fig. 7.12.

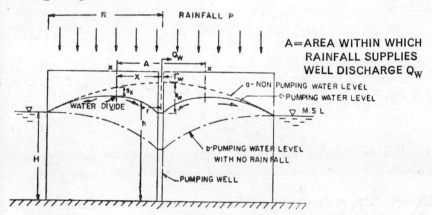

Fig. 7.12 Pumping island aquifer—unconfined and with rainfall.

From Darcy's law, discharge $Q = 2\pi\, rhK \dfrac{dh}{dr}$ (i)

From continuity principle, $dQ = -2\pi\, rdr\, P$

Integrating, $Q = -\pi r^2 P + C_1$

$r = r_w \approx 0,\, Q = Q_w.$ Therefore $C_1 = Q_w$

Therefore $Q = -\pi r^2 P + Q_w$ (ii)

Substituting (ii) in (i),

$$hdh = \frac{Q_w}{2\pi K}\frac{dr}{r} - \frac{P}{2K}rdr$$

$$2h\frac{dh}{dr} = \frac{Q_w}{\pi Kr} - \frac{Pr}{K}$$

At the water divide $\dfrac{dh}{dr} = 0,\, r = x,\, Q_w = \pi x^2 P$ (7.14)

The well discharge is thus equal to the recharge within the radius of influence.

Integrating, $h^2 = \dfrac{Q_w}{\pi K}\log_e r - \dfrac{P}{2K}r^2 + C_2$ (iii)

$r = R,\, h = H,$ therefore $C_2 = H^2 - \dfrac{Q_w}{\pi K}\log_e R + \dfrac{P}{2K}R^2$

Substituting this value of C_2 in (iii),

$$H^2 - h^2 = \frac{Q_w}{\pi K}\log_e \frac{R}{r} - \frac{P}{2K}(R^2 - r^2) \qquad (7.15)$$

which is the equation for the phreatic surface (due to pumping under a recharge rate of P), line c in Fig. 7.12.

If there is no pumping ($Q_w = 0$) the shape of the phreatic surface is given by

$$H^2 - h^2 = -\frac{P}{2K}(R^2 - r^2) \qquad (7.16)$$

indicating $h > H$, line a in Fig. 7.12.

Assuming pumping with no rainfall ($P = 0$), the phreatic surface (line b in Fig. 7.12) is given by

$$H^2 - h^2 = \frac{Q_w}{\pi K}\log_e \frac{R}{r} \qquad (7.17)$$

which is the same as the Dupuit's equation [Eq. (5.4)] for unconfined aquifer.

Example 7.3: A circular island of 600 m radius has an effective rainfall of 5 mm/day. A central well 30 cm in diameter is pumped at a constant

Sea Water Intrusion

rate of 30 m³/hr. K for the island aquifer is 25 m/day. The depth of sea around the island is 10 m. What is the drawdown in the well and at the water divide?

Solution: The phreatic surface during pumping is given by

$$H^2 - h^2 = \frac{Q_w}{\pi K} \log_e \frac{R}{r} - \frac{P}{2K}(R^2 - r^2)$$

$$100 - h^2 = \frac{30 \times 24}{\pi \times 25} \times 2.303 \log_{10} \frac{600}{r} - \frac{0.005}{2 \times 25}(600^2 - r^2)$$

$$100 - h^2 = \left(\frac{r}{100}\right)^2 + 21.14 \log_{10} \frac{600}{r} - 36 \qquad \text{(iv)}$$

At the well face, $r = 0.15$ m. Therefore $h = 7.71$ m.
Non-pumping water level in the well is given by putting $r = 0.15$ m in Eq. 7.16.

$$100 - h^2 = -\frac{0.005}{2 \times 25}(600^2 - 0.15^2)$$

$$h \text{ (non-pumping)} = 11.67 \text{ m}$$

Drawdown in the well

$$s_w = 11.67 - 7.71 = \mathbf{3.96 \text{ m}}$$

If the sea around the island is contributing, the hydraulic gradient line would be sloping downwards and inwards at every point, line b in Fig 7.12. If a water divide exists at a radial distance of x, then the entire pumpage is being contributed by rainfall. So the area contributing is obtained from Eq. (7.14)

$$Q_w = \pi x^2 \cdot P$$

$$30 \times 24 = \pi x^2 \times 0.005$$

Therefore $\qquad x = \mathbf{214 \text{ m}}$

which is well within the 600 m radius of the island.
Putting $r = 214$ m in (iv),

$$100 - h^2 = \left(\frac{214}{100}\right)^2 + 21.14 \log_{10} \frac{600}{214} - 36$$

$$h \text{ (pumping)} = 11.06 \text{ m}$$

Putting $r = 214$ m in Eq. (7.14)

$$100 - h^2 = -\frac{0.005}{2 \times 25}(600^2 - 214^2)$$

$$h \text{ (non-pumping)} = 11.45 \text{ m}$$

Drawdown at the water divide

$$s_x = 11.45 - 11.06 = \mathbf{0.39 \text{ m}}$$

Upconing

In coastal aquifers, where the fresh ground water is underlain by saline water, pumping in a well causes the fresh water-salt water interface to rise below the well in response to the pressure reduction on the interface due to the drawdown of the ground water table around the well. This is called 'upconing', Fig. 7.13. If the well discharge is relatively high or if the bottom of the well is close to the interface, the salt water cone may enter into the well causing the well discharge to be a mixture of fresh and salt water.

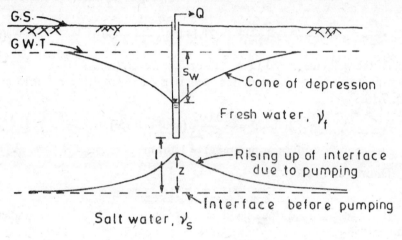

Fig. 7.13 Upconing of salt water beneath a pumped well.

Bear and Dagan (1968) developed an expression for the rise of salt water cone below the centre of the pumping well, and a simplified form for long duration pumping is

$$z = \frac{Q}{2\pi K'l} \qquad (7.18)$$

where $z =$ rise of cone and $l =$ depth of interface below the well bottom prior to pumping.

From Eq. (7.18) safe depths and pumping rates of wells to prevent entry of saline water, can be determined.

Example 7.4: A well screen 1 m in length is located 15 m below the ground water table in an unconfined aquifer having a permeability of 20 m/day. The fresh water-sea water interface exists at a depth of 36 m below the water table. What is the maximum discharge that can be sustained from the well without causing the salt water to intrude into the well.

Solution: The interface should not be allowed to rise more than one-third the distance between the bottom of the well and the original interface

Sea Water Intrusion

elevation. For this the maximum sustained discharge from the well is given by Eq. (7.18).

$$z = \frac{Q}{2\pi K'l}$$

$$1/3\,(36-16) = \frac{Q}{2\pi(0.025 \times 20)(36-16)}$$

Therefore $Q = 418$ m³/day or **290 lpm**

QUIZ 7

I. Select the correct answer(s):
   1. Sea water intrusion occurs in coastal aquifers
      a. when impermeable formations are exposed into sea water
      b. when permeable formations are exposed into sea water
      c. in (a) above, when the hydraulic gradient is towards the land
      d. in (b) above, when the hydraulic gradient is towards the land
      e. in (b) above, when the hydraulic gradient is towards the sea
      f. in all the above cases
   2. If follows from the Ghyben-Herzberg principle that a column of fresh water of 41 m will balance a column of salt water of 40 m.
      a. since sea water has a specific gravity of 1.025
      b. intrusion occurs at a depth of 40 m below m.s.. when the depth of fresh water is 1 m above m.s.l.
      c. intrusion occurs at a depth of 41 m below m.s.. when the depth of fresh water is 1 m above m.s.l.
      d. intrusion occurs at a depth of 80 m below m.s.l. when the depth of fresh water is 2 m above m.s.l.
      e. the ground water is entirely saline when the fresh water level is almost at m.s.l. or below m.s.l.
   3. As the fresh water extraction increases
      a. the ground water levels continue to decline
      b. the sea water wedge moves landward, the shape of interface being a parabola with the origin at the contact of m.s.l. with the land surface
      c. the sea water wedge moves landward, the shape of the interface being a straight line given by the equation $y = 40\,h_f$, where $y =$ depth of interface below m.s.l. $h_f =$ depth of fresh water above m.s.l.
      d. the wedge as a whole moves with a uniform velocity towards the land
      e. the toe of the wedge moves somewhat faster than the remaining portion thus the wedge becoming long and thin, i.e. it is simply stretched out;
      f. the toe of the wedge moves somewhat faster than the remaining portion and the wedge flattens out
      g. the fresh water flow into the sea decreases
   4. The movement of salt water wedge
      a. is a very slow process

b. the length of the intruded wedge is directly proportional to the rate of ground water extraction in the coastal aquifers
c. the wedge toe velocity is directly proportional to the porosity of the aquifer
d. the length of the intruded wedge is inversely proportional to the fresh water flow into the sea
e. the length of the intruded wedge linearly varies as the thickness of the aquifer
f. when the interface is not well defined (i.e. dispersed), the line of constant TDS say 1500-2000 ppm, can be assumed as the interface
g. the wedge front velocity is inversely proportional to the wedge length

5. Sea water intrusion can be controlled by
   a. increasing the ground water extraction, thereby minimising the loss of fresh water into the sea
   b. limiting the extraction of ground water, to maintain the fresh water surface appreciably above m.s.l.
   c. recharging by a line of recharge wells parallel to the coast
   d. intense pumping by a line of pumping wells along the coast
   e. constructing a subsurface dam or sheet piling parallel to the coast
   f. cultivation of paddy in the low lying coastal areas
   g. method (....) above is practicable and method (....) above is the best if a large quantity of fresh water resource could economically be utilised which otherwise runs waste into the sea
   h. all the above methods

6. The water supply from an oceanic island aquifer is
   a. mostly restricted to the effective rainfall
   b. mostly from infiltration galleries with lateral connections
   c. mostly from deep tubewells, resulting in high drawdowns
   d. from shallow tubewells pumped at such rates that the water divides exist within the island

PROBLEMS

I. Due to large scale ground water exploitation of a coastal aquifer, sea water intrusion is feared. From conductivity measurements in two wells located at 180 and 300 m from the shore (landward side) the 1500 ppm line was found to be located at depths of 18.75 and 23.60 m respectively, below m.s.l. The thickness of the aquifer is 30 m, permeability 50 m/day and porosity 10%. Determine: (a) the rate of fresh water flow towards the sea, (b) the length of the intruded wedge, and (c) the wedge front velocity.

If the fresh water flow towards the sea is doubled due to limits of extractions recommended by the G I Team, what will be the length of the intrusion of the wedge. Take specific gravity of sea water as 1.025.

(1.07 m^3/day/m width, 525 m,...,262.5 m)

II. Fresh water flow towards the sea from a coastal aquifer is at the rate of 40 m^3/day per metre length of coast. The depth of the homogeneous aquifer is 40 m below m.s.l. and has a permeability of 50 m/day. Calculate the position of the toe of the salt water wedge in the aquifer should ground water development in the coastal aquifer reduce the fresh water flow into the sea by 90%.

Hint: Use Eq. (7.7) (25 to 250 m)

III. A circular island of 800 m radius has an effective rainfall of 6 mm/day. A central well of 30 cm diameter is pumped at a constant rate of 600 lpm. K for the island aquifer is 20 m/day. The depth of sea around the island is 10 m. Determine the drawdown in the well and at the water divide.

(5.17 m, 0)

IV. The loss of fresh water from a coastal aquifer extending 6 km along the shore has been estimated to be 30000 m^3/day. The aquifer is underlain by an impervious layer at a depth of 50 m below m.s.l. Permeability of the aquifer is 40 m/day. Determine:
a. the depth of the interface below m.s.l. at 90 m from the shore inland
b. the location of the toe of the salt-water wedge
c. the width of the gap at the shore bottom through which the fresh-water escapes into the sea
d. how far the toe of the wedge will move if the loss of fresh-water is reduced by 80% by ground-water exploitation in the coastal aquifer?

(30 m, 250 m, 2.5 m, 1250 m)

V. The bottom of a well drilled in a coastal aquifer is 20 m above the fresh water-salt water interface. The permeability of the aquifer is 40 m/day. What is the maximum pumping rate so that the interface will not rise by more than 10 m?

(1257 m^3/day)

VI. A circular island of 1000 m radius has an effective rainfall of 8 mm/day. A central well of 0.4 m diameter is pumped at a constant rate of 800 lpm. The permeability of the island aquifer is 30 m/day. The depth of sea around the island is 12 m. Determine the drawdown in the well and at the water divide.

(3.5 m, 0.6 m)

VII. (a) When actually sea water intrusion takes place? How would you locate the fresh water-sea water interface?

(b) Indicate the practical methods to halt and abate sea-water intrusion in the Madras Coastal environs?

8

Ground Water Geophysics

Geophysics, in the past few years, has reached a place of vital importance to the scientific development and protection of the world's precious ground water supply. Geophysical investigations of the buried strata can be made either from the land surface or in a drilled hole in the formation.

Surface Geophysical Techniques

The surface methods include:
(i) Electrical resistivity.
(ii) Seismic refraction and reflection.
(iii) Other survey methods like soil temperature, magnetometer, gravity, remote sensing, etc.

Electrical resistivity—surface method

In the electrical resistivity method, the electrical resistance determined by applying an electric current (I) to metal stakes (outer electrodes) driven into the ground and measuring the apparent potential difference (V) between two inner electrodes (non-polarising d.c. type, i.e. porous pots filled with $CuSO_4$ solution, and metal stakes in a.c. type) buried or driven into the ground, Fig. 8.1, gives an indication of the type and depth of the subsurface material. Changing the spacing of electrodes changes the depth of penetration of the current and the apparent electrical resistivity ρ_a, obtained at different depths by measuring the resistance $R(= V/I)$, is plotted on a semi-log or log-log paper against the depth. The depth at which current enters a formation of higher or lower resistivity is signalled by a change in the resistivities recorded at the ground surface. By proper interpretation of the resistivity data from the field curves so obtained and matching them with standard curves available (Mooney and Wetzel, master curves, it is possible to identify the water bearing formations and accordingly limit the depth of well drilling.

There are mainly two common systems of electrode arrangement. In the

Ground Water Geophysics

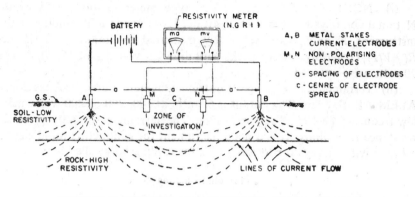

a. WENNER ELECTRODE ARRANGEMENT—EQUAL SPACING

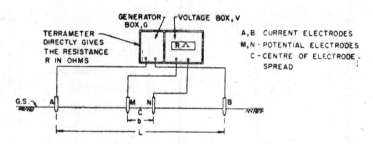

b. SCHLUMBERGER ELECTRODE ARRANGEMENT $= L \gg b$

Fig. 8.1 Electrical resistivity—surface method.

Wenner system, the electrodes are spaced at equal distances a, Fig. 8.1a, and the apparent resistivity ρ_a for a measured resistance $R(= V/I)$ is given by

$$\rho_a = 2\pi a R \tag{8.1}$$

and the field curve is plotted on a semi-log paper ρ_a versus a, ρ_a being in ohm-metres in logarithmic scale and a in metres in arithmetic scale. In Schlumberger system, Fig. 8.1b, the distance between the two inner potential electrodes (b) is kept constant for some time and the distance between the current electrodes (L) is varied. The apparent resistivity ρ_a for a measured resistance $R(= V/I)$ is given by

$$\rho_a = \pi \frac{(L/2)^2 - (b/2)^2}{b} R, \quad L \gg b \tag{8.2}$$

$$\rho_a = \frac{\pi L^2}{4b} R, \quad \text{if } L > 5b \tag{8.2a}$$

and the field curve is plotted on a log-log paper ρ_a versus $L/2$, ρ_a being in ohm-metres and $L/2$ in metres.

There are basically two types of instruments to conduct the electrical resistivity survey:

(i) NGRI resistivity meter, a d.c. type meter manufactured by the National Geophysical Research Institute, Hyderabad (South India). In this instrument V and I are separately measured to obtain the resistance $R(=V/I)$. Generally battery packs with different voltages of 15, 30, 45 and 90 volts are employed.

(ii) Terrameter, an a.c. type of instrument manufactured by Atlas Copco ABEM AB, Sweden. The output is 6 watts at 100, 200 or 400 volts using low frequency (1-4 Hz)* square waves. The terrameter directly gives the resistance R in ohms. It is a good instrument for conducting rapid electrical resistivity surveys for locating sites for drilling borewells.

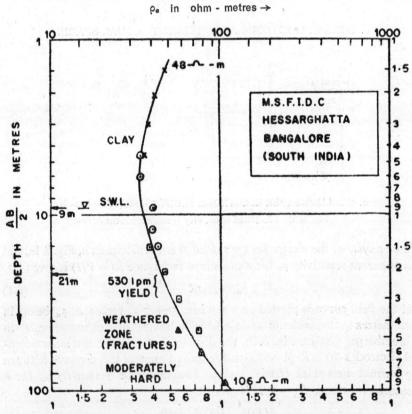

Fig. 8.2 Resistivity depth probe—Schlumberger field curve (Ex. 8.1).

Depth sounding and traversing: Two methods of investigation are generally employed in the electrical resistivity method of traversing.

(i) Resistivity depth probing or sounding to detect vertical changes. Here the centre of the electrode spread remains fixed and the spacing

---

*to avoid polarization (range provided in meter = 1, 4, 16, 64 Hz)

Ground Water Geophysics

between the electrodes is progressively increased until the maximum required depth is reached, Fig. 8.2.

(ii) Traversing or profiling method. Here the electrode separation is kept constant for two or three values (say $a = 10$ m, 15 m, or 20 m) and the centre of the electrode spread is moved from one station to another station (grid points) to have the same constant electrode separations. Traversing or profiling is used to detect subsurface changes in horizontal direction or the lateral spread. Profiling can be carried out along a series of parallel lines and a resistivity contour map of the area showing isoresistivity lines can be prepared. This will indicate areas of high resistivity and will be useful in identifying aquifer formations, Fig. 17.6.

Use of electrical resistivity method: Some of the geophysical investigations that can be done by the electrical resistivity method for ground water studies are:

(i) Correlating lithology and drawing geophysical sections.
(ii) Bed rock profile for subsurface studies.
(iii) Fresh water-salt water interface by constant separation profiling.
(iv) Contact of geological formations.
(v) Water quality in shallow aquifers and ground water pollution as in oil field brine pollution, pollution by irrigation waters and pollution by sea water intrusion, which cause change in electrical conductivity.

Resistivity data for the rock formations in Karnataka are given in Table 8.1.

Table 8.1 Resistivity data for rock formations in Karnataka.

Rock type	Resistivity (Ω-m)
Highly weathered and saturated gneiss/granite	<40
Weathered and saturated gneiss/granite	40–80
Weathered but less saturated	80–170
Unweathered granite/gneiss with water filled joints	170–400
Massive rock	⩾400
Vesicular basalts saturated with water	100–150
Highly weathered basalt saturated with water	5–10
Gravelly sands with fresh water	100
Shale and clay	1
Sea water	0.3
Brackish water	1

Example 8.1: The readings given in Table 8.2 were obtained from an NGRI resistivity meter. Plot ρ_a versus $L/2$ on log-log paper (Schlum-

Table 8.2 Resistivity depth probe—Schlumberger (Ex. 8.1) [Place—M.S.F.I.D.C., Hessaraghatta, Bangalore].

Distance of inner electrodes from the centre of the spread $\frac{MN}{2}$ (m)	Distance of outer electrodes from the centre of the spread $\frac{AB}{2}$ m	Metre readings		Resistance $R=\frac{mv}{ma}$ (ohm)	Factor $F(m)$ $=\frac{\pi}{MN}\left[\left(\frac{AB}{2}\right)^2-\left(\frac{MN}{2}\right)^2\right]$	Apparent resistivity $\rho_a = FR$ (ohm-m)
		Voltage mv	Current ma			
0.15	1.5	53.0	26	2.040	23.4	47.7
	2.1	20.6	23	0.895	46.0	41.2
	3.0	11.4	28	0.407	94.0	38.22
	4.5	3.8	22	0.173	212.0	36.66
0.75	4.5	17.0	20	0.850	41.2	35.0
	6.0	13.0	28	0.464	74.3	34.5
	9.0	7.7	32	0.240	169.0	40.6
	12.0	3.3	25	0.132	301.0	39.76
	15.0	2.6	28	0.093	470.0	43.2
3.0	15.0	9.5	28	0.339	113.0	38.3
	21.0	6.0	29	0.207	226.0	46.7
	30.0	2.3	20	0.115	466.0	53.6
	45.0	9.0	126	0.071	1054.0	75.0
7.5	45.0	18.0	130	0.138	412.0	56.9
	60.0	8.1	78	0.104	743.0	77.3
	90.0	7.3	116	0.063	1685.0	106.0

berger field curve) and interpret the results. ρ_a versus $AB/2$ is plotted on a log-log paper, Fig. 8.2. A 15 cm well drilled at the site to a depth of 30 m gave an yield of 530 lpm at a depth of 21 m.

Example 8.2: The following readings (Table 8.3) were obtained during a terrameter survey. Plot ρ_a versus a on a semi-log paper and intepret the results.

Table 8.3 Resistivity depth probe—Wenner (Ex. 8.2)
[Place: Tirumalai Samudram, Cauvery delta]

Electrode Spacing a (m)	Meter reading R (ohm)	Constant $K=2\pi a$	Apparent resistivity $\rho_a = K.R.$ (ohm-m)
2	3.3	12.6	41.5
4	1.2	25.1	30.0
6	0.469	37.7	17.7
8	0.311	50.3	15.6
10	0.258	62.9	16.0
15	0.179	94.3	16.9
20	0.163	126.0	20.2
25	0.147	158.0	23.2
30	0.137	187.0	25.6
35	0.135	220.0	29.8

ρ_a versus a is plotted on a semi-log paper and the Wenner field curve is shown in Fig. 8.3.

Master Curves for Layered Media

Resistivity data may be interpreted from master curves for a small number of earth layers assuming them as horizontal of uniform thickness and resistivity. They are prepared for particular electrodes configuration, like Wenner, Schlumberger, various thickness and resistivity ratios being assumed for the individual layers.

Resistivity curves for 3-layers are generally divided into four type as

$\rho_1 > \rho_2 < \rho_3$ high-low-high H-type
$\rho_1 < \rho_2 < \rho_3$ low-low-high A-type
$\rho_1 < \rho_2 > \rho_3$ low-high-low K-type
$\rho_1 > \rho_2 > \rho_3$ high-low-low Q-type

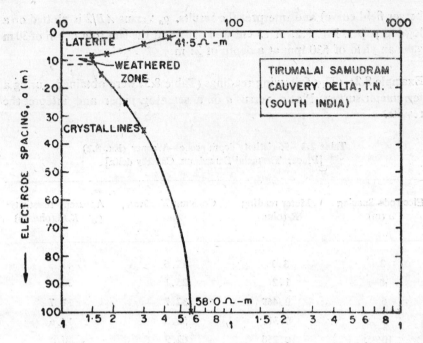

Fig. 8.3 Resistivity depth probe—Wenner field curve (Ex. 8.2).

Such ideal conditions like horizontal beds of uniform thickness and that the lowest bed extends in depth to infinity may not exist in the earth, but they are of help for comparison and interpretation of field curves obtained.

Two-layer Case: For a layer of thickness 'h' overlaying an infinitely thick homogeneous substratum of resistivity ρ_2, a family of curves is given by Tagg (Tagg master curves, 1934) as, Fig. 8.4,

$$\frac{\rho_1}{\rho_a} \text{ vs } \frac{h}{a} \text{ for } \rho_2 > \rho_1, \; k = \text{positive}$$

$$\frac{\rho_a}{\rho_1} \text{ vs } \frac{h}{a} \text{ for } \rho_2 < \rho_1, \; k = \text{negative}$$

where $\rho_1 = \rho_a$ as $a \to 0$ i.e. at small electrode spacings
ρ_a = resistivities for various electrode spacings by Wenner configurations

$$\text{resistivity coefficient } k = \frac{\rho_2 - \rho_1}{\rho_2 + \rho_1} \tag{8.3}$$

For particular value of a and ρ_1/ρ_a, the values of h/a are read from the 'master curves' for different values of k. Multiplying the h/a values by

Ground Water Geophysics 311

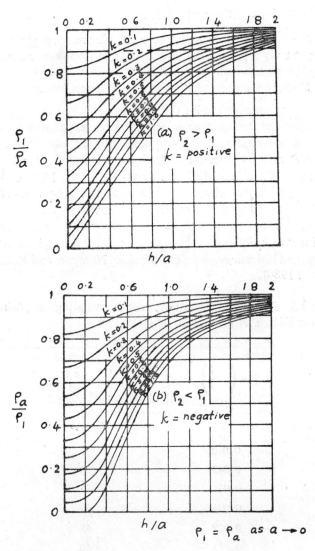

Fig. 8.4 Tagg's master curves for a two-layer case.

the corresponding a, h values are obtained. These are plotted as 'k vs h', (see Fig. 8.6 in Example 8.3).

If the curves for different electrode spacings 'a' intersect near a point, it can be assumed as a simple two-layer case, and the coordinates h and k of this point can be read. From this k, the resistivity of the substratum can be obtained from

$$\rho_2 = \rho_1 \frac{1+k}{1-k} \tag{8.3a}$$

and h = thickness of the surface layer.

The limitations of this method are:

(i) the value of ρ_1 obtained at small electrode spacings ($a \to 0$) may not truly represent the resistivity of the top layer unless it is homogeneous and isotropic.
(ii) it involves numerous steps and is time consuming.

It is now customary to plot the master curves and the field curve on a log-log paper and for the Wenner or Schlumberger configuration. The curve of best fit is one which is parallel to the relevant master curve.

Master curves for three- and four-layer configurations have been published by Moony and Wetzel (1956), by Orellana and Mooney (1966), and also by European Association of Exploration Geophysicists (1963) which allow greater flexibility in the choice of resistivity patterns.

New methods of plotting the field resistivity data by 'inverse slope' and 'direct slope' techniques have been developed for determination of absolute resistivity and thickness of layers (Sankar Narayan and Ramanujachari, 1967, Baig, 1980).

Example 8.3 Results of resistivity depth sounding at Adirampatnam (Thanjavur Dist. T.N.) are given below:

a (m)	$R\left(=\dfrac{V}{I}\right)$ (Ω)	a (m)	$R\left(=\dfrac{V}{I}\right)$ (Ω)
2	0.080	25	0.0165
4	0.049	30	0.0140
6	0.036	35	0.014
8	0.031	40	0.012
10	0.029	45	0.011
15	0.023	50	0.010
20	0.018		

Interpret the data
  a. by cumulative resistivity plotting
  b. by using Tagg's master curves.

Solution: The field data are processed as follows to obtain $\Sigma\, \rho_a$ for plotting the cumulative resistivity curve (ρ_a vs a), Fig. 8.5, and $\dfrac{\rho_1}{\rho_a}$ to obtain the field curves 'k vs h' for different values of k using Tagg's master curves, Fig. 8.6.

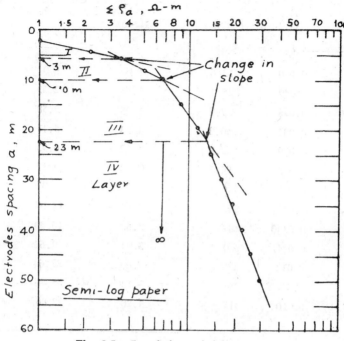

Fig. 8.5 Cumulative resistivity curve.

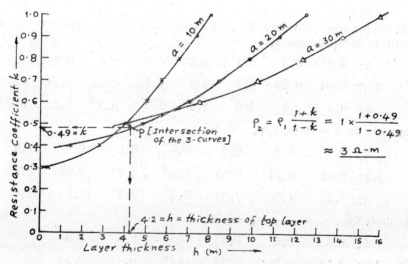

Fig. 8.6 Curves 'k vs h' for electrode spacings 'a' (using Tagg's master curves).

a (m)	R (Ω)	K ($=2\pi a$)	$\rho_a = KR$ (Ω-m)	$\Sigma \rho_a$	$\dfrac{\rho_1}{\rho_a}$
2	0.080	12.6	1.0	1.0	1.0*
4	0.049	25.1	1.23	2.23	0.81
6	0.036	37.7	1.35	3.58	0.74
8	0.031	50.3	1.53	5.11	0.65
10	0.029	62.9	1.82	6.93	0.55
15	0.023	94.3	2.18	9.11	0.46
20	0.018	126	2.26	11.37	0.44
25	0.0165	158	2.60	13.97	0.38
30	0.0140	187	2.62	16.59	0.38
35	0.014	220	3.05	19.64	0.33
40	0.012	251	3.00	22.64	0.33
45	0.011	284	3.12	25.76	0.32
50	0.010	315	3.20	28.96	0.31

*At $a=2$ m, at the top soil, $\rho_a \approx \rho_1$

Data for drawing 'k vs h' curves, for (b)

$a = 10$ m, $\rho_1/\rho_a = 0.55$; for this value, from Tagg's master curves, Fig. 8.4 for different values of k, h/a values are read and multiplied by 'a' ($=10$) to obtain h values:

k:	0.3	0.4	0.5	0.6	0.7	0.8	0.9	1.0
h/a:	0.04	0.28	0.4	0.5	0.57	0.65	0.73	0.80
h:	0.4	2.8	4.0	5.0	5.7	6.5	7.3	8.0

Similarly for $a = 20$ m, $\rho_1/\rho_a = 0.44$,

k:	0.4	0.5	0.6	0.7	0.8	0.9	1.0
h/a:	0.07	0.24	0.35	0.42	0.49	0.56	0.62
h:	1.4	4.8	7.0	8.4	9.8	11.2	12.4

Similarly for $a = 30$ m, $\rho_1/\rho_a = 0.38$,

k:	0.5	0.6	0.7	0.8	0.9	1.0
h/a:	0.14	0.25	0.34	0.41	0.47	0.53
h:	4.2	7.5	10.2	12.3	14.1	15.9

Interpretation

(a) The cumulative resistivity curve is drawn, Fig. 8.5.
Tangents are drawn to the curve and the values of 'a' at which the slope of the curve changes give the depths to the top of each layer. Thus

I layer, $a = 3$ m, thickness $= 3$ m (top soil)
II layer, $a = 10$ m, thickness $= 10 - 3 = 7$ m
III layer, $a = 23$ m, thickness $= 23 - 10 = 13$ m
IV layer, $a = 23$ m, thickness $=$ infinite.

It is a four-layer case.

(b) For $a = 10, 20, 30$ m, 'k vs h' curves intersect at nearly P, Fig. 8.6. The coordinates of P are $h = 4.2$ m and $k = 0.49$; from Eq. (8.3a).

$$\rho_2 = \rho_1 \frac{1+k}{1-k} = 1 \times \frac{1+0.49}{1-0.49} \approx 3m$$

Thickness of surface layer **h = 4.2 m**

Resistivity of the substratum $\rho_2 = 3\Omega m$.

It is assumed for two layers the top layer being of thickness 4.2 m, and the substratum extending to infinity below this layer.

Seismic Refraction Method

Seismic reflection methods provide information on the deep seated strata (750 m) while the seismic refraction methods cover only a few hundred metres below the ground surface. Hence the seismic refraction method is used in ground water investigation. The elastic waves caused by the detonation of explosives near the ground surface or a sledge hammer striking a metal plate on the ground, travel downwards into the various rock layers and are refracted back to the surface from the junctions between different rock layers. The waves are picked up at various points on the ground surface by a geophone, Fig. 8.7, and recorded. This record shows when the energy commenced and when it was picked up at the surface. By knowing the arrival times of different waves at different distances from the energy source, the velocity of propagation of the wave through each rock layer can be calculated. The velocities are characteristic of particular rocks in particular conditions, i.e., dry, jointed, saturated with water, weathered, etc. The refracted waves arrive at the surface only on the condition that the velocity of the propagation in the underlying layer is higher than that in the overlying area. Each layer through which the refracted wave travels horizontally must have a thickness that is great enough to permit transmission of the wave. The deeper a horizon is buried, the thicker it must be to properly refract the shock wave.

A time-travel curve (time versus distance from source to geophone) is drawn, Fig. 8.7 and by knowing the distance X_1 to the first point on the curve where a change in slope is indicated, the depth to the rock layer can be computed from the equation (Nettleton, 1940).

$$Z_1 = \frac{X_1}{2} \sqrt{\frac{VV}{V_1 + V_2}} \quad (8.4)$$

where V_1 and V_2 are the velocities of propagation through the earth and the rock layer respectively. Using the intercept time t_1, the depth Z_1 is given by the equation (Nettleton, 1940)

$$Z_1 = \frac{t_1}{2} \frac{V_2 V_1}{\sqrt{V_2^2 - V_1^2}} \quad (8.5)$$

The depth Z_2 of the second layer is given by

$$Z_2 = \left(\frac{t_2}{2} - Z_1 \frac{\sqrt{V_3^2 - V_1^2}}{V_3 V_1}\right) \frac{V_3 V_2}{\sqrt{V_3^2 - V_2^2}} \quad (8.6)$$

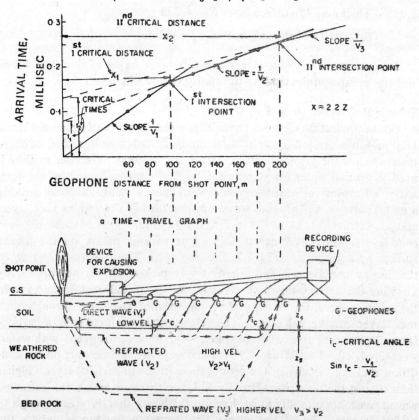

Fig. 8.7 Seismic refraction method.

The critical angle i_c is given by

$$\sin i_c = \frac{V_1}{V_2}$$

For angles of incidence greater than critical, there are no refractions into the deeper layers but the waves are totally reflected. Travel time is usually measured in milliseconds. The disturbances, commonly amplified, are recorded photographically on a moving film.

By moving the source of the shock wave along a line, a profile of the underlying bed rock surface (or other layer, such as the water table) can be obtained. The refraction method is faster and often finds application in

(i) Locating the ground water table.
(ii) Determining depth to bed rock or impermeable layer and configuration (volume of material).
(iii) Locating a buried stream channel (cut into bed rock).
(iv) Locating faults that could act as ground water barriers.

Generally water table delineation is confined to loose alluvium. In the Deccan the alternation of soft and hard layers, however, limits the method to obtaining depths and velocities down to the first high velocity layer.

The velocity of propagation varies from as low as 120 m/sec in dry top soil to more than 6000 m/sec in very dense rocks such as granite, limestone and basalt. The velocities in saturated strata are somewhat greater than in unsaturated strata. The average velocities of seismic waves in different formations are given in Table 8.4.

Table 8.4 Average velocity of seismic waves.

Rock formation	Range of velocities m/sec
Dry sand and loose soil	150—400
Alluvium	500—1500
Wet sand	600—1800
Clays	900—3000
Sandstone	2000—4300
Shale	2100—4000
Limestone	3000—6000
Deccan trap	4000—5000
Igneous and metamorphic rocks (compact)	4500—6500

Example 8.4: In a refraction shooting, nine geophones were placed along a straight line at distances of 40, 60, 80, 100, 140, 180, 220, 260 and 320 metres from the shot point. The seismic record gave the following data (Table 8.5). Draw the time-distance graph and determine the velocity of the shock wave and thickness of each layer.

Table 8.5   Seismic data

Geophone	Distance from shot point m	Time of first arrival milli-seconds
G_1	40	75
G_2	60	110
G_3	80	150
G_4	100	160
G_5	140	180
G_6	180	200
G_7	220	205
G_8	260	215
G_9	320	225

Solution: The time-distance graph is drawn as shown in Fig. 8.8. The velocity of shock wave (direct wave) in the top soil layer, i.e. reciprocal of slope

$$V_1 = \frac{80 \text{ m}}{0.15 \text{ sec}} = 533 \text{ m/sec}$$

The velocity of the shock wave (refracted wave) in the second layer (weathered rock)

$$V_2 = \frac{(180-80) \text{ m}}{(0.20-0.15) \text{ sec}} = 2000 \text{ m/sec}$$

The velocity of the shock wave (refracted wave) in the bottom hard rock layer

$$V_3 = \frac{(320-180) \text{ m}}{(0.225-0.200) \text{ sec}} = 5600 \text{ m/sec}$$

The thickness of the first layer using the critical distance formula,

$$Z_1 = \frac{X_1}{2}\sqrt{\frac{V_2-V_1}{V_2+V_1}} = \frac{80}{2}\sqrt{\frac{2000-533}{2000+533}} = 30.4 \text{ m}$$

Ground Water Geophysics

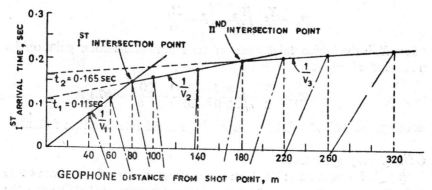

a. TIME - DISTANCE PLOT

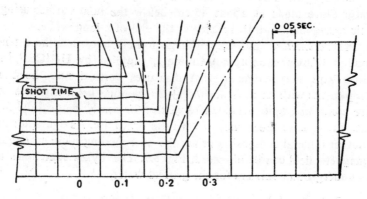

b. SEISMIC RECORD SHOWING SHOT INSTANT AND THE FIRST ARRIVAL TIMES

Fig. 8.8 Seismic refraction—travel-time plot (Ex. 8.3).

The thickness of the first layer using the intercept time formula,

$$Z_1 = \frac{t_1}{2} \frac{V_2 V_1}{\sqrt{V_2^2 - V_1^2}} = \frac{0.11}{2} \frac{2000 \times 533}{\sqrt{2000^2 - 533^2}} = 30.4 \text{ m}$$

The thickness of the second layer using the intercept time formula

$$Z_2 = \left(\frac{t_2}{2} - Z_1 \frac{\sqrt{V_3^2 - V_1^2}}{V_3 V_1}\right) \frac{V_3 V_2}{\sqrt{V_3^2 - V_2^2}}$$

$$= \left(\frac{0.165}{2} - 30.4 \frac{\sqrt{5600^2 - 533^2}}{5600 \times 533}\right) \frac{5600 \times 2000}{\sqrt{5600^2 - 200^2}}$$

$$= 54.8 \text{ m}$$

An approximate equation for Z_2 presented by Geophysical Specialities Company (1960), in modified form, is

$$Z_2 = \frac{X_2}{2}\sqrt{\frac{V_3-V_2}{V_3+V_2}} - \frac{Z_1}{6}$$

where X_2 is the horizontal distance of the second intersection point on the time-distance graph.

$$Z_2 = \frac{180}{2}\sqrt{\frac{5600-2000}{5600+2000}} - \frac{30.4}{6} = 57.1 \text{ m}$$

as compared with $Z_2 = 54.8$ m obtained from the intercept time formula.

Other Surveying Methods

(a) *Soil temperature*: The high specific heat of ground water can cause a shallow aquifer to act as a heat sink or heat source that influences the near-surface temperatures to a measurable extent. Measurements of soil temperatures are made at about 45 cm below the land surface using an electronic thermometer—a thermistor at the end of a long probe.

(b) *Magnetometer*: Basically, a magnetometer measures the intensity and direction of magnetic forces. 'Minimag', supplied by UNICEF, has a suspension wire magnetometer that measures vertical and horizontal field components with an accuracy of ± 20 gammas (γ) if tripod is used. They are useful in granite areas where vertical or nearly vertical dykes are frequent. In basaltic areas they are of little or no use. As dykes normally have another mineral composition than the surrounding rock, an anomaly in the magnetic field can be observed, Fig. 8.9. The dykes sometimes also serve as underground barriers for the ground water.

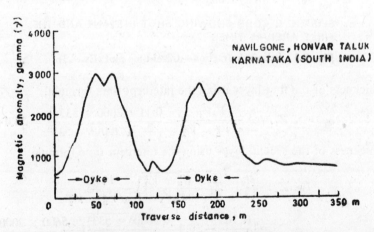

Fig. 8.9 Magnetic profile across dolerite dyke.

(c) *Gravity*: Gravity is directly related to the density and volume of earth materials beneath the point being measured. Gravity studies, because they are relatively insensitive to small changes in geology, have been used

Ground Water Geophysics

in ground water studies to map large, buried valleys. They have also been used to locate sink holes and caverns in limestone areas.

The most common type of gravity instrument is the gravimeter, which measures the direct effects of the pull of gravity on a mass suspended by a delicate spring. Changes in the length of the spring are related directly to the vertical intensity of the gravity field. Optical or electrical methods are used to amplify movements of the spring so that very slight changes can be measured. Gravity is measured in gals (1 gal = 1 cm/sec^2) or milligals 1 milligal = 0.001 gal).

(d) *Remote sensing*: This method includes infra-red photography, aeromagnetic surveys at low altitudes and other methods which use photographic techniques, conducted above the earth surface, either by aircrafts or satellites. Infra-red photography can detect temperature difference in water. Aerial magnetic surveys can delineate subsurface rock structures which might control the flow of ground water.

For ground water the most widely observable anomalies (i.e. changes in the properties of rocks due to the presence of the resource—water) are electrical conductance of the rock and velocity of sound through particular rock types. As such the resistivity and seismic methods are widely employed and are also relatively cheap. Electrical resistivity surveys are generally used for preliminary exploration of rather large areas. The interpretations based on such surveys should be confirmed by test drilling.

Borehole Geophysical Techniques

The geophysical techniques inside a drilled hole include

(i) Electric logging—electrical resistivity and spontaneous potential
(ii) Radioactive logging
(iii) Induction logging
(iv) Sonic logging
(v) Fluid logging—temperature logging, fluid resistivity logging, flow meter and tracer logging
(vi) Well construction and completion measurements—defining hole alignment, caliper logging, cement bond logging and water level measurements
(vii) Downhole photography

Electric Logging

A four electrode arrangement is commonly employed in measuring resistivity from bore holes similar to the four electrodes used in surface resistivity method. A current (I) is passed between the electrodes A and B while voltage is measured between electrodes M and N. One current electrode is always on the ground potential and its effect can be taken as negligible. Conventionally there are two systems of electrode arrangements called the 'normal' and 'lateral', Fig. 8.10. In the normal arrangement the distance MN is large compared to the distance AM. If AM is small, say 40 cm, it is called a 'short normal' and if it is longer, say 160 cm, it is called a 'long normal'. In the lateral arrangement MN is very small compared to the distance AM. If O is the midpoint of MN and AO is 1.8 m, then it is called 1.8 m lateral. Usually in electrical logging both normal and lateral devices are used to obtain maximum information. The lateral measures the resistivity of the formation beyond the zone of invasion, Fig. 8.11.

Invasion: During the drilling operations, the mud in the borehole is usually conditioned so that the hydrostatic pressure inside is greater than

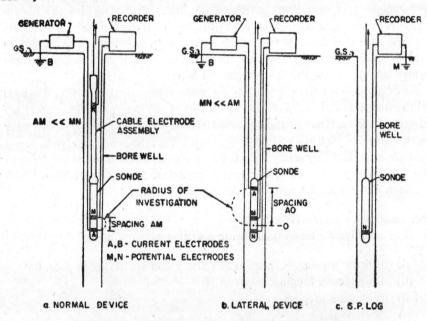

Fig. 8.10 Electrical logging—electrode arrangements.

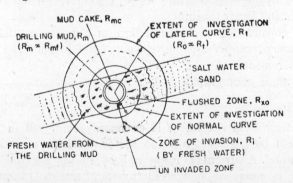

Fig. 8.11 Zones of invasion and their resistivities.

the pressure of the formations. The mud filtrate pushes the interstitial water of the formation up to a certain distance called 'flushed zone' though its influence is still further up to what is called the 'invaded zone', beyond which the formation is uncontaminated. The solid particles of the infiltrating fluid are deposited on the wall of the borehole forming a mud cake which considerably reduces the infiltration. The different zones and their resistivities are shown in Fig. 8.11.

In actual logging, the logging tool with three electrodes built up in it (one being on the ground surface) called the 'sonde' is connected to a multiconductor cable passing over a sheave and via winch and electronic recorder mounted on a truck to a power source. The sonde is lowered into the hole, Fig. 8 12, and the recorder with moving pen traces the electrical resistivity

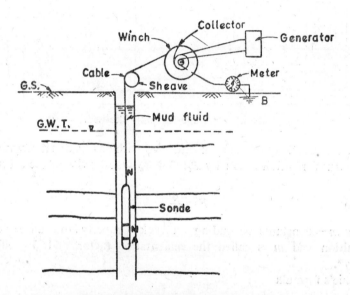

Fig. 8.12 Sonde lowered into the hole.

on graph paper continuously with depth as the sonde is withdrawn. This resistivity is the apparent resistivity R_a. Different sondes are used for electrical resistivity, S.P. (spontaneous potential) and radioactive logs. Oklahoma Logmaster, USA was used in UNDP investigations in South India and it recorded separate graphs for resistivity—short and long normal, lateral, SP and gamma logs. An idealised electric log is shown in Fig. 8.13 which will help to read the electric logs correctly and locate the different zones encountered down the drilled hole.

Formation Factor: The formation resistivity factor F depends on the lithology of the aquifer

$$F = \frac{R_0}{R_w} \tag{8.7}$$

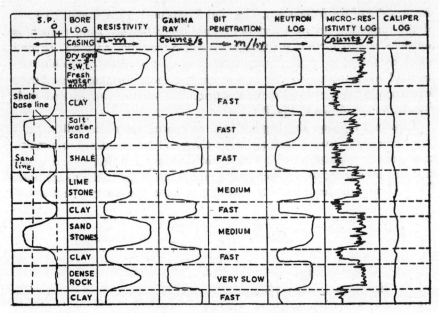

Fig. 8.13 Composite Log—idealised.

where R_0 = resistivity of the rock saturated with conducting fluid (assumed equal to true formation resistivity R_t) and R_w = resistivity of the saturating fluid.

Also
$$F = \frac{a}{\phi^m} \tag{8.8}$$

where a, m = constants depending on rock property and mineralogical composition and m is called the cementation factor and ϕ = effective porosity*.

Archie's formula
$$F = \frac{1}{\phi^m} \tag{8.9}$$

For sands
$$F = \frac{0.81}{\phi^2}$$

In compacted formations (limestone and dolomite)
$$F = \frac{1}{\phi^2}$$

For granular rock (Humble's formula)
$$F = \frac{0.62}{\phi^{2.15}}$$

---

*$\phi < n$, since pore water adheres to the particles even when drained. In gravels $\phi = n$; in finegrained soils, the difference is 5% or more.

Ground Water Geophysics

F is usually consistent for a given sedimentary unit within a depositional basin and may be determined by laboratory analysis of samples or from the Eq. (8.7), when both resistivity logs and formation water samples are available. Effective porosity may also be determined from the electric logs.

Example 8.4: The resistivity of a sample of formation water reduced to a standard temperature of 27°C is 15.2 ohm-m. If the formation resistivity read from the electric log is 131 ohm-m, determine the effective porosity of the formation, assuming a cementation factor of 2 in the Archie's formula.

Solution:
$$F = \frac{R_o}{R_w} = \frac{1}{\phi^m}$$

Assuming $R_o = R_t$

$$\frac{131}{15.2} = \frac{1}{\phi^2}$$

Effective porosity $\phi = 34.1\%$

Single point resistivity: In this method only two electrodes are employed, one in the tool and one on the ground. A current is passed between the two electrodes. The amount of current that will flow will be a function of the resistivity of the material close to the electrode in the borehole. Thus the measurement of current flow under a constant applied voltage will enable resistivity measurements to be made. The simplicity and economy of the equipment is an advantage of this method. The limitations of this method are

(i) it is not usually possible to determine the true resistivity accurately enough for quantitative interpretation in terms of porosity or lithology
(ii) the measured resistance is seriously affected by variations in the diameter of the well and the mud resistivity.

Spontaneous potential: The spontaneous or self potential (SP) in a drill hole is due to electrochemical and electrokinetic or streaming potentials. Electrochemical potentials are due to differences in concentrations of activities of the formation water and the mud filtrate, called the liquid junction potential, and membrane potential due to the presence of shale layers. The streaming potential is due to electro-filteration of the mud through the mud cake.

If the permeable formation is not shaly, the electrochemical potential (static SP or SSP)

$$E_c = -K \log \frac{a_w}{a_{mf}} \qquad (8.10)$$

where a_w, a_{mf} = chemical activities of the interstitial water and mud filtrate, respectively and K = coefficient proportional to the absolute temperature of formation ($K = 71$ at 25°C).

The chemical activity of a solution is related to the salt content and hence to the resistivity. The SP due to electrochemical activity may be written as

$$\text{SP} = -K \log \frac{R_{mf}}{R_w} \qquad (8.11)$$

where R_{mf} = resistivity of the mud fluid and R_w = resistivity of the formation water.

The SP curve generally provides the best logging approach to the determination of water quality.

The SP log is obtained by recording the potential differences, against depth, between a fixed surface electrode and a movable electrode in the borehole. Since the potentials associated with shales and clays are normally the least negative, the SP curve is a straight line called the 'shale baseline'. Opposite the permeable formations the SP curve shifts either to the left (negative) or to the right (positive) depending on the relative salinities of the formation water and the mud filtrate.

The shale baseline is drawn through as many deflection minima as possible. A sand line may then be drawn through negative deflection maxima and if fluid salinity is constant these lines will be parallel to each other and the zero baseline. The boundaries of the permeable formations are located at points of maximum SP slope rather than half amplitude as on many other logs.

The SP curve may be used to calculate formation water resistivity, locate bed boundaries, distinguish between shales and sandstone or limestone in combination with other logs, and for stratigraphic correlation. The SP log is affected by hole diameter, bed thickness, water or mud resistivity, density and chemical composition, and cake thickness, mud filtrate invasion and well temperature.

Static SP: The SP currents flow through the borehole, the invaded and the non-invaded part of the permeable formation and the surrounding shales. In each medium the potential along a line of current flow drops proportionately due to the resistance encountered by the SP current. If the SP currents could be prevented from flowing by insulation plugs, the potential difference in the mud equals the total e.m.f. The SP curve which would be recorded in such idealised conditions is called the static SP or SSP opposite clean formations.

Example 8.5: Electric and lithologic logs of Borehole No. 1008 at Pattukotai, Tamil Nadu are given in Fig. 8.14.

(i) Locate the main aquifer zones at their boundaries based on electric logs between 15 and 150 m depth.

Ground Water Geophysics 327

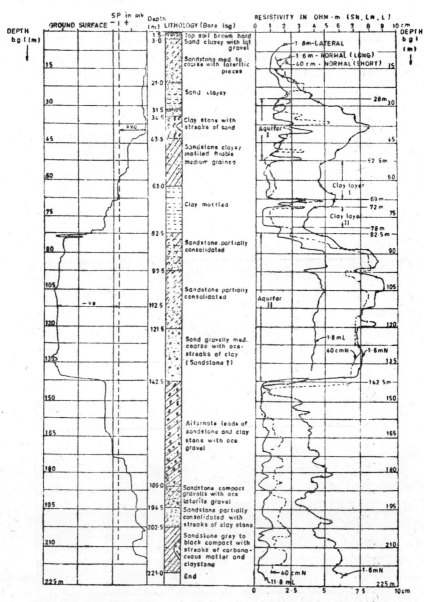

Date of electrical logs:
Logs horizontal scale:
40 cm N (normal): 1 cm=4 Ohm—m
1.6 m N (normal): 1 cm=10 Ohm—m
1.8 m L (lateral): 1 cm=20 Ohm—m
SP (Self potential): 1 cm=4 mV
Mud resistivity : 8.1 Ohm—m

Logs vertical scale: 1 cm=6m
SN—Short Normal, 40 cm
LN—Long Normal, 1.6 m
L—Lateral, 1.8 m

BH No. 1008 at Pattukkottai, Tamil Nadu

Fig. 8.14 Electric and lithologic logs (Ex. 8.5).

(ii) Locate the clay zones, if any, between the aquifers.
(iii) Determine the conductivity in micromhos/cm, TDS in ppm of the formation water and porosity of the formation material in the aquifers located.

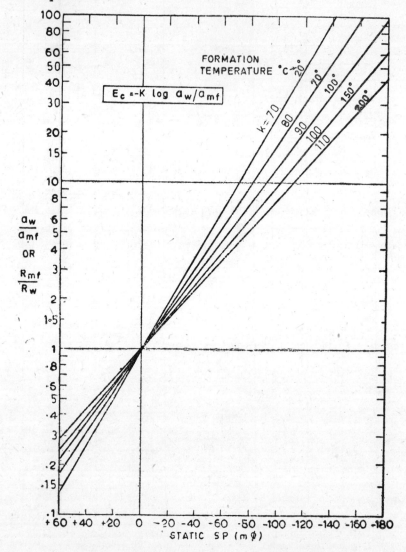

Fig. 8.15 R_{mf}/R_w versus SSP (Determination of R_w).

A graph showing R_{mf}/R_w against SSP is shown in Fig. 8.15. Assume a formation temperature 38°C and a cementation factor of 2 in the Archie's formula. The SP deflection may be taken as equal to SSP. Mud resistivity $(R_m) = 8.1$ ohm-m.

Solution: From a study of resistivity and SP logs between 15 and 150 m the following conclusion can be drawn.

(i) the main aquifer zones are located at half amplitudes on normal resistivity curves as
first aquifer between depths 28 and 52.5 m, maximum resistivity $R_{t1} = 77.5$ ohm-m
second aquifer between depths 82.5 and 142.5 m, maximum resistivity $R_{t2} = 82.5$ ohm-m

(ii) Clay zones are located at
first clay layer between 52.5 and 69 m
second clay layer between 72 and 78 m

(iii) Midpoint of the first aquifer zone is at $\dfrac{28 + 52.5}{2} = 40.25$ m. Maximum SP deflection at this point is $+8$ mv. Assuming SP $\approx$ SSP (static SP) from Fig. 8.15 $R_{mf}/R_w = 0.76$.

Resistivity of the formation water in the first aquifer,

$$\frac{8.1}{R_{w1}} = 0.76, \qquad R_m \approx R_{mf}$$

Therefore $R_{w1} = 10.7$ ohm-m

Midpoint of the second aquifer zone is at $\dfrac{82.5 + 142.5}{2} = 112.5$ m. Maximum SP deflection at this point is -17.2 mv and from Fig. 8.15 $R_{mf}/R_w = 1.8$. Resistivity of the formation water in the second aquifer

$$\frac{8.1}{R_{w2}} = 1.8, \qquad R_m \approx R_{mf}$$

Therefore $R_{w2} = 4.5$ ohm-m.

The electrical conductivity (EC) of the formation water in micromhos/cm is given by

$$EC = \frac{10,000}{R_w}$$

for the first aquifer,

$$EC = \frac{10,000}{10.7} = 935 \ \mu\text{mhos/cm}.$$

for the second aquifer,

$$EC = \frac{10,000}{4.5} = 2222 \ \mu\text{mhos/cm}.$$

The total dissolved salts (TDS) in ppm of the formation water is given by
$$TDS = 0.64 \times EC \text{ in micromhos/cm}$$

For the first aquifer,
$$TDS = 0.64 \times 935 = 600 \text{ ppm}$$
For the second aquifer,
$$TDS = 0.64 \times 2222 = 1422 \text{ ppm}$$

The formation factor
$$F = \frac{R_0}{R_w} = \frac{1}{\phi_m}, \; R_0 \approx R_t, \; m = 2$$

for the first aquifer,
$$\frac{77.5}{10.7} = \frac{1}{\phi^2}$$

porosity of the aquifer material, $\phi = 0.372$, or **37.2%**
for the second aquifer,
$$\frac{82.5}{4.5} = \frac{1}{\phi^2}$$

porosity of the aquifer material, $\phi = 0.233$, or **23.3%**

Radioactive Logging

Radioactive logs are of two general types—those which measure the natural radioactivity of formations (gamma ray log) and those which detect radiation reflected from or induced in the formations from an artificial source (neutron logs). Radioactive logs can be used in cased holes where most other types of logging will not work.

Gamma ray logs: The minerals in shales and clay emit more gamma rays than the minerals in gravels and sands. Thus gamma logs can be used to differentiate between sands, shales and clay. The probe is essentially a geiger counter or scintillometer and can be run in open or cased holes, Fig. 8.16a.

Gamma-gamma: The gamma rays from a source in the probe are scattered and diffused through the formation. Part of the scattered gamma rays re-enter the hole and are measured by an appropriate detector. The higher the bulk density of the formation, the smaller the number of gamma-gamma rays that reach the detector. The count rate plotted on a gamma-gamma log is an exponential function of bulk density. By knowing the bulk density, the porosity of the formation can be determined (uncased holes only) from the equation

$$n = \frac{\rho_g - \rho}{\rho_g - \rho_f} \quad (8.12)$$

where ρ_g = grain density; ρ_b = bulk density and ρ_f = fluid density.

Ground Water Geophysics

In a logging equipment used by Geological Survey of Canada 10 to 35 millicuries of cobalt[60] is used as a gamma source attached below a sodium iodide detector.

Neutron logging: Neutron rays are useful in determining the porosity of formations. A 'fast neutron' source is used to bombard the rock. When any individual neutron collides with a hydrogen ion (of a water molecule), some of the neutron's energy is lost and it slows down. A large number of slow neutrons, as recorded by a slow neutron counter, indicates a large amount of fluid, i.e. high porosity.

Each radiation produces a pulse in the circuit. The number of pulses per unit time is recorded. This can be done in cased or uncased holes, Fig. 8.16b. The gamma ray log does not indicate casing or presence of

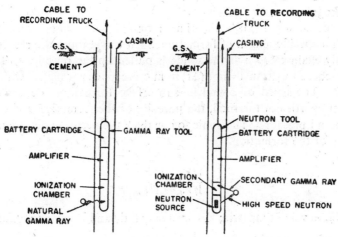

a. NATURAL GAMMA LOGGING   b. NEUTRON LOGGING

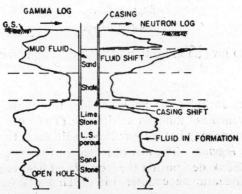

c. NEUTRON LOG INDICATES CASING AND FLUID (IN HOLE AND IN FORMATION) WHICH THE GAMMA LOG DOES NOT

Fig. 8.16 Radioactive logging.

fluid while the neutron log is sensitive to both casing and fluid in the hole as well as in the formation, Fig. 8.16c.

Induction Logging

Induction logging measures the conductivity (reciprocal of resistivity) of formations by means of induced alternating currents. Insulated coils (for induction), rather than electrodes, are used to energise the formations, and the bore hole may contain any fluid or be empty but the hole must be uncased. It is specially used to investigate thin beds because of its focussing abilities and its greater radius of investigation. It is a superior method for surveying empty holes and holes drilled with oilbased mud.

Sonic Logging

The sonic log records the time required for a sound wave to travel through a specific length of formation. Such travel times are recorded continuously against depth as the sonde is pulled up the bore hole. The sonic log is recorded as transit time (Δt) in microseconds per metre, with zero on the right. The speed of sound in subsurface formations depends on the elastic properties of the rock, the porosity of the formation and their fluid content and pressure. The sonic log enables the accurate determination of porosity of the formation.

$$n = \frac{1/V - 1/V_m}{1/V_f - 1/V_m} \qquad (8.13)$$

where V_m = velocity matrix; V_f = velocity fluid and V = velocity formation.

Since the transit time $\Delta t = \frac{1}{V}$,

$$n = \frac{\Delta t_{log} - t_{matrix}}{\Delta t_{fluid} - \Delta t_{matrix}} \qquad (8.14)$$

This log will also give an indication of rock type and fracturing.

Fluid Logging

Fluid logging includes the use of sondes to measure the temperature, quality and movement of fluids in a drill hole. (These characteristics of the fluid column may or may not truly reflect conditions in the aquifer system.)

Temperature logging: The rate of increase of temperature with depth (geothermal gradient) depends on the locality and heat conductivity of the formations. Temperatures encountered in drill holes are dependent not only on the natural geothermal gradient but also on the circulation of the mud. Temperature logs may be used to identify aquifers or perforated sections contributing water or gas to a well, to provide data on the source of water, as an aid in identifying rock types and for calculating fluid viscosity and specific conductance from fluid-resistivity logs.

Temperature logs can be used to distinguish moving and stagnant water in a well and identify the source of recharge or injected waste water. Temperature logging can also be used to verify that the cement on the outside of the casing has formed a proper bond because cement generates a great amount of heat as it sets, Fig. 8.17. Higher temperatures are usually recorded in caved sections where greater volumes of cement are deposited, permitting correlation with electric logs.

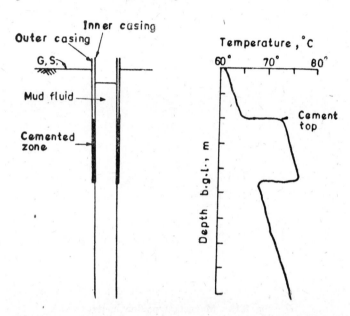

Fig. 8.17   Temperature log—temperature rise as cement hardens.

Fluid resistivity logging: Fluid resistivity logging is the measurement of resistivity of the fluid (water quality) between two closely spaced electrodes in the hole. The fluid resistivity log may be used to locate points of influx or egress of waters of different quality, to locate the interface between salt and fresh water, to correct head measurements for fluid density differences, to locate waste waters and to follow the movement of saline tracers. The resistivity of the fluid column is also important in interpreting SP, resistivity and neutron logs which may be affected by salinity changes.

Flow meter and Tracer logging: The devices used to measure vertical flow in water wells include the impeller flow meter, the radioactive tracer ejector-detector and the brine ejector-detector. Fluid movement from one aquifer to another can be measured by an impeller flow meter which records the number of impeller revolutions against time. Flow meters are useful for relatively high velocities. A magnetic-type flow meter for downhole current measurements is shown in Fig. 8.18.

Fig. 8.18 Magnetic-type flow meter.

Speed and direction of ground water flow can be detected by the use of dyes, soluble salts, radioactive tracers, electrical methods, heat dissipation and other means.

The common dyes are fluorescein, uranine and eosin. A common dye is sodium fluorescein, which can be detected in very low concentrations. Powdered fluorescein has a reddish-brown colour when dry; when dissolved in water it appears, by reflected light, a brilliant green. 1 part in 40 million can be detected by the naked eye. The tracing dye is placed in a central well at equal distances from which test wells have been sunk. The direction of flow is the direction from the central well to the well in which the reagent is first detected. The tracer is injected into the well at some point and a detector records the time it takes for the tracer to reach a second point. Very low velocity movements (of the order of metres per day) can be recorded with such a tracer set up.

Other in-hole tracers have been used to measure permeability. This includes insoluble radioactive tracers which are concentrated in the most permeable beds, and a single-well pulse technique relating to the recovery time under pumping conditions of a slug of tracer placed in the well.

Single well dilution technique was adopted in the Atomic Energy Establishment, Trombay, to study the amount and velocity of subsurface water flow. A radio-isotope solution was injected into a confined section of a well and measurement of the exponential dilution of the isotope solution with time as unlabelled water slowly moves into the well, was made. The velocity of flow (V) was calculated by the following equation (Moser, 1957)

$$V = \frac{2.3Q}{YAt} \log \frac{C}{C_o} \qquad (8.15)$$

where Q = volume of water in the well; A = cross-section of the well; t = time lapse after injection; C_o, C = initial and final (after time t) concentrations of tracer; Y = coefficient to take account of deformation of the hydro-dynamic field due to the presence of the well; $\left(Y = \frac{q}{q_w}\right)$; q = flow rate of water passing through the well and q_w = flow rate of water in the aquifer (formation) across the same cross-section of the well.

Radioactive tracers are very suitable for tracing the movement of ground water as against organic dyes since their mass concentrations as low as 10^{-6} to 10^{-18} can easily be measured by geiger counters and related circuits. Precautions should be taken against radiation effects. Radioactive tracers commonly used are Bromine[82], Calcium[45], Cobalt[60], Tritium (H[3]), Iodine[131], Phosphorous[32], Rubidium[86], and Iridium[192].

An ideal tracer

(i) Must be susceptible to quantitative determination in very low concentrations.
(ii) Should not be present in the natural water.
(iii) Must not react with the natural water to form a precipitate.
(vi) Must not be absorbed by the porous media.
(v) Must be cheap and readily available.

No tracer completely meets all the above requirements and a reasonably satisfactory tracer can be selected to fit the needs of a particular situation.

Dating of ground water: Tritium is produced in the atmosphere by cosmic radiation (and thermonuclear explosions) and is in abundance in rain. After the rain water infiltrates into the ground there is in no further addition of tritium and the tritium concentration diminishes exponentially. Thus, from ground water samples obtained, particularly from confined aquifers recharged from a single recharge area, the age of the ground water can be estimated. If several water samples are obtained from wells scattered over a basin, the direction and rate of ground water movement might be determined. The equipment required for measuring the radiation of extremely low levels of naturally occurring tritium is very expensive and time consuming which limit its application.

Focussed resistivity logs: The conventional sondes are 40, 160, 180 cm long and so tend to give average resistivity values unless the formations are fairly thick. To get over this focussed resistivity devices such as the laterolog, induction log, microlog and the microlaterolog have been developed.

A microlog or contact log is a resistivity log, measured with the electrodes spaced very closely (2.5 to 5 cm), in an insulating pad which is pressed against the walls of the drill hole, Fig. 8.19. For a 2.5 cm spacing, a depth of only 2.5 cm is investigated by the micro-resistivity study. The thickness of the mud cake has a significant major effect on the value of micro-resistivities. As the tool is lowered into the hole, the springs are clamped into the tool; upon reaching the bottom the spring arms are released. The average hole diameter and the micro-resistivities of each stratum exposed in the hole are recorded as the tool is drawn up the hole.

Microcaliper log: The microcaliper log records the average hole diameter and is run in conjunction with the microlog or contact log. The hole diameter will be equal to the size of the drilling bit, when a hard sandstone or limestone is traversed. Under normal conditions the well bore

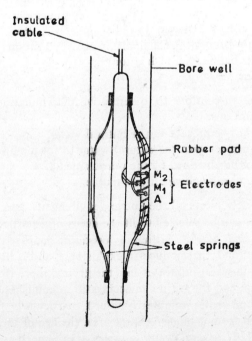

Fig. 8.19 Micro log or contact tool.

becomes enlarged in shale beds because the shales become wet with the mud fluid, slough off and cave into the hole. The microcaliper will indicate an enlarged hole up to the maximum spread of the caliper arms. Such information is useful for determining areas of formation caving, casing lengths, packers and perforations.

Cement bond logging: A cement bond log indicates whether or not cement is tight against the outside casing wall. This log is based on the principle that the signal strength of an acoustic signal travelling along the casing is greatly reduced where the cement is well bonded to the pipe, compared to no bonding or poor bonding.

Downhole photography—Drift indicator or photoclinometer: By means of a camera, pendulum and compass all housed in a probe, the inclination and direction of drill hole deviation (drift) can be determined.

The US Army Engineers developed the NX borehole camera at a cost of $ 80,000 to fit a 7.5 cm borehole and give a 360° scan of the borehole wall and used 8 mm colour movie film.

More recently, television cameras have been developed which can provide immediate and continuous visual inspection of a borehole wall— live and in colour. It can also be recorded on a video tape to be replayed later. They are usually less than 7.5 cm diameter and use 1,000 watts lighting apparatus. The camera or television transmitter is lowered down the hole, usually at constant rate. Depth is calibrated using a cable marked at intervals.

Photographs or television can be used to identify geologic formations in open holes, as part of well completion survey, to check damaged walls, to aid in removing foreign matter from a well, and to assist in development or well cleaning.

Geophysical logs are often used in conjunction with drilltime logs (bit penetration rate) and bore logs of the sample cuttings obtained from different depths during drilling to aid in identifying formation characteristics, Fig. 8.13. The record of bit penetration rate can be quantitative (cm/min) or even qualitative, termed as fast, slow or very slow.

Currently some sophisticated techniques that are being developed are the neutron life-time logging, several types of spectral logging, acoustic amplitude logging, nuclear magnetic logging and computer interpretation and collation of geophysical logs. Special sondes like the limestone sonde have been developed for specific purposes.

Geophysics is a specialised field by itself and information given here should serve only as an introduction for further study and interpretation of the results in ground water studies.

The Phenomenon of Dowsing

Dowsing is popularly known as divining. It is both an art and an empirical science; empirical in the sense that, the dowsers or diviners have not been able to put forth an acceptable explanation of their ability to detect ground water, minerals, oil deposits, etc. with so simple as instrument as a mere forked twig or a simple pendulum.

For the first time scientific explanation was attempted in the West, and recently in India also, for all the strange phenomena under the label 'radiesthesia'. The word radiesthesia has its origin in Egypt of bygone days. It connotes that it is a science dealing with the study of radiating energy of objects.

Radiesthesists believe that every object, cognate or incognate, radiates energy to all its surroundings, the result of which will be its impact on the structure and function of the surrounding objects. While each object behaves in this way with every other object in its proximal or distal vicinity relatively, dual polarity of the inherent energy becomes manifest. These poles are called 'attraction' and 'repulsion' and the interaction of these two is known as 'vibration'.

Rate, quality, intensity and relative susceptibility of these vibrations are the printed circuits of the radionic behaviour of objects. Objects find their specific ultrashort wavelengths in the ethereal media constantly during their life time. These radionic waves become omnifarious, omnipotent, and omnipresent.

It is man who is their variable gang condenser. He creates a susceptible medium through his little instrument which is nothing but a simple pendulum to receive, amplify rectify and screen distortions and transmit to the output section of the pendulum centre, and the dialcard which takes the pendulum to the correct pointer on the data chart.

Silent speaker: The 'silent speaker' of the chart speaks, indicating the answer you are entitled to receive. Here it is important to ignore the various factors affecting the radiesthetic divining. As all electromagnetic discharges in the atmosphere interfere with the electromagnetic waves of broadcasting stations, and as natural atmospheric electrical discharges of power (like heavy wind, gale, storm, thunder and lightning) cause heavy distortions, disturbances, shrills, whistles, and an array of unwanted sounds and noises, the radionic waves also similarly get affected by such agencies as the planetary positions, the influences, astral gems *mantras*, black magic, witchcraft, and such other less known practices, which are said to generate influential radionic vibrations.

One must bear in mind that these radionic vibrations are akin partly to electromagnetic waves in physical terms. They have their own paraphysical behaviours beyond the realm of physical electromagnetism. So also these waves are partly akin to physical sound waves in behaviour but yet they retain their paraphysical characters. This is why they are termed 'eloptic' in radiesthesia.

Dowsing could be employed to find out the condition of health, and the principle judiciously applied could be an aid to treatment. All that is required is a suspended mass of any material, to a band of strings held in between the right index finger and the thumb of the radiesthesist. Underneath the pendulum are placed what are known as data charts of probable answers to problems posed by the radiesthesist on behalf of the person whose photograph, signature, thumb impression, a specimen drop of his blood, urine, or a lock of hair, is either held in the left hand thumb and index finger or placed at a selected place on the reading base and at a distance from the data chart.

Sometimes the pendulum could even be retorted, instead of holding in the hand of the radiesthesist. A mere touch atop the retort would be enough to move the pendulum right on its job.

Neutral force: Placidity, benefic indifference, presence of mind and concentration will all go a long way to make perfect radiesthesist, who is said to be in short a 'neutral force', an imperative element in radiesthesia.

Radiesthestic map dowsing has been attempted by scientists in Soviet Russia during the 1940s and 50s, to detect mineral deposits.

QUIZ 8

I. Match the items of *A* and *B*:

1.

A	B
Depth sounding	Electrodes at equal distances
Seismic reflection	Electrical resistivity—variation with depth
Seismic refraction	Potential electrodes are very close
Wenner system	Low frequency a.c. type—quick
Schlumberger system	d.c. type—slow
NGRI resistivity meter	Depth to water table and bed rock
Terrameter	Large buried valleys, sink holes and caverns in limestones
Minimag	Granite areas with frequent dykes, nearly vertical
Gravity	Not used in ground water exploration
Master curves of Mooney and Wetzel	Radiesthesia
Dowsing	Interpretation of field resistivity data

2.

A	B
Electric log	Tritium
SP Log	Bore hole diameter
Gamma log	Identify foreign matter in a well

Neutron log   Electrical resistivity
Gamma-Gamma   Distinguish between clays and sandstones
Age of ground water   Porosity of formations
Caliper log   Investigation of thin beds
Borehole camera and television   Concentration of salts in formation water
Induction logging   Vertical flow in bore hole
Tracer   Speed and direction of ground water
Flow meter   Bulk density and porosity of rock types

II. Match the items in A, B and C:

A	B	C
Formation	*Resistivity*	*SP*
Fresh water gravel	low	negligible
Fresh water sand	moderate	low negative
Clay	high	moderate negative
Shale	very high	high negative
Dense formation		base line for SP
Saline water		
Brackish water		

III. State whether 'true' or 'false'; if false, give the correct statement:

1. The seismic refraction method can be used only where velocity of shock wave decreases with depth.
2. Due to intrusion of a dyke the magnetic profile decreases to nearly 3000–3400 gamma in the dyke area.
3. The gamma-gamma logging is the same as natural gamma logging.
4. Individual aquifer tests are conducted to know the exact quality and quantity of water from each aquifer and also the vertical leakage.
5. Since the shales offer least negative SP, the SP curve obtained is a straight line which is taken as the base line for measurement of SP.
6. Neutron log indicates casing and fluid in hole and in formation, which the gamma log does not.
7. Seismic reflection methods provide information on the deep seated strata (> 500 m) while the seismic refraction methods cover only a few hundred meters below the ground surface and hence the latter is used in ground water investigation.

IV. Select the correct answer(s):

1. An impeller flow meter measures
   a. velocity of flow of water in an aquifer
   b. volume rate of flow of water in an aquifer
   c. volume rate of flow of water from one aquifer to another
2. The total dissolved solids in formation water is given by
   $$\text{TDS in ppm} = \ldots\ldots\ldots \times \text{EC in micromhos/cm at } 25°C$$
   (1.2, 1.56, 0.64)

Ground Water Geophysics

PROBLEMS

I. The following data were obtained from a seismic refraction shooting:

Geophone No.	1	2	3	4	5	6	7	8	9
Distance from shot point, m	4	8	12	16	20	24	28	32	36
First arrival time, millisec	21	42	63	84	105	117	124	132	140

Estimate the depth to water table.

(Answer: 7.2 m)

II. Discuss the methodology of ground water investigation in a region, where adequate hydrologic data are not available, with the purpose of predicting the availability of ground water and its quantitative estimation.

III. The following readings were obtained from a Terrameter while conducting a resistivity depth probe by Wenner method in Thanjavur district, Tamil Nadu. Draw the resistivity curve and make interpretations for water well drilling.

Electrode spacing a (m)	Terrameter reading R (ohm)
10	0.13
15	0.12
20	0.11
25	0.105
30	0.095
35	0.0925
40	0.0915
45	0.0780
50	0.0755
60	0.0735
65	0.0720
70	0.0720
75	0.0725
80	0.0705
85	0.0700

IV. The following readings were obtained from an NGRI resistivity meter while conducting a resistivity depth probe by Schlumberger method. Draw the resistivity curve and make interpretations for water well drilling.

NGRI resistivity depth probe
Place: Dandipalya Village, Bangalore North district (South India)

Distance of inner electrodes from the centre of the spread $\frac{MN}{2}$ (m)	Distance of outer electrodes from the centre of the spread $\frac{AB}{2}$ (m)	Meter readings Voltage (mv)	Meter readings Current (ma)
0.15	1.5	46	25
	2.1	9	12
	3.0	3	9
	4.5	4	26
0.75	4.5	15	26
	6.0	3	9
	9.0	2	11
	12.0	1	14
	15.0	1	17
3.0	15.0	8	17
	21.0	3	11
	30.0	2	12
	45.0	1	16
7.5	45.0	3	16
	60.0	2	16
	90.0	1	17
15.0	90.0	2	17

A 15 cm well drilled at the site to a depth of 66 m gave an yield of 160 lpm between depths 37–65 m.

V. In a refraction shooting six geophones were placed along a straight line and the seismic record gave the following data:

Geophone	Distance from shot point m	Time of first arrival sec
G_1	100	0.15
G_2	200	0.10
G_3	300	0.15
G_4	400	0.19
G_5	600	0.23
G_6	800	0.28

Draw the time-distance graph and determine the velocities and thickness of layer(s) (2 layers, $V_1 = 1,943$ m/sec, $V_2 = 4,000$ m/sec, $Z_1 = 104.6$ m)

Ground Water Geophysics

VI. Following are the data obtained in a refraction shooting. Determine the depth to the water table:

Geophone No.	Distance from shot point (m)	First arrival time (milli sec.)
1	20	100
2	40	200
3	60	300
4	80	400
5	100	500
6	120	590
7	140	620
8	160	660
9	180	700

(Answer: 35.21 m)

VII. (a) Which is of utility in ground water investigation—Seismic reflection or refraction, and why?

(b) For ground water investigation of Karnataka which method you would prefer—Seismic or Electrical resistivity. Give reasons.

9

Geochemical Survey and Water Quality

The chemical composition of ground water is related to the soluble products of rock weathering and decomposition and changes with respect to time and space. Geochemical studies provide a complete knowledge of the water resources of a hydrological regimen. Sampling and testing in an area with some good quality and some poor quality water should serve to differentiate areas and aquifers of varying quality and on the results of this study recommendations can be made regarding different uses to which water in various areas and aquifers can be put.

Geochemical studies are also of value with respect to water use. They provide a better understanding of possible changes in quality as development progresses, which can in turn provide information about the limits of total development, or can permit planning for appropriate treatment that may be required as the result of future changes in the quality of water supply.

For analysis of chemical quality of water, tracing the movement of ground water is important. Induced tracers, including salt solutions, dyes such as fluorescein, and redioactive materials have been used as well as other techniques through the knowledge of hydrogeology.

Analysis of water samples for geochemical studies require a high degree of accuracy. The intensity of sampling should be gauged by the needs of the investigation and the severity of any water quality problems. The chemical characteristics of water are very important with respect to requirements for various uses.

To determine sea water intrusion, lines of piezometers are inserted at various depths at different distances from the sea coast. A 300 m cable may be used in a portable conductivity bridge or meter. A sudden increase in the BC (of the order of 50,000 μ mhos/cm) or chloride concentration (of the order of 19,000 ppm) at a particular depth, has sometimes indicated sea water intrusion. Fogged or connate water may also contribute to chloride concentration.

Geochemical Survey and Water Quality

Temperature measurements are usually made in ground water studies. These are particularly important in places where wide variations in the temperature are recorded. The depth of the source of ground water could be gauged from the temperature of the water (geothermal gradient $\approx$ 30 to 50 m/°C). Temperature results may lead to the discovery of an unsuspected source of pollution.

Water Sampling

Depth integrating samplers and point samplers are used in obtaining samples of water for estimating non-volatile constituents. The former consist only of a mechanism for holding and submerging the bottle. On lowering the bottle at a uniform rate, water enters throughout the vertical profile. Point samplers are used for collecting water at a specific depth below the water surface. In operation the bottle is worked and then thrown into the water with an excess of line. When the bottle comes to rest the suspension line is jerked, removing the cork. The bottle is allowed to fill in place and then stopped while being withdrawn from water. All these samplers are used with detachable weights which are necessary for submerging the bottle.

Special equipment and careful technique is necessary for the collection of dissolved gases and constituents affected by aeration.

Before a well water sample is taken, the well should be pumped for some time so that the sample will represent the ground water from which the well is fed. All bottles should be rinsed with water to be sampled before collecting the sample for analysis. If water sample is collected in glass bottles, sufficient air space may be provided, but if polythene bottles are used, they may be completely filled.

All particulars regarding the sample should be written in the field itself, immediately after sampling, and tagged to the sample bottle.

Polythene or glass bottles of 1 to 2 litres capacity are usually used for collection of samples. The sample number, date and other particulars are entered in the register in the laboratory.

Sample of water should be analysed as quickly as possible after collection. Special treatments may be given for preservation, fixation and handling of water samples before analysis. Otherwise the quality of water may change and many of the heavy metal ions normally present in small quantities in natural water may not remain in water till the sample is analysed. The sample should be freed of its sediment and acidified to about a pH of 3.5 with glacial acitic acid at the time of collection. A little formaldehyde (0.2 m/100 ml) is added to retard mold growth.

Seals of the bottle should be tightened before storage. High temperature should be avoided in the storage room.

Water quality

The water quality should satisfy the requirements or standards set for the specific use, namely, domestic, livestock, agricultural and industrial purposes.

The WHO drinking water standards 1963 are given in Table 9.1.

Table 9.1 World Health Organisation drinking water standards—International Standards 1963.

Characteristic	Limit of general acceptability (mg/l)	Allowable limit (mg/l)
Total solids	500	1500
Colour (°H)	5	50
Turbidity	5	25
Chloride	200	600
Iron	0.3	1
Manganese	0.1	0.5
Copper	1.0	1.5
Zinc	5	15
Calcium	75	200
Magnesium	50	150
Magnesium and sodium sulphate	500	1000
Nitrate (as NO_3)	45	—
Phenols	0.001	0.002
Synthetic detergents (ABS)	0.5	1.0
Carbon-chloroform extract	0.2	0.5
pH	7—8	min. 0.5 max. 9.2

The following gives the European drinking water standards 1970—

Toxic substances	Maximum concentrations (mg/l)
Lead	0.1
Arsenic	0.05
Selenium	0.01
Chromium	0.05
Cadmium	0.01
Cyanide	0.05

Geochemical Survey and Water Quality

The concentrations above which ill-effects may arise are:

Fluoride*	1.0-1.7 depends on climate (Fluorosis)
Nitrate	100 (Methaemoglobinaemia)
Iron	0.1
Chloride	200 (Taste)
Sulphate	250 (Gastrointestinal irritation)
Zinc	5 (Taste)

The upper limits of TDS in water for livestock (Department of Agriculture, Western Australia—1950) are as tabulated below.

Poultry	2860 mg/l
Pigs	4290 mg/l
Horses	6435 mg/l
Cattle (dairy)	7150 mg/l
Cattle (beef)	10000 mg/l
Adult sheep	12900 mg/l

Quality of Water for Industrial Uses

Recommended quality limits for water used for various industries as compiled from numerous sources are given in Table 9.2.

Rural Water Supply Requirement

Assuming a requirement of 50 to 70 litres per capita per day (lpcd) in the rural areas, discharge from a tubewell as 1000 lpm in alluvium and 250 lpm in hard rock areas, and 16 hours of pump operation per day, the rural population that can be served with water supply, at 70 lpcd, will be 14,000 and 3,500, respectively.

Requirement of Water for Livestock

Livestock	Water required (litres)
Per horse or mule	45
Per dairy cow (drinking only)	57
Per hog	15
Per sheep	7.5
Per 100 chickens	15
Per 100 turkeys	26.5

---

*Defluoridation of waters can be made by using Aluminium chloride and Aluminium Sulphate, alone or in combination [Bulusu, 1984]

Table 9.2 Quality of water for industrial uses—allowable limits in ppm.

Industry or use	Turbidity (Silica scale)	Colour (Standard Cobalt scale)	Odour and taste	Iron as Fe*	Manganese as Mn	Total solids	Hardness as CaCO₃	Alkalinity as CaCO₃	Hydrogen Sulphide	pH	Other requirements
Air conditioning											Cool, non-corrosive
Baking	10	10	Low	0.5	0.5				1.0		Moderately hard (calcium sulphate) water desirable
			Low	0.2	0.2				0.2		
Boiler Feed											
Pressure 0–10 kg/cm²	20	80				3000—500	80		5	8.0	Low dissolved oxygen and silica, non-corrosive
Pressure 10–17 kg/cm²	10	40				2500—500	40	Low	3	8.4	
Pressure 17–27 kg/cm²	5	5				1500—100	10		0	9.0	
Pressure >27 kg/cm²	1	2				50	2		0	9.6	
Brewing and distilling	10		Low	0.1	0.1	500—1000		75–150	0.2	6.5–7.0	NaCl 275
Canning	10		Low	0.2	0.2		25–75 for peas & other legumes		1.0		Pure water (potable)
Carbonated beverages	2	10	Low	0.2	0.2	850	250	50–100	0.2		Pure water (potable)
Confectionery			Low	0.2	0.2	100			0.2	7.0	Requirements differ for different candies
Cooling	50			0.5	0.5						cool, non-corrosive
Cotton bandages	Low		Low	0.2	0.2		50		5		
Food (general)	10		Low	0.2	0.2						Pure water (potable)
Ice	5	5	Low	0.2	0.2	1300					SiO₂ 10 (potable water)
Laundering				0.2	0.2		50				

Contd.

Geochemical Survey and Water Quality

Use								Remarks
Plastics (clear)	2	2	0.02	0.02	200			
Paper and pulp								
Ground wood	50	20	1.0	0.5		180		No grit or corrosiveness
Kraft pulp	25	15	0.2	0.1	300	100		
Soda and sulphite pulp	15	10	0.1	0.05	200	100		No slime formation
High-grade light papers	5	5	0.1	0.05	200	50		
Rayon (Viscose)								
Pulp production	5	5	0.05	0.03	100	8		OH 8, Al_2O_3 8, SiO_2 25, Cu 5
Starch manufacture	0.3		0.0	0.0		55	7.8–8.3	OH 8
Tanning	20	10–100	0.2	0.2		50–135	135	
Textiles	5	20	0.25	0.25	200			Constant composition; Al_2O_3 0.5

*As iron alone or iron and manganese

For a compound A_mB_n, ppm of ion $A = \dfrac{m\,(\text{atomic wt. } A)}{\text{molr. wt. } A_mB_n} \times \text{ppm } A_mB_n$

Ex. certain water has

35 ppm $CaCl_2$, $\quad$ ppm Cl $= \dfrac{2 \times 35.5}{111} \times 35 = 22.4$

Typical Industrial Water Demands

Industry	Water demand (m^3/ton product)
Baking	4
Chemicals	up to 1100
Coal-mining	5
Laundering	45
Milk-processing	4
Paper making	90
Steel production	45
Sugar refining	8
Synthetic fibres	140
Vegetable canning	10

Irrigation Water Quality

The quality of irrigation water is judged by the following characteristics.
 (a) Total dissolved solids (TDS)
 (b) Relative proportion of sodium to other cations
 (c) Concentration of certain specific elements
 (d) Residual carbonates

(a) *Total dissolved solids (TDS)*: If the salt concentration in the water increases, it is difficult for plants to extract water. Experiments have indicated that under osmotic pressure of 15 to 20 atm, plants wilt permanently. The relationship between osmotic pressure and concentration of a solution is given by

$$P = iRTC \tag{9.1}$$

where P = osmotic pressure, atm; i = Vonthoff factor; R = gas constant, litre-atm; T = absolute temperature and C = concentration, moles/l.

In very dilute solutions i may be identified with the number of ions per molecule. For example in 0.1% NaCl, $i = 2$, $C = 1$ gm/l = 1/58.5 moles/l (atomic weight of Na + Cl ions = 58.5); taking $RT = 22.4$ litre-atm,

$$P = \frac{2 \times 22.4}{58.5} = 0.766 \text{ atm.}$$

$$\text{In } 0.1\% \text{ Na}_2\text{SO}_4, \quad i = 3, \quad P = \frac{3 \times 22.4}{142} = 0.470 \text{ atm}$$

$$\text{In } 0.1\% \text{ CaCl}_2, \quad i = 3, \quad P = \frac{3 \times 22.4}{111} = 0.605 \text{ atm}$$

$$\text{In } 0.1\% \text{ CaSO}_4, \quad i = 2, \quad P = \frac{2 \times 22.4}{136} = 0.329 \text{ atm}$$

Thus the chlorides are more toxic than sulphates. The toxicity due to a given salt content increases with temperature.

Some important relations
1. TDS in ppm = $0.64 \times$ EC in μ mhos/cm
2. Osmotic pressure in atm = $0.00036 \times$ EC in μ mhos/cm
3. 1 mho/cm = 1,000 millimhos/cm = 10^6 μ mhos/cm
4. Concentration of ions expressed as

$$\text{Milliequivalent per litre, me/l} = \frac{\text{concentration of salt in mg/l}}{\text{equivalent weight}}$$

$$\text{equivalent per million, epm} = \frac{\text{concentration of salt in ppm}}{\text{equivalent weight}}$$

Since mg/l $\approx$ ppm, me/l = epm

5. The logarithm of the negative reciprocal of the hydrogen ion concentration is called pH

$$\text{pH} = -\log_{10}[\text{H}^+] \tag{9.2}$$

A solution with a pH < 7 is acidic, pH > 7 is alkaline and pH = 7 is neutral. For natural water pH = 6 to 8.

6. Most of the hardness of water is due to the presence of Ca^{++} and Mg^{++}. Total hardness (TH) is expressed as ppm of $CaCO_3$

$$\text{TH} = \text{Ca} \times \frac{CaCO_3}{Ca} + \text{Mg} \times \frac{CaCO_3}{Mg} \tag{9.3}$$

ppm ppm ratio of ppm ratio of
equivalent   equivalent
weights   weights

$$\text{TH} = 2.497\,\text{Ca} + 4.115\,\text{Mg} \tag{9.3a}$$

where all the constituents are expressed in ppm.
Total hardness in ppm = sum of epm of Ca and Mg $\times$ 50 (significant amounts of iron and manganese, if present, are included). (9.3b)

7. Noncarbonate hardness (NCH) in ppm = [sum of epm of Ca and Mg – sum of epm of CO_3 and HCO_3] $\times$ 50. When the difference is negative NCH = O

8. TDS in ppm = sum of the determined constituents in ppm except that the bicarbonate concentration should be multiplied by 30.10/61.02 = 0.49.

9. EC in μmhos/cm = total cations or anions in epm $\times$ 100

10. Per cent reacting value (PRV) of any constituent is the concentration of the constituent in epm expressed as a percentage of the total cations or anions in epm.

11. Salt index (SI) = total Na – 24.5 – 4.85 (total Ca – Ca in $CaCO_3$) where all the quantities are expressed in parts per 100,000 parts of water. Salt index is negative for waters suitable for irrigation and positive for those unsuitable.

12. The characteristics of some major constituents in water are given in Table 9.3.

Table 9.3 Values for major constituents.

Constituents	Atomic weight	Valency	Equivalent weight*
Cations		+	
Ca	40.08	2	20.04
Mg	24.32	2	12.16
Na	23.00	1	23.00
K	39.10	1	39.10
Anions		−	
CO_3	60.01	2	30.00
HCO_3	61.02	1	61.02
SO_4	96.06	2	48.03
Cl	35.46	1	35.46
NO_3	62.01	1	62.01
F (Fluoride)	19.00	1	19.00

* Equivalent weight $= \dfrac{\text{Atomic weight}}{\text{Valency}}$

13. When TH $\leqslant$ alkalinity, all hardness is carbonate hardness. When TH $>$ alkalinity, carbonate hardness = alkalinity, NCH = TH − alkalinity. The range of hardness recommended by USGS is given in Table 9.4.

Table 9.4 Recommended range of hardness.

Class	Range of hardness (mg/l) as $CaCO_3$	Remarks
Soft	0–55	Require little or no softening
Slightly hard	56–100	
Moderately hard	101–200	Require softening
Very hard	201–500	

(b) *Relative proportion of sodium to other cations*: While a high salt concentration in water leads to formation of a saline soil, a high sodium leads to development of an alkali soil. USDA has defined an alkali soil as having a pH of 8.5 or more with a Na-saturation of 15% or more. An alkali soil has an unfavourable structure, puddles easily and restricts the aeration. Further the high sodium saturation directly causes calcium deficiency. Irrigation water with a low sodium absorption ratio (SAR) is desirable. SAR is defined as

$$\text{SAR} = \frac{Na^+}{\sqrt{\frac{Ca^{++} + Mg^{++}}{2}}} \quad (9.4)$$

where the concentrations are expressed in me/l.
The sodium percentage is calculated as

$$Na\% = \frac{Na + K}{Ca + Mg + Na + K} \times 100 \quad (9.5)$$

where all ionic concentrations are expressed in me/l.
USDA has constructed a nomogram to read exchangeable sodium percentage (ESP).

(c) *Concentration of certain specific elements*: Elements such as Selenium, Molybdenum and Flourine are tolerated by plants but are toxic to animals that feed on them; elements such as Boron and Lithium are toxic to plants. Subsurface waters are richer in Boron than surface waters. Traces of Boron > 0.5 ppm are injurious to citrus, nuts and deciduous fruits; cereals and cotton are moderately tolerant to Boron while alfalfa, beets, asparagus and dates are quite tolerant (1 to 2 ppm). Boron is present in many soaps and thus may become a critical factor in the use of sewage in irrigation.

(d) *Residual carbonate (RC)*: When the sum of carbonates and bicarbonates is in excess of calcium and magnesium, there is almost complete precipitation of the latter.

$$RC = (CO_3^{--} + HCO_3^-) - (Ca^{++} + Mg^{++}) \quad (9.6)$$

where all the concentrations are expressed in me/l.
Tungabhadra water in Karnataka is said to be of excellent quality and has the following constituents:

TDS	100 ppm
ESP	2%
Boron	0.07 ppm
RC	0-0.2 me/l

Water Quality Plots

Graphical representations of the concentrations of different ions in a water sample have been developed from time to time. Such graphical patterns make understanding easier and quicker. Some of them are bar graphs, patterns drawn on radial coordinates, Stiff's pattern, cumulative percentage composition, Schoeller's logarithmic plotting and trilinear diagrams. These are described in the following.

Ionic concentration diagrams: The Collin's bar chart is used by USGS (Hem, 1960). Here the total concentration of anions or cations in epm is represented by the height of the vertical bars and the concentration of anions and cations by horizontal breaks, Fig. 9.1. Usually there are six subdivisions but more can be used. Cations are plotted on the left bar and anions on the right. Some times the total hardness ($CaCO_3$) is shown by a dark band on one side.

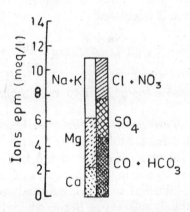

Fig. 9.1 Collin's bargraph.

Stiff's polygon of epm: Stiff's (1951) system uses four parallel horizontal axes and one vertical axis. Four cations are plotted on one side and four anions on the other side of the vertical axis. The vertices of the polygon are connected to give a shape characteristic of the type of water, Fig. 9.2.

Pie (sector) diagram: Here the total concentration is represented by the area of the circle which is divided into sectors to give the percentage composition, Fig. 9.3.

Radial Vectors: Maucha's (1940) systems of plotting uses radial vectors, Fig. 9.4, the length of each vector represents the epm of the constituent. This system is simple and convenient. Some systems indicate only the percentages of epm, Fig. 9.5.

Piper's trilinear diagram: In The Piper's (1953) diagram consists of two lower triangular fields and a central diamond-shaped field, Fig. 9.6a. All the three fields have scales reading in 100 parts. The percentage reacting values of the cations and the anions are plotted as a single point (according to the trilinear coordinates) at the lower left and right triangles,

Geochemical Survey and Water Quality

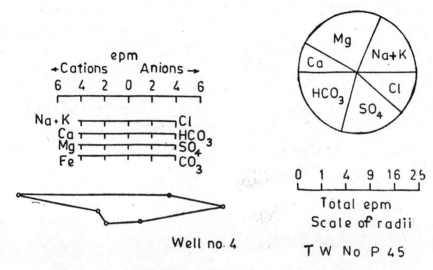

Fig. 9.2 Stiff's pattern.

Fig. 9.3 Pie (sector) diagram.

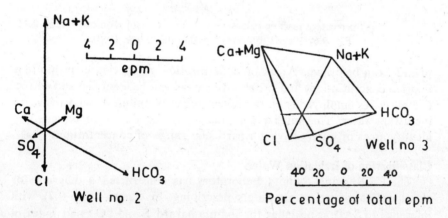

Fig. 9.4 Maucha's radiating vectors of epm.

Fig 9.5. Radial patterns.

respectively. These are projected upwards parallel to the sides of the triangles to give a point (P) in the rhombus. The point is represented by a circle whose area is proportional to the absolute concentration (actual ppm) of the water. The water quality types can be quickly identified by the location of P in the different zones of the diamond-shaped field as shown in Fig. 9.6b.

Water quality maps: Water quality of ground water bodies in a region can be studied by plotting the result of analysis of water samples on maps, by means of symbols, a shading, a bar graph, a small pie circle, etc. Lines of equal concentration of dissolved solids (e.g. chloride-isochlor), equal

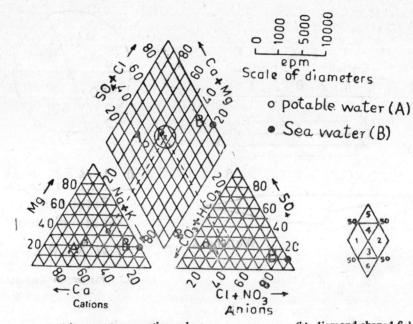

(a) percentage reacting values (b) diamond-shaped field
Fig. 9.6 Piper's trilinear diagram (after Piper, 1953).

pH, etc. can be drawn. Areas of concentration of specific elements (e.g. fluoride concentration) and problematic areas can be identified and corrective measures employed (say, to prevent tubewell failures) in corrosive or incrusting water. The water quality maps may be shaded or coloured to identify areas of waters having a particular range of concentration.

Classification of Irrigation Water

The US Regional Salinity Laboratory has constructed a diagram for classification of irrigation waters describing 16 classes (Fig. 9.7), with reference to SAR as an index for sodium hazard S and EC as an index of salinity hazard C. The quality classification of irrigation water is given in Table 9.5.

The chart shown in Fig. 9.8 has been developed by Doneen and is based on permeability index (PI) given by

$$PI = \frac{Na + \sqrt{HCO_3}}{Ca + Mg + Na} \times 100 \tag{9.7}$$

where all the ions are expressed in me/l.

In general water is good if:
(i) its position in the U.S. salinity diagram is within the zone of good or moderate waters
(ii) it belongs to class I or II in the Doneen's chart,

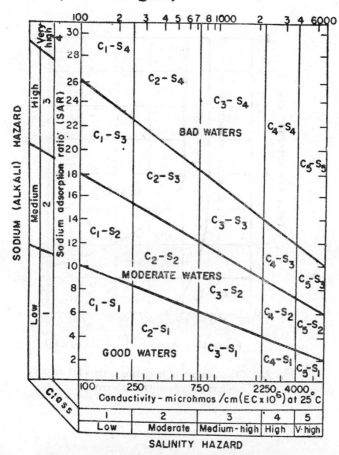

Fig. 9.7 Classification of irrigation waters (USDA).

Table 9.5 Quality classification of irrigation water*

Water Class	Salinity hazard		Alkali hazard SAR	RC in me/l
	EC in µ mhos/cm at 25°C	Salt concentration in me/l		
Excellent	<250	<0.25	upto 10	<1.25
Good	250–750	0.25–7.50	10–18	1.25–2.50
Medium	250–2250	7.50–22.50	18–26	>2.50 poor
Bad	2250–4000	22.50–40.00	>26	
Very bad	>4000	>40		

*IS 2296—1963: Cl <600, SO_4<1000 ppm
ESP<15, B<1 EC<250 µ mhos/cm

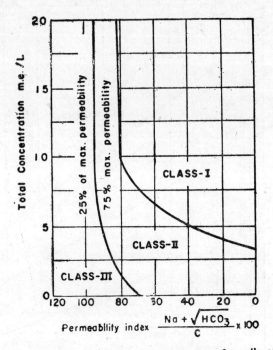

Fig. 9.8 Classification of irrigation waters for soils of medium permeability (Doneen)

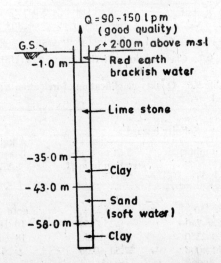

Fig. 9.9 Bore log-borewell 1 km from sea shore.

Geochemical Survey and Water Quality

(iii) TDS ≯ 1000 ppm. This limit can be increased to 1700 ppm if Ca forms 25% of the total bases (Na+Ca),

(iv) its salt index value is negative.

The greatest reliance is on the first two criteria. Waters other than these are generally (but not invariably) either unfit for irrigation or have restricted use depending upon the soil type, nature of crops, and drainage conditions etc.

Example 9.1: The chemical analysis of a water sample, obtained from a medium depth borewell, Fig. 9.9, in Thanjavur district, Tamil Nadu at a distance of 1 km from the sea shore, is given above. Report the results qualitatively.

Constituents	Ca	Ma	Na	HCO_3	CO_3	SO_4	Cl	TDS	TH as $CaCO_3$
Value (mg/l)	56	16	85	256	24	43	82	440	205

pH=7.5
EC at 25°C=700 mhos/cm

Solution:

	Cations				Anions		
Constituent	mg/l	eq. wt.	me/l	Constituent	mg/l	eq. wt.	me/l
Ca	56	20	2.75	HCO_3	256	61	4.20
Mg	16	12.2	1.32	CO_3	24	30	0.80
Na	85	23	3.70	Cl	82	35.5	2.31
K	—	39.1	—	SO_4	43	48	0.89
	Total		7.77				8.20

For the cations to balance the anions, their average value may be taken as 8 me/l.

(i) Total hardness

$$TH = (Ca + Mg) \, 50 = (2.75 + 1.32) \, 50$$
$$= 203.50 \text{ mg/l as } CaCO_3$$

which is very near the value determined experimentally.

(ii) Non-carbonate hardness

$$NCH = [(Ca + Mg) - (CO_3 + HCO_3)]\,50$$
$$= [(2.75 + 1.32) - (0.80 + 4.20)]\,50$$
$$= 0, \text{ since the value is negative}$$

(iii) $\quad TDS = 56 + 16 + 85 + 256\,(0.49) + 24 + 43 + 82$
$$= 431.5 \text{ mg/l}$$

which is very near the value determined experimentally.

(iv) Sodium percentage

$$Na\,\% = \frac{Na + K}{Ca + Mg + Na + K} \times 100$$

$$= \frac{3.70}{8.00} \times 100 \quad \text{(after balancing)} = 46.2\%$$

This is also the PRV of Na.

(v) $\quad EC = 8 \times 100 = 800\mu \text{ mhos/cm}$

which is near to the value experimentally determined.

(vi) $\quad SAR = \dfrac{Na}{\sqrt{\dfrac{Ca + Mg}{2}}} = \dfrac{3.70}{\sqrt{\dfrac{2.75 + 1.32}{2}}} = 2.58$

(vii) $\quad RC = (CO_3 + HCO_3) - (Ca + Mg)$
$$= (0.80 + 4.20) - (2.75 + 1.32) = 0.93$$

(viii) Permeability Index

$$PI = \frac{Na + \sqrt{HCO_3}}{Ca + Mg + Na} \times 100$$

$$= \frac{3.70 + \sqrt{4.20}}{2.75 + 1.32 + 3.70} \times 100 = 74\%.$$

From the US salinity diagram, Fig. 9.7, the water is classified as C_3–S_1, i.e. medium-high salinity and low sodium, which is good for irrigation. Even from Doneen's classification chart, Fig. 9.8, the water is classified as class-I, which is good for irrigation. Also from WHO Drinking Water Standards, Table 9.1, the water is good for drinking. In general, the water is of good quality even though it is very near the sea coast.

Crops should be selected on the basis of their salt tolerance, and the salt content of the irrigation water and the soil. Salt tolerance of some crops are given in Table 9.6.

Recommendations for the use of water must take into account soil types, crops grown, drainage and management practices. Good water

Table 9.6 Salt tolerance of crops

High tolerant EC=8,000—16,000 μmhos/cm Boron=2—4 ppm	Semi-tolerant EC=4,000—8,000 μmhos/cm Boron=1—2 ppm	Salt-sensitive EC=2,000—4,000 μmhos/cm Boron=0.3—1 ppm
Barley	Rye	Field beans
Beets	Wheat	Ladino clover
Cotton	Sorghum (*jowar*)	Radish
Bermuda grass	Flax	Celery
Date	Alfalfa	Green beans
Asparagus	Tomato	Orange
Spinach	Lettuce	Grape fruit
Turnip	Carrots	Lemon
Sugar beet	Grapes	Plum
Sugarcane	Fig	Peach
Tabacco	Olive	Apricot
	Potato	Pear
	Sunflower	Cherry
	Rice	Walnut
	Maize	Peas
	Cabbage	Grams
	Onion	
	Banana	

management and proper use of amendments may make it possible to use successfully some of the poor waters for irrigation.

Saline and Alkaline Soils

Some of the characteristics of saline and alkaline soils are given in the following.

Saline soils: Conductivity of the saturation extract $>4m$ mhos/cm at 25°C, ESP<15, pH<8.5.

White alkali: White crusts of salts on the surface.

Saline-alkali soils: Conductivity of the saturation extract $>$ 4 m mhos/cm at 25°C, ESP* $>$ 15, pH $\not>$ 8.5, the particles remain flocculated and sometimes contain gypsum.

Alkali soil: ESP $>$ 15, conductivity of the saturation extract $<$ 4 m mhos/cm at 25°C, pH = 8.5 to 10.

Black alkali soils: Slick spots. Dispersed and dissolved organic matter present in the soil solution of highly alkaline soils may be deposited on the soil surface by evaporation.

The main causes leading to the development of saline and alkaline soils are

(i) Arid climate
(ii) High subsoil water table
(iii) Poor drainage
(iv) Irrigation with water containing soluble salts
(v) Inundation of sea water
(vi) Saline nature of the parent materials.

The phenomenon of salts coming up in solution due to a rise in ground water table and forming a thin crust on the surface after evaporation of water is called salt efflorescence. The following measures are adopted to reclaim the salt-affected land.

1. Artificial drainage to lower the water table by 1 to 1.25 m below surface by open or tile drains.

2. Better irrigation practices and lining of canals to prevent seepage.

3. By leaching, i.e. by applying 5-30% more water depending on salt content of water and type of soil to leach out the salts.

$$\text{Leaching requirement LR} = \frac{\text{EC of irrigation water}}{\text{EC of drainage water}} = \frac{\text{Depth of } i_w}{\text{Depth of } d_w}$$

$$\text{LR} = \frac{\text{EC}_{iw}}{\text{EC}_{dw}} = \frac{D_{dw}}{D_{iw}} \tag{9.8}$$

For crops which can tolerate an EC_{dw} of 8000 micromhos/cm and if EC_{iw} = 2000 micromhos/cm, LR = 2000/8000 = 25%, which is a maximum value. Basins are made and water is filled in.

4. Growing salt tolerant crops like rice, sweet clover and bermuda grass. They also cover the land surface and reduce evaporation.

5. Use of soil amendments: Gypsum ($CaSO_4$) as a source of soluble Ca is widely used for reclamation of black alkali soils. When gypsum is not available, sulphur is used. Sulphur is oxidised to sulphuric acid and

---

*Exchangeable sodium percentage, ESP

$$= \frac{\text{Exchangeable Na, me/100 gm soil or me/l}}{\text{Total exchangeable cations, me/100 gm soil or me/l}}$$

reacts with $CaCO_3$ to form gypsum. Waste lime from beet sugar refineries is sometimes used. Other amendments are Al_2SO_1 and $FeSO_4$. After the chemical amendments have converted the Na-soil into Ca-soil, the soluble products of the reaction, like sodium bicarbonate and sodium sulphate, have to be removed by leaching through flooding.

$$CaCO_3 + H_2SO_4 = CaSO_4 + H_2O + CO_2$$
$$2(Na\text{-soil}) + CaSO_4 = Ca\text{-soil} + Na_2SO_4 \text{ (soluble)}$$
$$\uparrow$$
$$\text{leached out by flooding}$$

$$CO_2 + H_2O = H_2CO_3$$
$$CaCO_3 + H_2CO_3 = Ca(HCO_3)_2$$
$$2(Na\text{-soil}) + Ca(HCO_3)_2 = Ca\text{-soil} + 2Na\,HCO_3 \text{ (Soluble)}$$
$$\uparrow$$
$$\text{leached out by flooding}$$

The quantity of amendment required depends on the soil depth of reclamation and the amount of exchangeable sodium initially present. While the salt content of 1:5 saturation extract is only 0.2 to 0.4%, the ex-Na may account for more than 40% of the exchangeable cations.

6. Applications of manure and organic matter improves tilth and permeability. Decomposition of organic matter liberates CO_2 which dissolves in water forming carbonic acid which increases solubility of $CaCO_3$ in soil.

7. Salt balance can be kept at the desired level if there is a tile drainage. Additions of salt from irrigation water and losses from tile flow are measured. The balance is on the soil.

Example 9.2: A soil has a total exchangeable capacity of 40 me/100 gm and exchangeable sodium of 15 me%. Determine the quantity of gypsum needed to reclaim 1 ha of such a soil to a depth of 0.5 m so as to reduce ex-Na to 2 me% (ESP = 5%).

Solution: $15 - 2 = 13$ me of ex-Na will have to be replaced from 100 gm of soil. Assuming quantitative replacement, 13 me of gypsum ($CaSO_4$, $2H_2O$) are needed; i.e. $13\,(20 + 48 + 4 + 16) = 1144$ mg or 1.144 gm of gypsum per 100 gm of soil. To reclaim 1 ha of soil to a depth of 0.5 m, theoretical quantity of gypsum needed

$$= \frac{1.144}{100} \frac{(0.5 \times 10^4)(1.8 \times 1,000)}{1,000} = 103 \text{ tons}$$

Criteria for Incrustation and Corrosion

Incrustations formed due to precipitation of bicarbonates of calcium and magnesium (decomposed to form insoluble carbonates and release of carbon dioxide) are soft and can be easily removed by acids or other chemicals. Hard incrustations are formed due to sulphates and silicates of calcium and magnesium which being insoluble in acids or other chemicals, cannot be removed.

The following general criteria have been adopted:

(i) Waters carrying > 400 ppm of bicarbonates may cause soft type incrustations.

(ii) Waters carrying > 100 ppm of sulphates or 40 ppm of silicon may cause hard type incrustations.

(iii) Waters with pH < 7 are corrosive.

(iv) Waters with EC $> 1,500$ μ mhos/cm may cause corrosion of iron and steel.

(v) Presence of chlorides > 500 ppm may indicate corrosive property of water.

(vi) Presence of > 2 ppm of iron or 1 ppm of manganese may cause precipitation of hydroxides and oxides of iron and manganese (incrustations).

Sea Water Contamination in Ground Water

Figure 9.10 a, b, c indicate a typical sample of a normally good ground water, slightly contaminated ground water and sea water respectively. Since

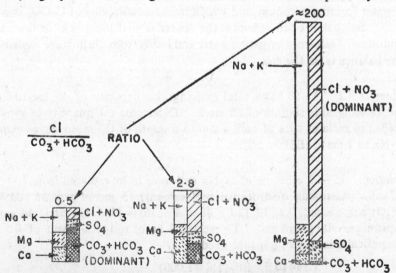

Fig. 9.10 Sea water contamination.

a temporary increase of TDS may lead to misconception of salt water contamination, Revelle (1941) recommended the chloride-bicarbonate ratio to identify sea water intrusion. Chloride is the dominant ion in sea water and it is only available in small quantities in ground water while bicarbonate which is available in large quantities in ground water occurs only in very small quantities in sea water. The chemical analysis of a sea water sample from the Bay of Bengal, Fig. 9.10c, gave the following results:

Constituent	Ca	Mg	Na	K	HCO_3	Cl	SO_4	EC
Concentration in ppm	600	1,190	11,150	480	140	19,740	3,120	50,000 (μ mhos/cm)

It can be seen that while the chloride-bicarbonate ratio $= \dfrac{19,740/35.46}{140/61.02} = 243$, the bicarbonate $HCO_3 = 140/61.02 = 2.3$ me/l. Sea water intrusion can be detected by plotting a graph of chloride-bicarbonate ratio against distance from the sea coast, Fig. 9.11.

The results of chemical analysis on a water samples are entered in a standard form as shown in the following.

Water quality determination—chemical analysis

Water sample:	A	B
Bottle no.	A	B
Field no.	A	B
Date of collection:	A	B
Date of receipt:		
Lab. ref. no.		
Location:		
Source:		
Owner:		
Water-bearing formation	from	to
Temperature:	°C	
Water use:	Domestic/Irrigation/Agriculture/Industry	
Appearance:	Clear/coloured/turbid/sediments, etc.	
Remarks:		

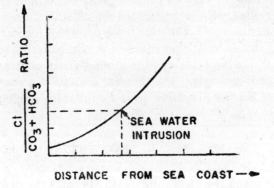

Fig. 9.11 $\dfrac{Cl}{(CO_3+HCO_3)}$ Ratio versus distance from sea coast.

Chemical analysis (laboratory)

Sample: Date: Constituent	A			B		
	mg/l	me/l	me %	mg/l	me/l	me %
Ca						
Mg						
Na						
K						
Si						
Fe						
Mn						
B						
CO_3						
HCO_3						
SO_4						
Cl						
NO_3						
F						
Turbidity						
Colour						
Carbondioxide						
Hydrogen sulphide						

Geochemical Survey and Water Quality

	A	B
TH (as $CaCO_3$) mg/l		
EC micromhos/cm		
TDS mg/l		
pH		
SAR		
Na %		
PI (Doneen)		

	A	B
Classification—USDA		
—Doneen		
—WHO Drinking Water Standards		
Date:		Chemist

Example 9.3: What is the specific resistance in ohms of a water sample having a conductivity of 4000 units.

Solution: $EC = 4000$ μ mhos/cm $= 4000/10^6 = 0.004$ mhos/cm

Specific resistance $R = \dfrac{1}{\rho} = \dfrac{1}{0.004} = $ **250 ohms/cm**

Example 9.4: In a titration test, 4.5 cc of silver nitrate solution was used for a water sample. If 1 cc of the silver nitrate solution is equivalent to 0.001 gm of the chloride ion and if 100 cc of water was used, what is the chloride content of water in mg/l?

Solution: For determination of chloride ion in the water sample, titrate with silver nitrate solution.

4.5 cc of $AgNO_3$ is equivalent to $0.001 \times 4.5 = 0.0045$ gm of Cl

Chloride content per litre $= 0.0045 \times \dfrac{1000}{100} = 0.045$ gm, or **45 mg/l**

QUIZ 9

I. Match the items in A and B:

A	B
i. $EC \approx 50{,}000$ μ mhos/cm $Cl \approx 19{,}000$ ppm	a. Permeability index, PI (Doneen)
ii. ≈ 50 m/°C	b. Leaching requirement, LR

iii. TDS < 1,500, pH 6.5-8
iv. 50-70 lpcd
v. $\log \dfrac{1}{H^+}$
vi. pH<7, Cl>500 ppm, CO_2>50 ppm, EC>1,500 micro-mhos/cm, TDS>1,000 ppm, H_2S 1-2 ppm, DO
vii. EC up to 750, SAR up to 18
viii. $\dfrac{Na+\sqrt{HCO_3}}{Ca+Mg+Na} \times 100$ all ions in me/l
ix. $\dfrac{EC-\text{irrigation water}}{EC-\text{drainage water}} \times 100$
x. Fe>2 ppm, Mn>1 ppm TH>300 ppm, pH>8
xi. Chloride-bicarbonate ratio versus distance from sea coast
xii. Orthotolidine test
xiii. Fluoride concentration of 19 ppm in Sagalia village in Sirohi district of Rajasthan

c. Incrustation
d. Plot to detect sea water intrusion
e. Available chlorine in water
f. People crippled with bent backs, deformed limbs and mottled teeth

g. Drinking water
h. Rural water supply requirement

i. pH

j. Corrosive water

k. Geothermal gradient
l. Sea water intrusion
m. Good quality irrigation water

II. State whether 'true' or 'false':
(a) Colour discs are used in the Jackson candle turbidimeter.
(b) Alkacid pH test papers are used to estimate important dissolved metal ions in water, quickly.
(c) Before sampling ground water for chemical analysis from a well, the well should be pumped for some time.
(d) The turbidity in ground water sample can be quickly estimated by a flame photometer.
(e) Point samplers are used for collecting water at a specific depth below the water surface.
(f) If water samples are collected in glass bottles, sufficient air space may be provided but if polythene bottles are used they may be completely filled.

($\times$, $\times$, $\checkmark$, $\times$, $\checkmark$, $\checkmark$)

III. Choose the correct ions given in brackets:
...... is the dominant ion in sea water while ...... is the dominant ion in fresh ground water.
(CO_3, HCO_3, SO_4, Cl, NO_3)

QUESTIONS

I. (a) How do you determine the EC of a ground water sample?
(b) If a water sample has a TDS of 1152 mg/l and EC of 1800 units, calculate the specific resistance in ohms of another water sample which has a TDS of 6,400 mg/l.
Hint: $K \times EC = TDS$, where K is a constant

(R = 100 ohms/cm)

Geochemical Survey and Water Quality

II. Define pH. How do you determine the pH value of a water sample in the laboratory?
What do you infer if a water sample has a pH value of 8.5?
How will the pH value of this water change
(a) if ten drops of a strong acid are added
(b) if 5 cc of distilled water is added
(acidic — pH<7, pH=8.5)

III. What is the procedure for determining TDS in a water sample?
Determine the TDS in mg/l of a ground water sample from the following data:

Weight of evaporating dish	54.5505 g
Volume of water evaporated	250 cc
Weight of dish and evaporated dry material	54.7565 g

(824 mg/l)

IV. The chemical analysis of ground water from a well in Adyar, Madras is given in the following. Classify the water with respect to salinity and alkali hazards and its suitablity for irrigation use,

Constituent	SiO_2	Ca	Mg	Na	HCO_3	SO_4	Cl	TDS
Concentration in mg/l	60	200	43	495	300	366	812	1700

Total hardness as $CaCO_3$=670 mg/l, pH=7.8.
Also plot the bar and Stiff diagrams for the water Sample.

(C_4—S_2, fair for irrigation)

V. The chemical analysis of a ground water sample gave the following results. Report the results qualitatively.
Cations: Ca=Mg=Na=50 ppm
Anions: HCO_3=138 ppm, SO_4=Cl=NO_3=100 ppm

(C_3S_1, good water)

10

Water Well Design

A water well has to be designed to get the optimum quantity of water economically from a given geological formation. The water requirements for the particular scheme—rural water supply, agricultural or industrial needs, has to be carefully determined. The choice of open wells or bore-wells (tube-wells) and the method of well design depends upon topography, geological conditions of the underlying strata, depth of ground water table, rainfall, climate and the quantity of water required. A water well design involves selection of proper dimensions like the diameter of the well and that of the casing, length and location of the screen including slot size, shape and percentage open area, whether the well has to be naturally developed or a gravel pack is necessary, design of the gravel pack, selection of screen material, etc. Screened wells in unconsolidated formations involve consideration of more design details when compared to wells in consolidated rock formations. Good water well design aims to ensure an optimum combination of performance and long service life at reasonable cost. For instance it is not economical to design a well to yield 2,000 lpm to serve a suburban home requiring 70 lpm. On the other hand, the use of correct sizes of well casing and well screen, choice of materials of good quality, and strength and proper development of the well, will reduce long term power costs due to higher rates of pumping and maintenance costs and increase the useful life of the well.

Well Diameter

The size of the well should be properly chosen since it significantly affects the cost of well construction. It must be large enough to accommodate the pump that is expected to be required for the head and discharge (yield) with proper clearance (of at least 5 cm around the maximum diameter of the bowl assembly) for installation and efficient operation. Also the diameter must be chosen to give the desired percentage of open area in the screen (15 to 18%) so that the entrance velocities near the screen do not exceed

Water Well Design

3 to 6 cm/sec so as to reduce the well losses and hence the drawdown, to exclude the finest particles of sand from migrating near the slots and prevent incrustation and corrosion at the strainer slots. In deep wells which have both high static and pumping water levels, the well diameter can be reduced below the level of the lowest anticipated pump setting during dry weather, particularly in artesian aquifers where the artesian head is relatively high.

It can be seen from the Dupuit's equation [Eq. (5.7)] for steady state flow conditions (constant drawdown), that the yield of the well is

$$Q \propto \frac{1}{\log_{10} \frac{R}{r_w}}$$

where R is the radius of influence and r_w is the radius of the well. For $R=300$ m, a 60 cm well will yield only 25% more than a 15 cm well and 12% more than 30 cm well, which shows that drilling a large diameter well will not necessarily mean proportionately large yields. Recommended well diameters for various yields are given in Table 10.1.

Table 10.1 Recommended well diameters

Anticipated well yield lpm	Nominal size of pump bowl* cm	Size of well casing (ID)	
		minimum, cm	optimum, cm
400	10	12.5	15
400–600	12.5	15	20
600–1,400	15	20	25
1,400–2,200	20	25	30
2,200–3,000	25	30	35
3,000–4,500	30	35	40
4,500–6,000	35	40	50
6,000–10,000	40	50	60

*Allow if possible 10 casing diameters between the top of the screen and the bowl setting.

Well Depth

The depth of a well and the number of aquifers it has to penetrate is usually determined from the lithological log of the area and confirmed from electrical resistivity and drilling time logs. An experienced driller can decide the depth at which drilling can be stopped after being advised by the hydrogeologist who analyses the samples collected during the drilling. The well is usually drilled up to the bottom of the aquifer so that the full aquifer thickness is available, permitting greater well yield. The poor quality aquifers are backfilled or sealed so that this water will not migrate upward when the well is pumped.

Design of Well Screen

The design of the well screen consists of the length of the screen, its location, percentage open area, size and shape of the slots and selections of the screen material.

Screen length: The optimum length of the well screen is chosen in relation to the aquifer thickness, available drawdown and stratification of the aquifer. In homogeneous artesian aquifer about 70 to 80% of the aquifer thickness is screened. The screen should best be positioned at equal distance between the top and bottom of the aquifer. In the case of non-homogeneous artesian aquifer, it is best to screen the most permeable strata. In the case of homogeneous water table aquifer, the well screen is positioned on the bottom portion of the aquifer, since the upper part is necessarily unwatered to form a hydraulic gradient for flow into the well. Selection of screen length is something of a compromise between two factors—a higher specific capacity can be obtained by using as long a screen as possible, while more available drawdown results by using as short a screen as possible. Theory and experience have shown that screening the bottom one-third of the aquifer provides the optimum design. The principles of design in a non-homogeneous water table aquifer are the same as in the case of non-homogeneous artesian aquifer.

Slot size: The size of the slots depends upon the gradation and size of the formation material, so that there is no migration of fines near the slots and all the fines around the screen are washed out to improve permeability. In the case of naturally developed wells the slot size is taken as 40 to 70% of the size of the formation material. If the slot size selected on this basis becomes smaller than 0.75 mm, then it calls for an artificial gravel pack. Artificial gravel pack is required when the aquifer material is **homogeneous** with a uniformity coefficient less than 3 and effective grain size less than 0.25 mm. The pack-aquifer ratio, i.e. the ratio of the 30 or 50% size of the gravel pack material to the 30 or 50% size of the formation material, is kept at 4:1 if the formation is fine and uniform, and 6:1 if the formation is coarse and non-uniform. The gravel-pack material should have a uniformity coefficient less than 2.5. The design procedure of selecting the gravel material is to determine the point D_{30} of the gravel pack which is equal to 4 to 6 times the D_{30} of the aquifer material obtained from the mechanical analysis of the aquifer material, and then drawing a smooth curve through this point (corresponding to D_{30} of the gravel pack) representing a material with a uniformity coefficient of 2.5 or less. This is the gradation of the gravel pack to be used. The slot size of the strainer (well screen) is kept at 10% size (D_{10}) of the gravel pack material, to avoid segregation of fine particles near the strainer openings. The width of slots ranges from 1.5 to 4 mm and the length 5 to 12.5 cm. A ratio of 5 of the 50% sizes of the gravel pack and aquifer material has been successfully used in water wells. The thickness of the gravel pack should be between 10 to 20 cm. The gravel pack material should be clean, rounded, smooth and uniform, consisting mostly of siliceous rather than calcareous material for which the allowable limit is usually up to 5%. Particles of shale, anhydrite and gypsum are also undesirable in the pack material. The maximum grain size of the pack material should be less than 10 mm. Usually the size of the pea gravel varies from 4 to 8 mm.

Screen diameter: After the length of the screen (depending upon the aquifer thickness) and the slot size (based on the size and gradation of the aquifer material) have been selected, the screen diameter is determined so that the entrance velocities near the well screen will not exceed 3 to 6 cm/sec to prevent incrustation and corrosion and to minimise friction losses. The entrance velocity is calculated by dividing the expected yield of the well by the total area of openings in the length of the screen chosen.

Selection of screen: The mineral content of the water, presence of bacterial slimes and strength requirements are some of the factors which govern the selection of the screen material. The screen material should be resistant to incrustation and corrosion and should have strength to withstand the column load and collapse pressure. The principal indicators of corrosive ground water are (Anon, 1966) low pH, presence of dissolved oxygen, $CO_2 > 50$ ppm, $Cl > 500$ ppm. The principal indicators of incrusting ground water are total hardness > 330 ppm, total alkalinity > 300 ppm, iron content > 2 ppm, and pH > 8. Slime producing bacteria are often removed with chlorine treatment. This is followed by acid treatment to redissolve the pricipitated iron and manganese. The selection of the screen material also depends on the quality of ground water, diameter and depth of the well and the type of strata encountered. Some of the commonly used types of screens are shown in Fig. 10.1.

The continuous-slot type of well screen is made by winding cold-drawn wire, approximately triangular in cross-section, spirally around a circular array of longitudinal rods. The V-shaped openings facilitate the fine particles to move into the well during development without clogging them. This type has the maximum percentage of open area per unit length of screen, and the slot openings can be varied by adjusting the spacing of the wires wrapped. These screens are being made of metal such as GI, steel, stainless steel and various types of brass.

The Louver-type of screen has openings in the form of shutters. There is a tendency of the openings being blocked by the fine particles during development. This type of screen is, therefore, best used in artificially gravel-packed wells.

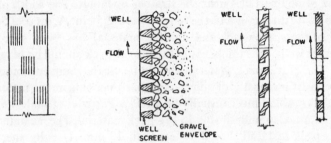

a. Slotted screen—vertical slots
b. V-shape continuous slot (no bridigng of soil grains)
c. Louver-type screen
d. Rectangular slots as in slotted pipe

Fig. 10.1 Well screens.

The rectangular slot or the slotted pipe screen is produced by cutting slots, vertical or horizontal, with a sharp saw, oxyacetylene torch, or by punching with a chisel and die or, a casing perforator. Some of the limitations of the slotted type well screens are wide spacing from the strength point, resulting in a low percentage of open area, lack of continuity and uniform size of the openings; the slots or perforations made in the steel pipe may be more readily subject to corrosion at the jagged edges and surfaces, and the chances of blockage of such openings are high. This type of screen is least expensive. Slotted PVC pipes are finding increasing use since they are light and easy to handle and are not subjected to corrosion. The use of slotted PVC pipes is generally limited to small diameter wells because of their relatively low strength and difficulty in providing proper fittings.

The pipe-base well screen or metallic filter point is made by using a perforated steel pipe. A wire mesh is wrapped around the perforated pipe and is in turn is covered by a brass perforated sheet. The percentage of open area in this type is usually low and the perforations are blocked by incrustation. This type of screen is relatively inefficient.

In the Cauvery delta the coir-rope screen is sometimes employed as an inexpensive substitute for other types of screens. Coir rope is wrapped tightly around a circular array of steel flat or rods. The life of the coir ranges from 7 to 8 years and can be increased by treating the coir with cashew shell oil. Hand boring sets are used for constructing coir rope screen wells. Coir rope screen is lowered into the casing pipe and the outside casing. Coir rope screen does not need gravel packing and development, but at the same time gives very good supply. Coir screen is used in shallow wells where the depth generally does not exceed 12 to 15 meters.

The best type of opening is the V-shaped slot that widens towards the inside of the screen, i.e. openings bevelled inside. Regarding the choice of the screen material, steel has good strength but it is not corrosion resistant. Brass has fair to good resistance to corrosion but has only half

Water Well Design

or less the strength of steel. However, for most situations, the strength of a well-made brass screen is adequate. Stainless steel has excellent strength and is highly resistant to most corrosive conditions. Well screens of corrosion resistant alloys such as Everdur metal, type 304 stainless steel and silicon red brass should be used in all except temporary installations. Metals used in fabricating screens and their resistance to corrosion are given in Table 10.2.

Table 10.2 Corrosion resisting metals

Name of metal	Analysis	Cost factor	Colour of finish	Suitability
Monel metal	70% nickel 30% copper	1.5	Bluish silver	High NaCl and DO as in sea water, not usually used for drinking water.
Super nickel	70% copper 30% nickel	1.2	Bright nickel	—do—
Everdur metal Silicon-bronze	96% copper 3% silicon 1% manganese	1.0	Rich copper red	High TH, NaCl (with no DO), Fe; usually used for municipal and industrial production wells. Highly resistant to acid treatment.
Stainless steel	74% l.c. steel 18% chromium 8% nickel	1.0	Dark silvery steel	Water containing H_2S, DO, CO_2. Fe, bacteria; in municipal and industrial production wells.
Cupro nickel	70% opper 29% nickel 1% arsenic		Bright nickel	
Silicon red brass	83% copper 16% zinc 1% silicon			Resistant to acid and corrosion

Open Wells Versus Borewells

In choosing the type of well the following factors have to be considered:

(i) Availability of space.
(ii) Hydrogeologieal characteristics of the subsurface strata.
(iii) Seasonal fluctuation of water levels.

(iv) Cost of well construction including provision of water lifting appliances.
(v) Economics and ease of water lifting operation.

Some of the advantages and disadvantages of the open wells and tube-(bore)wells are given in the following for making a choice of a particular type for a given situation.

Open wells—Advantages

(i) Storage capacity of water is available in the well itself.
(ii) Does not require sophisticated equipment and skilled personnel for construction.
(iii) Can be easily operated by installing a centrifugal pump at different settings for low and high water levels.
(iv) Can be revitalised by deepening by blasting or by putting a few vertical bores at the bottom, or horizontal or inclined bores on the sides to intercept the water bearing fractures.

Open wells—Disadvantages

(i) Large space is required for the well and for the excavated material lying on the surface like a big mound.
(ii) Construction is slow and laborious.
(iii) Subject to high fluctuations of water table during different seasons.
(iv) Susceptibility to dry up in years of drought.
(v) High cost of construction as the depth increases in hard rock areas.
(vi) Deep seated aquifers cannot be economically tapped.
(vii) Uncertainty of tapping water of good quality.
(viii) Susceptibility for contamination or pollution unless sealed from surface water ingress.

For the same area of cross-section, the perimeter of the well and so the area of the exposed surface (for the seepage of water into the well from fractures, fishures and cracks), is the least in a circular well, larger in a square well and larger still in a rectangular well. Hence a rectangular well is preferred in hard rock areas but in unconsolidated formations, a circular well will be the most economical as the cost of construction of the well and its steining will be the minimum as compared to the other shapes wherein the remaining wells of larger cross-section will involve a high cost.

A well sunk in loose and unconsolidated formations requires steining or retaining wall. In soft soils where problems of caving occur, wells are

Water Well Design

constructed by sinking precast RCC rings or by constructing at site a circular RCC curb and raising brick masonry in cement mortar over the curb, with alternate bands laid dry (i.e. without putting cement mortar for joints), for the seepage of water. When the RCC curb sinks due to the weight of the brick masonry constructed above it. The soft soil inside is scooped out and the water inside is bailed out either manually or by a centrifugal pump. Additional heights of brick masonry are constructed as the well sinks, gradually reducing the thickness at certain heights, as per design requirements against earth pressure, and the well is sunk to the required depth. The brick masonry above the high water table may completely be plastered inside with cement mortar, Fig. 5.40.

Thickness of well steining for open (dug) wells 2.4 to 6.0 m dia. is given in Table 10.3.

Table 10.3 Thickness of well steining for open wells

Depth below ground level (m)	Thickness of steining	
	Brick masonry (cm)	Stone masonry (cm)
up to 3	33	30
3–10	46	38
10–12	53	46
12–14	61	54

When the strata is stable or hard, the well can be excavated without having to construct steining. Wells have to be sealed by backfill in order to keep out contaminated surface water. Where side springs are encountered week holes should be provided in the steining or retaining wall with a back fill of coarse gravel or pebbles.

The depth to water table mainly depends upon the topography, geology and rainfall of the area. The water level in the adjoining wells will be a good indicator. It is advisable to excavate or sink a well to the bottom of the water-yielding strata or to the extent of the weathered portion of the rock with fissures and cracks exposed in order to tap most of the springs.

Tube (Bore) Wells—Advantages

(i) Do not require much space.
(ii) Can be constructed quickly.
(iii) Fairly sustained yield of water can be obtained even in years of scanty rainfall.
(iv) Economical when deep-seated aquifers are encountered.

(v) Flowing artesian wells can sometimes be struck.
(vi) Generally good quality of water is tapped.

Tube (Bore) Wells—Disadvantages
(i) Requires costly and complicated drilling equipment and machinery.
(ii) Requires skilled workers and great care to drill and complete the tube wells.
(iii) Installation of costly turbine or submersible pumps is required.
(iv) Possibility of missing the fractures, fissures and joints in hard rock areas resulting in many dry holes.

Selection of site should be made by careful observation of the outcrops in the area, geologic interpretation of the trial bores, yields of the existing borewells in the area, hydrologic information combined with pumping and recuperation tests, supplemented by an electrical survey of the area.

After conducting a hydrogeological study of the area and ascertaining the purpose and the quantity of water required, the type of well best suited can be determined. Wells located at the lowest levels in valleys generally have greater chances of striking water and also yield larger amount of water than those on slopes or ridges. Similarly, wells located nearer to rivers and streams or canal banks, or within the influence of water bodies like tanks and reservoirs will have assured supply

Regarding the quality of water, generally, the ground water from igneous rocks will be acidic in nature and low in mineral contents. The water will be hard and brackish in basalts and shales and in the alluvium in deltaic areas close to the sea. Good quality of water can be expected from sandstones and river alluviums.

Example 10.1: A borewell log is shown in Fig. 10.2 and an yield of around 900 lpm is expected. Design all the components of the water well both for naturally developed and artificially gravel packed cases, assuming (a) ground water occurs under artesian conditions with piezometric head 6 m below ground level, (b) ground water occurs under water table conditions with

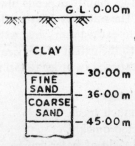

Fig. 10.2 Borewell log, Example 10.1.

Water Well Design

water level 30.6 m below ground level. A mechanical analysis data of the sample obtained between depths 36 to 45 metres is given in Table 10.4.

Table 10.4 Results of mechanical analysis

IS sieve aperture dimension	Weight ratained in each sieve (gm)	Cumulative Weight retained (gm)	Cumulative % retained	Cumulative % passing
2.80 mm	57.4	57.4	14.4	85.6
2.00 mm	112.2	169.6	42.4	57.6
1.40 mm	84.8	254.4	63.6	36.4
1.00 mm	59.6	314.0	78.5	21.5
710-micron*	43.2	357.2	89.3	10.7
Bottom pan	41.3	398.5	100.0	0
Total	398.5			

* 1 micron = $\frac{1}{1000}$ mm

Design of Water Well

Case (a)—Confined aquifer: From Table 10.1, for an anticipated well yield of 900 lpm a 20 cm well is recommended. The screen may be located in the aquifer which lies between depths 36–45 m. The thickness of the aquifer is 9 m which has a grain size mostly in the range 0.6–2 mm and is classified as coarse sand as per IS scale.

$$\text{Length of screen } l = \frac{3}{4} \times 9 = 6.75 \text{ m}$$

Keeping 15% open area, the entrance velocity (V_e) to obtain an yield of 900 lpm is given by

$$\frac{900 \times 1000}{60} = 0.15 \, (\pi \times 20 \times 6.75 \times 100) V_e$$

$$V_e = 2.358 \text{ cm/sec which is permissible}$$

The screen of 6.75 m length may be centrally located in coarse sand aquifer. From the mechanical analyis data for the aquifer sample, a grading curve is plotted on a semi-log paper, Fig. 10.3. From the grading curve, the effective size $D_{10} = 0.69$ mm and the uniformity coefficient $C_u = 2.94$. Artificial gravel pack is not required since $D_{10} > 0.25$ mm and $C_u > 2.5$ and the well may be naturally developed when the slot size should be kept at D_{50} or $D_{60} = 1.75$ to 2.03 mm, say, 2 mm.

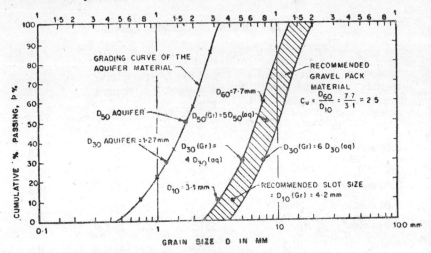

Fig. 10.3 Design of gravel pack, Example 10.1.

Case (b)—Water table aquifer: Recommended well diameter = 20 cm

Length of screen $l = \frac{1}{3} \times 9 = 3$ m located at the bottom one-third of the coarse sand aquifer. Assuming 15% open area for the screen, the entrance velocity (V_e) to obtain an yield of 900 lpm is given by

$$\frac{900 \times 1000}{60} = 0.15(\pi \times 20 \times 3 \times 100)V_e$$

$$V_e = 4.78 \text{ cm/sec}$$

This is slightly on the higher side. To bring this below 3 cm/sec, adopt a screen length of 4.5 m with a maximum of 18% open area, when V_e is given by

$$\frac{900 \times 1000}{60} = 0.18 (\pi \times 4.5 \times 100)V_e$$

$$V_e = 2.95 \text{ cm/sec, which is permissible}$$

Artificial gravel pack is not required and the slot size may be kept at 2 mm as in case (a).

Gravel-pack design: If an artificial gravel pack is desired since $C_u < 3.0$ for the coarse sand aquifer, then

D_{30} (gravel pack material) = 4 to 6 times D_{30} of aquifer material

= 4 to 6 times 1.27 mm, as read from the grading curve

= 5.08 mm to 7.62 mm

Water Well Design

With these points for D_{30} for the gravel pack material smooth curves are drawn such that C_u for the gravel pack material is 2.5. The shaded area in Fig. 10.3 shows the recommended gravel pack material; clean pea gravel of size 3 to 10 mm may be used. The slot size is kept at D_{10} of the gravel pack material which is 4.2 mm. The thickness of the artificial gravel pack may be 15–20 cm.

Check for drawdown

$$K = CD_{10}^2 \text{(Allen Hazen equation)}$$
$$= 100(0.069)^2 = 0.4761 \text{ cm/sec}$$

Since Allen Hazen's formula overestimates K, taking 2/3 of the avove value

$$K = \frac{2}{3} \times 0.4761 = 0.32 \text{ cm/sec.}$$

Transmissibility

$$T = Kb = 0.0032 \times 9 = 0.0288 \text{ m}^3/\text{sec/m}$$

Specific capacity

$$= \frac{T}{1.4} \times \text{Efficiency of the well (60\%)} = \frac{0.0288}{1.4} \times 6.0$$
$$= 0.01235 \text{ m}^3/\text{sec/m}$$

Probable drawdown

$$S_w = \frac{0.9}{60 \times 0.01235} = 1.215 \text{ m which is permissible.}$$

Example 10.2: Preliminary test shows that a tubewell can yield 1,800 lpm when the drawdown is limited to 10 m from an aquifer situated at a depth of 90–110 m below ground level. The corresponding radius of influence is estimated as 300 m. The static water level in the well is about 12 m b.g.l. The aquifer soil has $D_{10} = 0.23$ mm, $D_{50} = 0.60$ mm and $D_{60} = 0.67$ mm. Determine the diameter, length of strainer, slot size and size of the gravel pack required.

Assuming an average daily consumptive use of 3 mm for the cropping pattern followed in the agro-climatic region, allowing 30% more water to take care of the peak demand at plant flowering stage, and an irrigation efficiency of 70% with better water management and intensive irrigation with double and triple cropping (complete crop cover), what will be the total area that can be brought under the tubewell irrigation with 12 hours of pumping per day? Assuming the soil as sandy loam with moisture holding capacity of 3.6 cm per 30 cm depth of soil layer, the average depth of effective root zone for the type of crops as 1 m and a seasonal consumptive use of 40 cm for the crops, determine the depth of irrigation, frequency and the number of waterings.

Assuming that the highest portion of the irrigation area is about 8 m above the ground level near the tubewell site and 360 m away from the well, determine the hp of the pump set required and the monthly electric bill assuming 15 ps per unit (kWhr) consumed. Also calculate the line current of a 440-volt delta connected induction motor assuming a power factor of 0.85.

Assuming such schemes cost Rs 75,000, work out the economic viability of the scheme and the basis of irrigation assessment.

Tubewell design: Diameter of the strainer is given by

$$Q = p\pi Dl\, V_e$$

Assuming a minimum entrance velocity of 2.5 cm/sec and percentage open area of the strainer as 15%

$$\frac{1800 \times 1000}{60} = 0.15\, \pi Dl\, (2.5)$$

$$Dl = 25,500 \text{ cm}^2$$

$$K = CD_{10}^2 \qquad \text{(Allen Hazen equation)}$$

$$= 100\, (0.023)^2$$

$$= 0.053 \text{ cm/sec}$$

Since Allen Hazen's formula overestimates the value of K, take 2/3 of this value

$K = 0.053 \times \frac{2}{3} = 0.0353$ cm/sec, say 0.036 cm/sec,

Dupuit's formula

$$Q = 2.72\, kb\, \frac{H - h_w}{\log_{10} \frac{R}{r_w}}$$

Assuming the yield as roughly proportional to the length screened $l = 25,500/D$

$$\frac{1800 \times 1000}{60} = 2.72\, (0.036)\, \frac{25,500}{D} \times \frac{10 \times 100}{\log_{10} \frac{300 \times 100}{D/2}}$$

$$D \log_{10} \frac{60,000}{D} = 83.4$$

Try $D = 25$ cm, LHS $= 84.5$

A diameter of 25 cm for the strainer can be adopted which is also the correct choice from practical considerations (Table 10.1).

Length of strainer

$$l = \frac{25,500}{25} = 1,020 \text{ cm or } 10.2 \text{ m}$$

Water Well Design

The strainer should be centrally located in the aquifer.

Design of gravel pack: Since for the aquifer soil $D_{10} = 0.23$ mm < 0.25 mm and $C_u = \dfrac{D_{60}}{D_{10}} = \dfrac{0.67}{0.23} = 2.91 < 3$, an artificial gravel pack is required. Assuming a pack-aquifer ratio of 5,

$$D_{50} \text{ (gravel pack)} = 5 + D_{50} \text{ (aquifer)} = 5 \times 0.60 = 3 \text{ mm}$$

With this point for D_{50} for the gravel pack, a smooth curve is drawn such that $C_u = 2.5$ (Fig. 10.4) which is the grading curve of the gravel pack

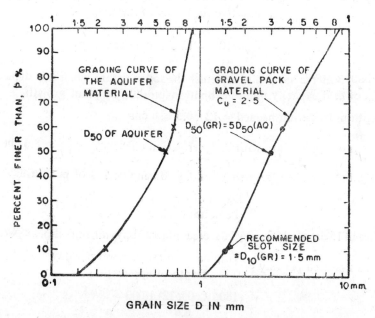

Fig. 10.4 Design of gravel pack, Example 10.2.

material required. Clean pea gravel of size 1 to 9 mm may be used. The slot size of the strainer is kept at D_{10} of the gravel pack material which is 1.5 mm. The openings may be bevelled inside to avoid clogging of the fine particles during development. The thickness of the artificial gravel pack may be 15–20 cm.

Check for drawdown

$$\text{Specific capacity} = \frac{T}{1.4} \times \text{Efficiency of the well (60\%)}$$

$$= \frac{0.036\,(20 \times 100)}{1.4} \times 0.6 = 30.8 \text{ ccs/cm}$$

$$= \frac{30.8 \times 60 \times 100}{1000} = 184.8 \text{ lpm/m}$$

Probable drawdown

$$S_w = \frac{1800}{184.8} = 9.75 \text{ m} < 10 \text{ m}$$

The design is suitable since the drawdown is within the permissible limit with a well efficiency of 60%.

Area under tubewell irrigation: Average daily consumptive use = 3 mm Peak use at plant flowering stage = $3 + 3 \times 0.3 = 3.9$ mm

Depth of irrigation = moisture holding capacity of the soil × depth of effective root zone of the crop

$$= \frac{3.6}{30} \times 100 = 12 \text{ cm (field capacity)}$$

But usually water is applied to replenish the soil moisture when it is depleted to 50% of field capacity due to evapotranspiration by plant growth.

Depth of irrigation needed = $12 \times 0.5 = 6$ cm

Depth of application with 70% irrigation efficiency = $\frac{6}{0.7} = 8.6$ cm

Irrigation interval (frequency) in days during period of peak use

$$= \frac{\text{Depth of irrigation}}{\text{Peak use per day}} = \frac{6}{0.39} = 15.4$$

say once in 15 days, which may be increased to 20 days during off-peak period.

$$\text{Area irrigated per day} = \frac{\text{Volume of water pumped per day}}{\text{Depth of application}}$$

$$= \frac{(1800 \times 60 \times 12)\, 1{,}000}{8.6}$$

$$= 1.51 \times 10^8 \text{ cm}^2 = 1.51 \text{ ha}$$

Total area that can be irrigated = area irrigated per day × irrigation interval

$$= 1.51 \times 15 = 22.65 \text{ ha, say 23 ha}$$

$$\text{Number of waterings} = \frac{\text{Seasonal consumptive use for the crop}}{\text{Depth of each irrigation}} = \frac{40}{6}$$

= 6.6, say 6–7 waterings of 8.6 cm depth

The irrigation interval and the number of waterings have to be adjusted according to the rainfall, if any, during the growing season.

Power of the pumpset: A suitable concrete pipe has to be selected for conveyance of irrigation water to the farm 360 m away. From Table B-2 (Appendix) for a flow rate of 1800 lpm a 15 cm concrete pressure pipe

Water Well Design

(P2-class) gives a friction loss of 3.5 m per 100 m length and a 20 cm concrete pipe gives a friction loss of 0.8 m per 100 m length, while the increase in the cost for 20 cm pipe is about Rs 12 per meter. But a 15 cm concrete pipe gives a friction loss of $\frac{3.5}{100} \times 360 = 12.6$ m which means an additional power $P = \rho g Q h_f = 1,000 \times 9.81 \, (0.03) \, 12.6 = 3710$ W or 3.71 kW and the monthly power cost becomes $3.71 \times 12 \times 30 \times 0.15 \approx$ Rs 200. If the 20 cm pipe is selected, the monthly power cost is only $\left(\frac{0.8}{3.5}\right) \times 200 =$ Rs 45.70 and the monthly interest on the additional capital cost of Rs 12×360 at an interest rate of 12% p.a. $=$ Rs $\frac{(12 \times 360 \times 0.12)}{12} =$ Rs 43.20. The net savings per month by using 20 cm pipe $=$ Rs $200 -$ (Rs $45.70 +$ Rs $43.20) =$ Rs 111, which means an annual saving of Rs 1332. Hence a 20 cm P2-class RCC Hume pipe will be used.

Total head acting on the pump (H) is calculated as follows.

Depth to static water level in the well	12 m
Drawdown during pumping	10 m
Delivery head (from tubewell site to the highest portion of the irrigated land)	8 m
Friction loss in the 20 cm RCC Hume Pipe @ 0.8 m per 100 m for 360 m, say	3 m
Minor losses due to bends and elbowes, velocity head at the delivery end, etc., say	1 m
Total head	34 m
Add 10% extra	3.4 m
Total head (H)	37.4 m

The total head acting on the pump, allowing for the seasonal fluctuation of ground water table, may be taken as 37.5 m, when the concrete pipe has to withstand a pressure of 3.75 kg/cm^2 (hence P 2-class RCC Hume pipe can be used).

Power of the pumpset assuming an overall efficiency of 60% for the pump and the efficiency of the electric motor and drive as 85%,

$$P = \frac{\rho g Q H}{\eta_p \, \eta_m} \text{ watts} = \frac{1,000 \times 9.81 \, (0.03) \, 37.5}{0.60 \times 0.85}$$

$$= 21,600 \text{ W, or } 21.6 \text{ kW}$$

and the monthly electric bill

$$= 21.6 \text{ kW} \times 12 \text{ hr} \times 30 \text{ days} \times 0.15 \text{ Re} = \text{Rs } 1170$$

Line current in a 3-phase delta connected system.

Input power in watts $P = \sqrt{3}\ EI \cos \phi$

$$21.6 \times 1000 = \sqrt{3} \times 440 \times I \times 0.85$$

Line current $I = 33.2$ amp.

From the pump-catalogues available with the prospective pump manufacturers the characteristic curves of different pump makes should be compared and the one with 21.6 kW under 37.5 m head yielding 1800 lpm under maximum efficiency should be selected. It should be driven by a 21.6 kW motor operating on 440 V and 33.2 amp.

By charging for the water supplied (by measuring over the V-notch or by noting the power units consumed) and lining the water courses by any cheap material locally available like 10-15 cm semicircular baked clay tiles, laterite sheets, cuddapah slabs with joints finished with cement mortar 1:4 or soil cement, or brick in cement mortar, the economic use of water is achieved. It is clear from this example that a tubewell 25-30 cm diameter, tapping 20-30 m of good aquifer and pumping 10-15 hours per day can bring a total area of 20-50 ha under intensive irrigation with double and triple cropping. The economic returns from such schemes are worked out in the following.

Economic viability of 23 ha of tubewell irrigation:

I. Capital costs

Cost of ground water survey for locating the tubewell, drilling, casting, and strainer, development (gravel pack), conducting step drawdown and yield tests, installation of pumpset (deep-well or submersible turbine), panel board, switch and starter, pumphouse, etc., say   0.75 lakh

$$\text{Capital cost per ha} = \frac{0.75 \text{ lakh}}{23 \text{ ha}} = \text{Rs. } 3{,}260$$

II. Operating costs

(i) Depreciation @ 8% (average), repairs and maintenance of equipment @ 3%, interest on capital investment @ 12% (flat)—total 23% of capital cost   Rs 17,250

(ii) One pump operator @ Rs 500 p.m.   Rs 6,000

(iii) Electric power charges—21.6 kW×12 hr×0.15 ps+270 days for triple cropping (excluding rainy days)   Rs 10,500

(iv) Making channels and bunds and their maintenance @ Rs 100 per ha   Rs 2,300

(v) Labour @ Rs 50 per ha   Rs 1,150

(vi) Cost of cultivation like ploughing, sowing, seeds, manures and fertilisers, plant protection, intercultivation, transporting and market @ Rs 750 per ha (average) for 3 crops in a year = $750 \times 3 \times 23$ Re 51,700

                                                                Rs 88,900

Operating cost per ha under triple cropping = $\dfrac{88,900}{23}$ = Rs 3,860

III. Gross revenue per ha

    (i) First crop (monsoon) of maize yielding 50 qn/ha @ Rs 120 per qn   Rs 6,000

    (ii) Second crop of wheat (winter) yielding 35 qn/ha @ Rs 120 per qn   Rs 4,200

    (iii) Third crop of millets or pulses yielding 25 qn/ha @ Rs 100 per qn   Rs 2,500

               Total revenue per ha from 3 crops   Rs 12,700

IV. Net annual income per ha

    = Rs 12,700 − Rs 3860 = Rs 8840.

V. Repayment of loan and economic status of farmers
Assuming the capital has to be repaid in 8 years, annual repayment of loan per ha = $\dfrac{0.75 \text{ lakh}}{8 \times 23}$ = Rs 408

The net annual savings of the farmers holding 1 ha each

    = Rs 8840 − Rs 408 = Rs 8432 (or Rs 703 p.m.)

Thus the poor farmers in the area holding 0.5–1 ha each can have a net savings of Rs 350–700 p.m. apart from the labour provided for them from cultivation of their lands, under the tubewell co-operative irrigation schemes. They have also some money (64% of the capital investment) put aside towards the depreciation of the tubewell and pumpset. They can hope for prosperity after the complete repayment of loan, i.e. from the 9th year onwards when they can have a net savings of Rs 737 p.m./ha which can be further increased up to Rs 900 p.m. by including crops like sunflower, soyabeans, chillies, potato, tomato, etc.

From the above design example, the following factors apply to tubewell irrigation schemes:

1. Capital cost per ha   Rs   3,000–3,500
2. Operating cost per ha

   (i) cost of irrigation (see item II, i–v) per crop   Rs   500–600
   (ii) cost of cultivation per crop   Rs   650–800

3. Electric power cost per crop per ha   Rs   150
4. Net annual income per ha (under 2–3 crops)   Rs   7,000–10,000
5. m. hp required per ha (for lifting)   1–1.1
6. Additional m.hp, required if sprinkler irrigation is adopted, i.e. another 40 m of head for the pressure to be maintained at the base of the nozzle and frictional losses in the main and laterals   1–1.2
7. Cost per cubic meter of water supplied (volumetric sale)

$$= \frac{\text{Rs } 37{,}200 \text{ (cost of item II, i–v)}}{(1{,}800 \times 60 \times 12 \times 270)/1{,}000} \quad \text{Rs } 0.106$$

or 10.6 ps/m^3

Cost of irrigation per ha-cm = Rs 0.106×100 = Rs 10.60

Assuming a seasonal consumptive use of 40 cm for wheat (assuming no rainfall in the growing season) and an irragation efficiency of 70%, depth of water required = $40/0.7 = 57$ cm. The cost of irrigation per hectare of wheat = Rs 10.60×57 = Rs 604. Thus the irrigation assessment is made on.

(i) Volumetric basis 10–12 ps/m^2 or Rs 10–12 per ha-cm
(ii) Crop-area basis Rs 400–600 per crop-ha of wheat

Economic use of water is achieved by assessment on volumetric basis, and on tubewells it has been possible to arrange volumetric sale of water.

In the above design example the two principles of irrigation, namely, 'how much to apply?' and 'when to apply?' have been illustrated. The third principle, namely, 'how to apply', i.e. the efficient method of application for the maximum utilisation of water resources, will be treated in chapter 17.

Economics of open wells in hard rock areas of North Arcot district, Tamil Nadu is given in the following. Open wells of size $6 \times 4.5 \times 12$ m deep are feasible.

(i) Cost of excavation and steining   Rs 15,000
(ii) Cost of pump and motor, pipes and fittings   Rs 4,000
(iii) Construction of pumpshed and electrical connection   Rs 2,000

Total cost   Rs 21,000

Under the cropping pattern followed, about 1.6 ha of land comes under an open well with a net income of Rs 5000.

Water Well Design

Example 10.3: A 30 cm tube well was drilled in an area for which the bore log is given below. The GWT varies between 10 m in monsoon to 15 m in summer, bgl. A preliminary test showed that the well can yield 2500 lpm with a drawdown of 5 m. The average permeability of the sandy strata may be taken as 30 m/day. Determine the length of the strainer required and its location. Assume a radius of influence of 300 m.

The Well bore log

Depth (m)	Strata
0–5	Clay with shingle
5–20	very fine sand
20–35	clay with kankar
35–50	coarse sand
50–60	clay
60–80	medium sand
> 80	clay

If the water have to be lifted to a height of 30 m agl what is the power of the pump required. Assume total losses of 5 m and pump efficiency of 60%. What is the monthly electricity bill at 20 ps per kW hr assuming a motor efficiency of 85% and 12 hr of pumping per day?

Solution:
$$Q = C\pi D\, l p V_e$$

Allowing an entrance velocity of 25 mm/s, assuming 16% open area for the screen, and 50% of the open area is clogged (clogging coefficient $C = 0.5$),

$$\frac{2500}{1000 \times 60} = 0.5\, \pi \times 0.3\, l \times 0.16 \times 0.025$$

Length of screen $l = 22.2$ m

Check for drawdown, assuming the yield as roughly proportional to the length screened,

$$Q = \frac{2.72\, Kl\, s_w}{\log \frac{R}{r_w}}$$

$$\frac{2500}{1000 \times 60} = \frac{2.72 \left(\frac{30}{86400} \times 22.2\right) s_w}{\log \frac{300}{0.15}}$$

$$s_w = 6.6\, \text{m}$$

which is higher than 5 m due to partial screening. If the drawdown is to be limited to 5 m, the well has to be pumped at

$$Q \approx 2500 \times \frac{5}{6.6} = 1900 \text{ lpm}$$

The well screen may be located in the coarse sand and medium sand strata, each of 10.1 m length, centrally.

Power of the pump required

$$P = \frac{\rho g Q H}{1000 \, \eta_o} \text{ kW}$$

$$H = \underset{s_w}{\underset{\text{bgl}}{6.6}} + \underset{\underset{\text{agl}}{\text{swl}}}{15} + \underset{\text{height}}{30} + \underset{\text{losses}}{\underset{\text{total}}{5}} = 56.6 \text{ m}$$

$$P = \frac{1000 \times 9.81 \times (2.5/60) \, 56.6}{1000 \times 0.60} = 38.5 \text{ kW}$$

which is the power to be delivered to the pump shaft.

Power of the motor $= \dfrac{P}{\eta_m} = \dfrac{38.5}{0.85} = $ **45 kW**

Monthly electricity bill $= 45 \text{ kW} (30 \times 12) \, 0.20 = $ **Rs 3260**.

Design of Collector Wells

To get large irrigation supplies from permeable alluvial aquifers with a permanent source for continuous recharge, like gravel formations from banks of lakes or perennial streams or from water-logged areas irrigated by big canal systems, radial water collectors are employed. A radial water collector is essentially a large diameter, normally 4-6 m, shallow well from which horizontal strainers protrude, radially near the bottom, into the permeable aquifers. The central well is a vertical concrete caisson, about 4 m in diameter and a wall thickness of 45 cm. The cylindrical caisson is precast in RCC rings on the ground surface preferably with steel form work. Each ring in lowered under the weight of the next as the earth material is excavated from inside. When the required depth is reached, the bottom is sealed by pouring a thick concrete plug heavy enough to overcome the buoyancy. When the caisson is in place, lateral pipes, fabricated from heavy steel plate 10 mm thick and slotted to have 15-18% of open area with their leading ends protected by streamlined nose cap, are driven horizontally into the water bearing formation by special hydraulic jacks installed at the bottom of the caisson through the precast port holes (left in position during casting of the caisson according to the information available from boring data) to form a radial pattern of laterals. Laterals or screens are usually available in 2.5 m lengths and the successive lengths are buttwelded till the required length is reached. The diameters vary from anything between 15 to 50 cm depending upon the local conditions and the yield required. The maximum number of radials are limited to 16 in one level and a second level of radials is possible, as may be desired if many clay lenses inhibit free vertical water movement in the vicinity of the well. The laterals need not be equally long and some may be driven with a provision for some more to be driven when growing requirements warrant an increase in yield. A common situation is a caisson on a river bank with radials only along 90° of the circumference. The maximum length of a radial so far has not exceeded 135 m. The total length of the laterals may range from 120 to 900 m depending upon the diameter and yield required. The arrangement of laterals on river Avon at Somerdale Works in England for an yield of 3,415 m^3/day is shown in Fig. 10.5. Seven laterals with slotted pipes of 20 cm diameter and total length 126.5 m were driven. Provision was made for driving three more laterals in case growing requirements should demand a greater yield.

There are different patents developed long back such as by Ranney, Fehlmann, Preussag, etc. In the Ranney method, the slotted pipe itself is jacked out with a digging head with large holes in front for the removal of the aquifer material and the slotted pipe once packed in remains as the radial collector. In the Fehlmann method, a blank casing is installed after which the perforated pipe is put inside and the blank casing withdrawn. In both the types, the fine particles are washed out by flushing so that natural gravel packs are formed around the perforations. The water should percolate into the strainers with an entrance velocity not exceeding 6–9 mm/sec to prevent sanding and incrustation. The location of laterals near the bottom of the well ensures removal of water from the entire water bearing formation above them. This also increases the effective radius of the well or its circle of influence, resulting in higher yields. Normally caissons operate at depths ranging from 10–25 m. In waterlogged canal irrigated tracts, this depth is not likely to exceed 15 m. The top of the caisson is raised to an elevation above the maximum flood level. The intake gate valves, operated from top, are installed both for low and high water operations. The well diameter should be large enough to accommodate the travelling screens and hydraulic jacks for driving operations. The initial cost of a radial water collector exceeds that of a vertical tubewell but the large yields obtained under low pumping heads and low maintenance costs lower its cost per cumec of water lifted. Over 300 such installations are already in operation in USA besides many others in France, Germany, and other parts of Europe with discharges varying from 2,000 to 60,000 lpm. A collector well located adjacent to a surface water body like a stream or lake induces infiltration of surface water through the bed of the water body into the well and thus large supplies of good quality water (since it gets filtered through the natural river bed) are obtained for supply to municipalities, industries, etc. Recently, a collector well has been constructed, in the Vaigai river bed for water supply to Madurai city. Two such wells have been constructed at Koyali near Baroda below the bed of the river Mahisagar

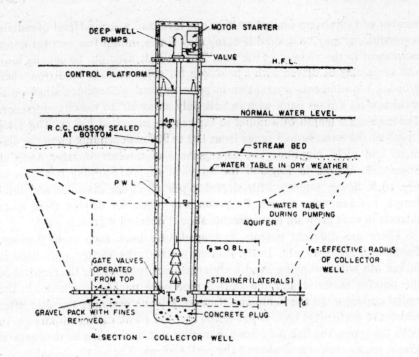

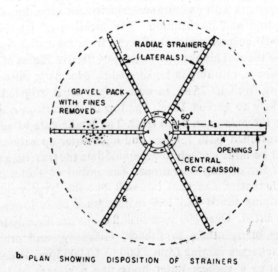

Fig. 10.5 Radial water collector.

for water supply to the Gujarat Refinery. Each well has a diameter of about 6 metres and yields about 50,000 m³/day. The quality of water is very good.

In the design of collector wells, the maximum drawdown may be calculated with reasonable accuracy assuming the radial collector well as an

Water Well Design

ordinary vertical well with an effective radius of 75 to 85% of the individual lateral lengths spaced around the entire circumference of the caisson and having equal length laterals.

Example 10.3: A radial water collector is to be designed for extraction of 15,000 m³/day from the ground water stored in the Palar river bed which has an aquifer extending up to 13 m below the bed, with the width limited to the actual river bed itself, which is 600 m wide. A long duration pump test indicates a transmissibility of 3.6×10^6 lpd/m for a saturated thickness of 12 m when the water table is 1 m below the river bed and a storage coefficient of 33% for the aquifer as confirmed by the laboratory tests. The water table goes down by 4.2 m over a period of 8 months (exactly 250 days) during which period of the summer rainfall amounts to 40 cm.

Solution: Let the collector well be located in the middle of the river. The water level goes down by 4.2 m during a period of 250 days. Due to the natural recharge during this period by summer rainfall, the water table rises by $40/0.33 = 121$ cm or 1.21 m. Hence the permissible drawdown could be $4.20 + 1.21 = 5.41$ m over a 250 days period. Although the well would recover from 5.41 m drawdown including 1.21 m recovery in the dry season by occasional showers, it is a good practice to design the collector well for a maximum drawdown of 4.20 m only allowing the extreme dry periods of no rainfall at all in the dry season.

Since the river bed is a narrow stripped aquifer and the collector well is located in middle of the river bed at 300 m from both the banks which are impermeable boundaries, the drawdown due to both image wells located at 300 m beyond each bank or each at 600 m away from the collector well, has to be included.

From the Theis non-equilibrium equation

$$u = \frac{r^2 S}{4Tt} = \frac{600^2 (0.33)}{4(3.6 \times 10^3) 250} = 0.033$$

for $u = 0.033$, $W(u) = 2.78$, from Theis type curve

$$s = \frac{Q}{4\pi T} W(u) = \frac{15 \times 10^3 \times 2.27}{4\pi(3.6 \times 10^3)} = 0.92 \text{ m}$$

The cumulative effect of both image wells is 1.84 m and the total drawdown allowable for the real collector well is only $4.20 - 1.84 = 2.36$ m. From the Theis equation

$$\frac{0.92}{2.36} = \frac{2.78}{W(u)}$$

$W(u) = 7.14$, for the real collector well

For $W(u) = 7.14$, $u = 0.0004$, from Theis type curve

$$\frac{0.033}{0.0004} = \frac{600^2}{r^2} \quad r = 66 \text{ m, the effective radius of the collector well}$$

Since the effective radius of the collector well is equal to 75 to 85% of the individual lateral lengths L_s (circular array),

$$0.8 L_s = 66$$

Therefore

$$L_s = 82.50 \text{ m, length of each lateral}$$

Let n be the number of laterals spaced equally around the circumference of the caisson. Assuming 16% open area, 40% clogging of the slots, entrance velocity of 6 mm/sec, the number n of laterals of 20 cm diameter is given by

$$Q = (\pi \, dn L_s p) \, V_e \qquad (10.1)$$

where p = effective per cent open area

V_e = entrance velocity

$$\frac{1.5 \times 10^4}{24 \times 60 \times 60} = (\pi \times 0.20) n \times 82.50 \, (0.16 \times 0.60) \, 0.006$$

Therefore $n = 5.8$, say, 6 laterals spaced uniformly at 60° around the circumference of the caisson, Fig. 10.5. The diameter of the caisson may be taken as 4 m and may be sunk up to 13 m after which the bottom is sealed by pouring a thick concrete plug. The laterals may be placed at about 1.5 m above the bottom of the aquifer to ensure proper functioning. After the well is installed, it should be properly developed usually by compressed air. Its yield should also be tested for supplying data to the pump manufacturers. Details of some radial water collectors installed in India are given in Table 10.4.

Infiltration Gallery

Infiltration gallery is a horizontal perforated or porous pipe with open joints, surrounded by a gravel filter envelope laid in a permeable aquifer with a high water table and a continuous recharge with a perennial flow. Infiltration galleries are usually laid parallel to river beds at depths of 3 to 6 m for intercepting and collecting ground water by gravity flow. The horizontal pipes may be of vitrified clay, brick, or concrete of 0.5 to 1.5 m diameter, set in a trench across the aquifer well below the lowest permissible water table, and packed in granite chips around the concrete pipes to fine gravel against the aquifer material, Fig. 10.6, so that clear water percolates with low entrance velocities. Manholes may be provided at intervals of about 100 m for inspection and maintenance. The gallery is laid with a longitudinal slope leading to a central collecting shaft from where the water is pumped to the surface for use.

Table 10.4 Details of some radial water collectors installed in India

Sl. No.	Name of river	Owner of collector well	Length of laterals m	Saturated thickness of aquifer m	Permeability m/day	Width of slots mm	Discharge m³/day
1.	Mahisagar (Baroda)	Gujarat Refinery	400	12	432	6	45,000
2.	Mahisagar (Baroda)	Gujarat Refinery	200—top 200—lower	27	432	6	31,500
3.	Mahisagar (Baroda)	Baroda Municipal Corporation	550	17.5	336	6	54,000
4.	Tapti (Surat, Gujarat)	Baroda Rayons	600	16	384	6	40,500
5.	Jamuna (Delhi)	Delhi Municipal Corporation	600	16	60	3	18,000
6.	Vagai (Madurai, Tamil Nadu)	Tamil Nadu Water Board	600	6	120	3	22,700
7.	Nagavalli (Rayagada, Orissa)	J.K. Industries	600	6	360	6	22,500
8.	Sabarmati (Ahmedabad)	Ahmedabad Municipal Corporation	550 (2 tiers*)	13.5	131	6	36,000

*For a length of 550 m assuming 10 radial directions, the length of lateral will be ≈ 55 m, which may start buckling during driving. On the other hand if the laterals are driven in two tiers in staggered rows, the force required for driving (a shorter lateral) will be less and also the hydraulic gradient and hence the flow efficiency will improve.

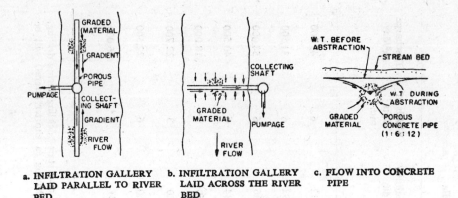

a. INFILTRATION GALLERY LAID PARALLEL TO RIVER BED
b. INFILTRATION GALLERY LAID ACROSS THE RIVER BED
c. FLOW INTO CONCRETE PIPE

Fig. 10.6 Infiltration gallery.

Design of infiltration gallery: For the data given for the design of a radial water collector in the Palar river bed, as an alternative, an infiltration gallery is desired to be constructed for a supply of 15×10^6 lpd with a maximum permissible drawdown of 4.2 m after a 250 days period, at a

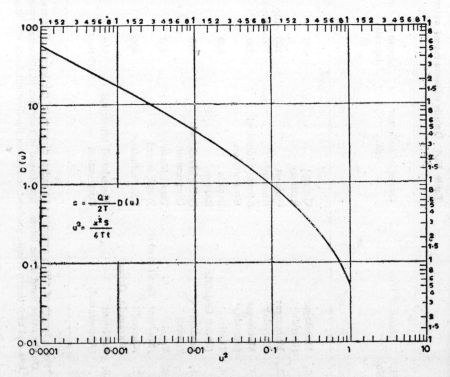

Fig. 10.7 Ferris drain function.

distance of 3 m from the infiltration gallery. Boundary conditions need not be taken into account if the gallery is planned across the river bed as the flow is essentially two-dimensional in the vertical plane in the upstream-downstream direction. The drawdown pattern away from the gallery is given by Ferris (1950) as

$$s = \frac{Qx}{2T} D(u) \tag{10.2}$$

$$u^2 = \frac{x^2 S}{4Tt} \tag{10.3}$$

where x = distance from drain to the point of observation of drawdown and $D(u)$ = 'drain function'; values of $D(u)$ for values of u^2 are given in Table 10.5 and Fig. 10.7.

In the design example for $x = 3$ m

$$u^2 = \frac{3^2 (0.33)}{4 (3.6 \times 10^3) 250} = 8.25 \times 10^{-7}$$

for $u^2 = 8.25 \times 10^{-7}$, $D(u) = 625$ from Table 10.5.

For an allowable drawdown of 4.2 m,

$$4.2 = \frac{Q \times 3}{2(3.6 \times 10^3)} \times 625$$

$$Q = 16.13 \text{ m}^3/\text{day/m of gallery}$$

For a supply of $15 \times 10^3 / \text{m}^3 / \text{day}$,

$$\text{length of gallery required} = \frac{15 \times 10^3}{16.13} = 940 \text{ m}$$

But as the river bed aquifer is only 600 m wide, the yield that could be obtained

$$= \frac{600}{940} (15 \times 10^3) = 9.6 \times 10^3 \text{ m}^3/\text{day}$$

If, however, 15×10^3 m³/day is required, two galleries should be designed. Optimum spacing of the galleries can be found, if each gallery is allowed to influence the neighbouring gallery to the extent of 0.3 m only. Then if Q is known, the distance x for $s = 0.3$ m can be found from Eq. 10.2.

From the above designs of radial water collector and infiltration gallery, it can be seen that the collector well is more sophisticated and expensive but has higher capacities than the infiltration gallery. Hence choice should be made by the required yield followed by economic aspects.

Table 10.5 Values of D(u) and u^2 for drain function

u^{2*}	$D(u)^*$	u^2‡	$D(u)$‡
0.0025	10.32	1×10^{-7}	1783
0.0036	8.468	2×10^{-7}	1261
0.0049	7.109	4×10^{-7}	891.1
0.0064	6.130	7×10^{-7}	673.2
0.0081	5.331	1×10^{-6}	563.1
0.010	4.714	$2 + 10^{-6}$	398.0
0.013	4.008	4×10^{-6}	281.1
0.016	3.532	7×10^{-6}	212.1
0.020	3.079	1×10^{-5}	177.4
0.025	2.657	2×10^{-5}	125.2
0.030	2.354	4×10^{-5}	88.2
0.035	2.109	7×10^{-5}	66.42
0.040	1.943	1×10^{-4}	55.42
0.050	1.658	2×10^{-4}	38.92
0.060	1.441	4×10^{-4}	27.20
0.070	1.282	7×10^{-4}	20.36
0.090	1.049	1×10^{-3}	16.90
0.110	0.8810	2×10^{-3}	11.67
0.130	0.7598	4×10^{-3}	7.99
0 160	0.6284	7×10^{-3}	5.84
0.190	0.5324	1×10^{-2}	4.698
0.230	0.4384		
0.280	0.3517		
0.330	0.2895		
0.380	0.2434		
0.440	0.2008		
0.500	0.1837		
0.580	0.1345		
0.660	0.1094		
0.760	0 0864		
0.900	0.0623		
1.000	0.0507		

* (Ferris J.G., 1950)
‡ (Shanmugam C.T., 1968)

Water Well Design

QUIZ 10

I. Match the items in A and B:

Group 1

	A		B
i.	Entrance velocity	a.	Shallow wells in Cauvery delta and filter point tubewells in Orissa
ii.	Slot size in naturally developed wells	b.	V-shaped, widening towards inside
iii.	Slot size gravel packed wells	c.	15–16%
iv.	Pack-aquifer ratio	d.	D_{60}/D_{10}
v.	Thickness of gravel pack	e.	4:1 to 6:1
vi.	Uniformity coefficient	f.	≈ 2.5 cm/sec
vii.	Percentage open area in slotted pipe screen	g.	In relation to aquifer, 3–8 mm
viii.	Coir-rope screen	h.	10–20 cm
ix.	Pack gravel size	i.	D_{40} to D_{70} of formation
x.	Best type of the slot opening	j.	D_{10} of gravel pack material

Group 2

	A		B
i.	Drawdown pattern in the vicinity of infiltration gallery	a.	Large irrigation supplies from highly permeable aquifers within depths of 10–25 m with permanent source of continuous recharge like banks of lakes or streams
ii.	Effective radius of collector well	b.	Municipal supplies from perforated or porous pipes with open joints surrounded by a gravel filter envelope usually laid parallel to river beds below the lowest permissible water table, usually at depths of 3–6m below river beds, for collecting ground water by gravity flow
iii.	Permeability, K cm/sec	c.	Ferris 'drain function'
iv.	Radial water collectors	d.	$100\, D_{10}^2$, D_{10} in cm.
v.	Infiltration galleries	e.	$\approx 80\%$ of the length of individual

II. State whether 'true' or 'false'; if false give the correct statement:
1. In cable tool and straight rotary drilling, doubling the diameter of the well doubles the cost of drilling, while it may cost slightly more with reverse rotary.
2. Well screen should be made of a bimetal to avoid any galvanic corrosion.
3. Gravel packed wells should strictly adhere to aquifer-gravel-slot size relationship.
4. The cost of a gravel well is more than a naturally developed well.
5. Doubling the diameter of a well doubles the yield.
6. Usually in a water table aquifer about three-fourths of the aquifer thickness is screened while in an artesian aquifer the bottom one-third is screened.

7. The gravel pack material should be angular and well graded, mostly of calcareous material rather than siliceous.
8. Artificial gravel pack is required when the aquifer material is homogeneous with $C_u<3$ and $D_{10}<0.25$ mm.
9. Allen Hazen's formula underestimates permeability.
10. In an open well sunk in fractured and weathered rock the bottom portion of steining (if at all provided) should be completely plastered with cement mortar.
11. Cost of tubewell irrigation in unconsolidated formation is around Rs 10 per ha-cm, and if irrigated by sprinklers, it will be around Rs 30 per ha-cm, which may further go up to Rs 35 per ha-cm for spinkler irrigation from borewells in hard rock areas.
12. The cost of a standard tubewell in alluvial areas is around Rs 1 lakh while it may be around Rs 25,000 in hard rock areas and the yield from the tube or bore well is also in the same ratio.
13. While the cost of a standard open well (for irrigation) in an alluvial tract is in the range of Rs 8000–12000, it is in the range of Rs 15000–20000 in hard rock areas.   ($\checkmark$, ×, $\checkmark$, $\checkmark$, ×, $\checkmark$, ×, $\checkmark$, ×, ×, $\checkmark$, $\checkmark$, $\checkmark$)

III. Select the correct answer (s):
1. The well depth is selected considering the factors
    a. maximum number of aquifers should be tapped so that the discharge is maximum
    b. aquifer of poor quality should be avoided
    c. there should be no partial penetration
2. The well diameter is selected considering the factors
    a. cost of well
    b. adequate annular space to provide proper thickness of artificial gravel pack
    c. large enough to accommodate a suitable pump unit
    d. to obtain maximum yield
    e. all the above factors
3. The selection of the well screen material depends on
    a. water quality
    b. strength of material
    c. corrosion resistance
    d. cost
    e. all the above factors
4. For a partially screened well
    a. the discharge is proportional to the length screened
    b. the drawdown is proportional to the length screened
    c. the specific capacity is proportional to the length screened
    d. all the above relations
    e. none of the above relations

Water Well Design

QUESTIONS

1. The results of sieve analysis of an aquifer material are given in the following. Determine:
   a. effective size and uniformity coefficient of the aquifer material
   b. the gradation requirements of the gravel pack you recommend around the well screen
   c. the slot size for the well screen

   Results of sieve analysis of aquifer material

Size mm	Cumulative % finer by weight
0.8	95
0.59	90
0.5	85
0.4	80
0.38	60
0.32	40
0.29	20
0.26	11
0.2	6

   (0.24, 1.5 mm; 1.2—1.8 mm, $C_u=2.5$; 1.2 mm)

II. (a) What is the reason for partial screening? Indicate some good screening materials and how you would make a choice.

   (b) From a preliminary test, it is expected that a tubewell can yield 1,300 lpm under a drawdown of 4 m from a confined aquifer of 12 m thick, struck at a depth of 25 m b.g.l. The static water level is 10 m b.g.l. The aquifer soil has $D_{10}=0.46$ mm, $D_{30}=0.77$ mm, $D_{50}=1.08$ mm, $D_{60}=1.19$ mm, Determine the diameter, length of strainer, slot size gradation and thickness of gravel pack if required.

   ($D_w=20$ cm, $L=9$ m, Slot=3 mm, 10 cm of gravel pack (optional) of average size 4–6 mm)

III. (a) Explain the criteria of design for
   (i) Artificially gravel-packed wells
   (ii) Naturally developed wells

   (b) A sample of aquifer material has the grain size distribution given in the following table. If the overlying material is fairly firm determine the proper slot size:

Size of sieve opening (mm)	Weight retained (gm)
1.15	65
0.82	41
0.57	73
0.40	87
0.30	46
0.20	45

(0.8 mm)

IV. (a) Under what circumstances can a radial collector well be most advantageously used?

(b) How do you determine the length and number of laterals for a proposed radial collector well?

11

Water Well Drilling

Wells may be dug, bored, driven, jetted or drilled. Simple drilling methods like drive point, jetting, and hand boring can be adopted in favourable conditions for construction of shallow wells up to 25 cm diameter and 45 m deep. Selection of drilling equipment depends upon the hydrogeology of the formation, diameter and depth of the production well, availability of funds, maintenance and spares, production capacity, volume of work, operating crew and easy movement of the rig.

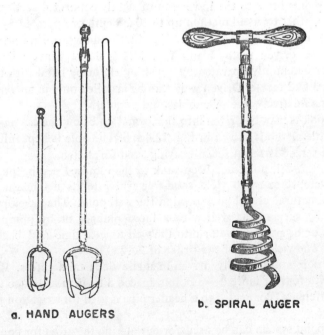

a. HAND AUGERS   b. SPIRAL AUGER

Fig. 11.1 Well boring with augers.

Boring

Boring of small diameter wells up to 15 m in clay, silt and sand may be done with hand-turned or power-operated augers, Fig. 11.1. Wells deeper than 5 m may require the use of a light tripod with a pulley at the top or a raised platform, so that the auger shaft can be lowered and removed from the hole without dissembling all shaft sections. The spiral auger is used to remove stones or boulders encountered. In caving formations boring is done by lowering the casing to the bottom of the hole.

When the casing pipe reaches the desired depth, well pipe with screen is lowered inside the casing pipe and the annular space between the well pipe and casing pipe is filled with gravel. The casing pipe is gradually pulled out with a winch and the pipe is rotated simultaneously. The well is developed by surging or by instant pumping or by using a compressor, till sand-free water is obtained.

Depending upon the diameter, depth of the well and water table, the well can either be pumped by suction lift or by installing a deep well pump. One or more wells can be coupled to get greater discharge. In the Cauvery delta in Tamil Nadu (south India) hand boring sets are extensively used to drill filter point wells up to 18 m and deep tube wells; coir rope and coir filters are used for filter points.

Driving

Well points fitted to the lower end of tightly connected sections of pipe are driven either by hand method up to 10 m depth or operated by a drilling rig for depths of 15 m and above. Fig. 12.4 a shows the assembly for a purely hand driven method and Fig. 12.4 b shows heavier drive block assemblies commonly operated by a drilling rig or by hand with the aid of of a tripod and tackle. Driven wells can be installed only in unconsolidated formations relatively free of cobbles and boulders.

Well points may be driven into the formation below the casing in drilled wells by the methods shown in Fig. 12.4 a, b. The hole is kept full of water while the screen is being set in heaving sand formations.

When the well point has been sunk to the required depth, the well has to be developed so as to yield sand-free water to its maximum capacity, forming a natural gravel pack around the well point. The development can be done by surging the well with a loose plunger or by pumping intermittently or by pouring water into the well time to time and backwashing the well. The yields are of the order of 90-250 lpm.

Well points are usually driven in shallow coastal aquifers. Well point system with four or more drive points spaced 8 to 16 m apart (so that there is no interference) connected to a header pipe is used for irrigation purposes, Fig. 11.2.

Well point system can be used for dewatering the areas for construction

Water Well Drilling

works with a closer spacing of 1.5 to 6 m (so that their areas of influence overlap), Fig. 11.3.

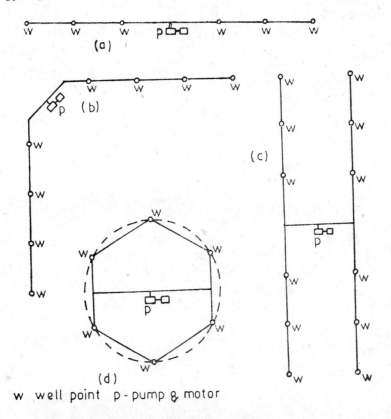

w well point   p - pump & motor

Fig. 11.2 Multiple well-point systems (Centrally located pump equalises suction lift).

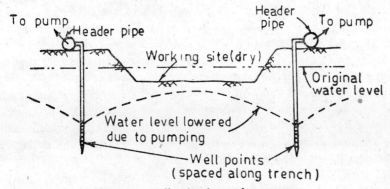

Fig. 11.3 Well-point dewatering system.

Well points are available in sizes of 3 to 8 cm diameter and 0.5 to 2 m lengths.

Cavity Wells

If there is a relatively thin impervious formation or stiff clay layer available at a shallow depth underlain by a thick alluvial stratum, it is an excellent situation for a cavity well. A hole is drilled using the hand boring set and the casing pipe is lowered to rest firmly on the stiff clay layer, Fig. 11.4. A hole of small cross-sectional area is drilled into the sand formation and is developed into a big hollow cavity by pumping at a high rate or by operating a plunger giving a large yield.

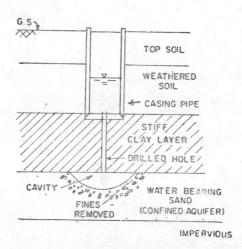

Fig. 11.4 Cavity well.

Jetting

The force of a high velocity stream or jet of fluid loosens the subsurface materials and transports them upward and out of the hole. Drilling may be a combination of chopping and turning of the drill bits shown in Fig. 11.5. The spudding percussion action can be imported to the bit by means of a hoist or by quickly releasing the rope. A combination of jetting and driving casing, washing out samples, can give information quickly. A tripod and pulley, winch and a small pump of approximately 680 lpm at a pressure of 3.5 to 5 kg/cm^2 is used to force the drilling fluid (very often plain water) through a hose on to the drill pipe and bit. Small truck mounted jetting drills are manufactured. They can drill the hole, install the casing and screens, develop the well and install the pump.

The jetting method is suitable for unconsolidated formations for holes up to 15 cm diameter. Hard clays and boulders pose problems.

Water Well Drilling

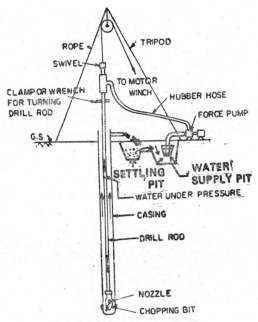

Fig. 11.5 Water jet method.

Core Drilling

A ring with black diamond bits or steel teeth is attached to a drill rod which is rotated. Water is circulated to remove cuttings. As the cutter

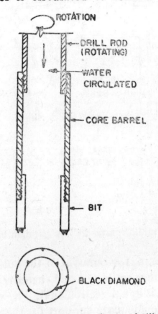

Fig. 11.6 Diamond core drilling.

advances, a core rises inside the ring, which is broken off time to time by feeding chilled steel shots with the circulating water. In diamond core drilling, the diamond bit is used instead of the shot bit and calyxite. Diamond bits are used for cutting extremely hard and abrasive formations, loosely cemented or consolidated. These bits have the diamonds set in a hardened alloy steel matrix which will retain the diamonds under critical conditions. For diamond core drilling a fairly high rotating speed is required to get good, smooth and unbroken core samples, Fig. 11.6.

Rotary Drilling

The equipment and working of the hydraulic or straight rotary method is very similar to that of a calyx drill, except that they are designed to drill large diameter holes to a greater depth at a much faster rate. This method consists of a rotating drill bit, Fig. 11.7, for cutting the borehole with a continuously circulated drilling fluid, forced through the hollow drill pipe

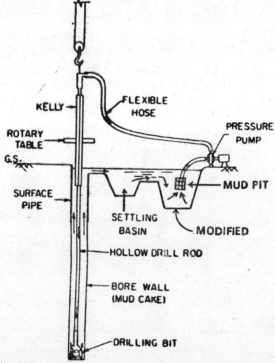

Fig. 11.7 Hydraulic rotary drilling (straight circulation).

on to the bit by a mud pump, for removal of the cuttings, which flow to a settling pit where the cuttings settle out and then overflow to a storage pit from where the mud fluid is again recirculated. The properties of the mud fluid are such as to provide adequate support for the wall of the hole and

Water Well Drilling

are usually viscous mixtures of water and natural or commercial clays such as bentonite. The driller can control the fluid characteristics by checking the density of the mud by a balance and its viscosity by a Marsh funnel shown in Fig. 11.8. The balance has a cup at one end and a sliding weight on the other portion of its beam. The cup is filled with the drilling fluid and the weight is moved until it balances the filled cup. The density of the fluid is then read on the balance arm; a fluid density of about 9 N per litre is usually satisfactory. The viscosity of the fluid is determined by closing the lower end of the Marsh funnel by a finger and filling the funnel to the proper level (about 1,500 cc) and then removing the finger to allow the fluid to discharge from the funnel. The time in seconds to drain 1,000 cc of the fluid is defined as the Marsh-funnel viscosity, expressed in seconds, which is usually in the range of 30–40 seconds. The driller has to keep the density and viscosity within the above limits by periodically adding water and/or clay to the drilling fluid.

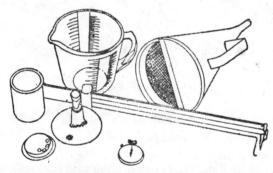

Fig. 11.8 Balance for determining mud weight, stop watch, Marsh funnel for measuring mud viscosity and 1 litre measuring cup.

The drilling mud forms cake on the wall of the borehole which seals the pores to prevent loss of fluid into permeable formations and prevents caving. The mud cake has to be effectively removed during development for an efficient well. Hydraulic rotary method is suitable for drilling 15–45 cm diameter holes 100–150 m deep in soft rock and unconsolidated formation. For large diameters small holes are sunk and reamed to large size. This method is not suitable for boulder formation and requires more water (450–900 lpm), repairs and maintenance. The driller must not only be able to operate the drill but also should have a knowledge of mud technology. The accurate sampling and logging of the formations penetrated may be difficult.

The volume of the settling pit should be at least three times the volume of the hole being drilled. A settling pit of 2 m long × 1 m wide × 1 m deep and a storage pit 1 m square × 1 m deep may be suitable for the drilling of 10 cm wells (15 cm holes), 35 m deep. A system of baffles may also be used to provide extra travel time in the pit to improve the settling.

The basic parts of a conventional rotary drill are—a mast and hoist, a power operated revolving table that rotates the drill stem and bit, a mud pump and a power unit or engine. Drag bits of either the fish tail or three way design, Fig. 11.9 a, b, are best suited in an unconsolidated clay and sand formations and roller type bits, Fig. 11.9 c, in coarse gravel and boulder formations.

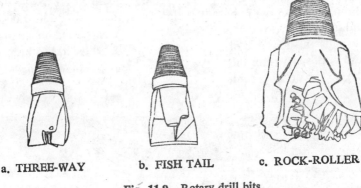

a. THREE-WAY   b. FISH TAIL   c. ROCK-ROLLER

Fig. 11.9 Rotary drill bits.

The speed of rotation of bit in the bore hole is 30 to 60 rpm. The drilling mud is essentially bentonite clay and the density of the mud fluid is 1.02 to 1.14 g/cc. The upward velocity of flow in the bore hole is 0.7 to 1 m/s. In unconsolidated materials, the drilling rate is around 100 m/day.

Air Rotary

The drilling fluid may be any fluid lighter than water like air, foam, etc.

Air drilling is used in fractured rocks. Air rotary is specially suitable for limestones. Air foam is used to remove cuttings. With dry air upward velocities are in the range of 10 to 30 m/s in the annular space between the drill pipe and hole. In foam type fluids the ratio of air to liquid is of the order of 200 : 1. Stiff foam is preferred for drilling in unconsolidated materials.

Reverse Circulation (Rotary) Drilling

In this system the direction of flow is reversed, i.e., from the annular space between the drill pipe and the wall of the hole through the bit into the hollow drill pipe upwards and discharged by the pump into the settling pit. The clear fluid returns to the borehole by gravity flow, Fig. 11.10. Relatively high velocity of the fluid in the drill pipe enables the cuttings to be carried to the surface without the deliberate use of clay or other additives

Water Well Drilling

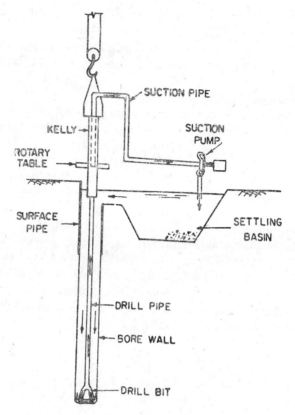

Fig. 11.10 Hydraulic rotary drilling with reverse circulation (reverse rotary).

to increase the viscosity. The boring is done without a casing and hydrostatic pressure is used to support the walls of the bore-hole during construction. Water level in the bore hole is about 2 m above natural level or at ground level. The settling pit is about three times the volume of the material expected to be removed from the bore hole. Circulation of large amount of make-up water (2500–5000 lpm) by the use of a large capacity suction pump or by an air lift is required and minimum development is required to remove the light filter cake formed unlike the impervious mud cake in the direct rotary. Reverse rotary is best suited to drill large diameter wells 45—60 cm, 75—90 m deep in 8—12 hours, in soft unconsolidated formations. Drilling in coarse dry gravels poses the greatest difficulty.

The diameter of the hole is large in relation to the drill pipe in order that the velocity of the descending water in the annular space is low (30 cm/sec or less) and the bit and drill pipe are rotated at speeds varying from 10 to 40 rpm. The large diameters favour completion of the wells by artificial gravel packing. The suction end of the rig pump is connected through the swivel to the kelly and drill pipe and suction head of the pump limits the drill pipe lengths to 3 m. 15 cm drill pipes are commonly used so that cuttings of size up to 13 cm can be brought up through the pipe.

Cable-tool Percussion Drilling

The cable-tool percussion method consists of a tool string, comprising the drill bit, drill stem, drilling jars and rope socket, suspended by a cable from a walking beam (truck mounted) or operated from a diesel engine, which lifts and drops the tool string, Fig. 11.11. Water forms a slurry with the pulverised material which is bailed out at intervals. When drilling in dry formations, water must be added to the hole to form the slurry. Tools for drilling and bailing may be carried on separate lines or cables spooled on independent hoisting drums. The casing closely follows the drilling bit, adding lengths of casing as the hole is deepened, to prevent caving. The top of the casing is fitted with a drive head which serves as an anvil. Periodic checks should be made with a plumb bob or carpenter's level to ensure that a straight and vertical hole is being drilled. Usually drilling is started with a large diameter and the diameter is reduced telescopically after drilling certain depths.

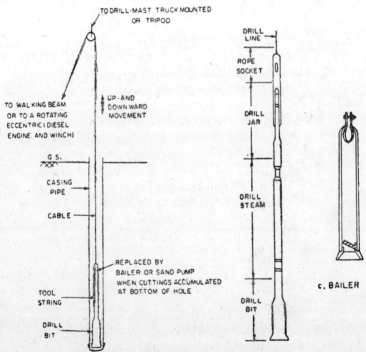

Fig 11.11 Cable tool percussion drilling.

Water Well Drilling

This method is suitable for rock, medium hard, soft and boulder formation to drill 20–45 cm and 75–180 m holes; drilling is relatively slow and casing has to be provided as the drilling progresses. Some of the advantages of the percussion method are:

(i) Reasonably accurate sampling of the formation material can be readily achieved.
(ii) Rough checks on the water quality and yield from each water-bearing stratum can be made as drilling proceeds.
(iii) Minimum water is required for drilling, a point for consideration in arid regions.
(iv) Any encounter with a water-bearing formation is readily noticed as the water seeps into the hole.
(v) Water-bearing strata of poor quality may be grouted or sealed off.
(vi) Bits can be dressed in the field.

Fishing of lost or stuck parts may be difficult and utilises of many fishing tools.

Rotary-cum-Hammer Drilling (Down-the-Hole Hammer)

In this method a pneumatic hammer operated at the lower end of the drill pipe is used. It combines the percussion effect of cast tool drilling and the rotary action of rotary drilling. The bit is an alloy steel hammer with heavy tungsten carbide inserts which help in faster rate of penetration in hard rock areas. The diameter and depth of the hole is limited by the volume of air that can be exhausted through the hammer to remove the cuttings. Compressed air must be supplied at pressure of 750 to 1350 kN/m^2 (to remove the cuttings effectively) and free air supply of at least 9–10 m^3/min for drilling 15 cm holes. The upward velocity in the space outside the drill pipe should be about 900 m/min. Proper rotation speed is from 15 to 50 rpm and 10 to 20 blows per second to bit. Reduced speed in best in harder and more abrasive rock. Down-the-hole hammer (DHD) drilling is the fastest drill suitable for hard rock areas to drill 15–20 cm holes, 120 m deep in 10–15 hours. It uses compressed air and a foaming agent is injected into the air line under pressure by means of an injection pump to lubricate the bit and dust control. A flush pump is used for flushing the hole and bringing the cuttings to the surface. Air compressor, pump and prime mover are all mounted on one truck. Some of the DHD drilling rigs used in the hard rock areas are Ingersoll Rand, RMT, Halco 625, etc.

Logging and Sampling

A well log denotes the characteristics of the materials penetrated by a bore hole along a vertical line from the ground surface. It may be a geophysical log like the electrical resistivity, SP, gamma ray, sonic, temperature and caliper logging or photograph logging by a deep well camera which gives a 360° picture of the bore hole. The direct methods of logging are drill cutting log or sample log, core log, drilling time log, drilling bit behaviour log, mud characteristics log, etc.

Samples of subsurface material obtained from different depths during the process of drilling operations are, in most cases, the best source of geologic information. The principle object of test drilling is to obtain samples that reveal the character, depth and thickness of various strata. In consolidated rocks the cores obtained serve as representative samples. The drilling time log is an accurate record of the time required to drill each meter depth; and the penetration rate represents the character of the material being penetrated, Fig. 8.10.

The drilling bit behaviour during drilling is indicative of the type of subsurface material encountered. In the case of rotary drilling, the drilling action may be smooth and even, indicating a homogeneous formation, or crunchy if gravel and boulders are present. In the cable tool method, the bit may stick in clay formations and tend to retard the rebounce, whereas the reverse is true if hard rock is encountered.

Mud loss, mud thinning, mud thickening, water loss, etc. should be carefully observed and recorded against depth. This gives useful information regarding the nature of the formations encountered. Sudden and persistent mud loss may be due to cavernous limestones, major joints or faults, dry sand, etc. whereas the mud thickening is due to clay formation. When artesian aquifers are encountered mud thinning may take place. It is also a good practice to have a complete record of mud pressure, mud density and viscosity, pressure on drilling bit, etc.

Samples of subsurface materials obtained during the process of drilling operations are, in most cases, the best source of geologic information. Samples are collected at regular intervals from different depths during drilling and analysed by the well site geologist and a driller's log prepared. Better sampling results are obtained in cable tool drilling than in other methods. The resultant log a more dependent on the well site geologist and driller than on the drilling method. Representative samples of all the formations penetrated have to be collected for geological study, mechanical analysis and preservation for correlation and verification purposes. The samples are kept packed in well labelled sample sacks of cloth. Good logs and samples are a valuable asset.

All the various types of direct and indirect logs available should be in the same vertical scale and put side by side to form a composite log Fig. 8.10. Then the characteristics of the various logs should be carefully studied and analysed critically to interpret the lithology and stratigraphy to obtain the final interpreted log.

Water Well Drilling

Drilling Programme

If a commission is given for execution of a ground water development project, the three essential documents to be prepared are:
(i) the construction schedule
(ii) the equipment and materials schedule
(iii) the costs schedule

The construction schedule is quite the most important as this shows exactly the method of approach to the problem and provides the basis for both the other schedules and the smooth field operation of all phases of the project.

A typical construction schedule is given in Chart 11.1. For economy the work pattern must resemble as closely as possible the assembly line in a large factory. Machinery must be maintained in operation, if possible, 24 hours a day performing the job for which it is designed.

Well Revitalisation

Existing open dug wells may be revitalised by boring, deepening and rock blasting. Vertical bores at the bottom of the existing open wells are recommended when it is found from hydrogeological studies that the ground water table in the vicinity is at a fairly higher level and water is under pressure at the bottom of the well and rises up if a trial bore is drilled at the bottom of the well. Horizontal and inclined bores may also be drilled to intercept the water-bearing fractures to augment the yield into the open

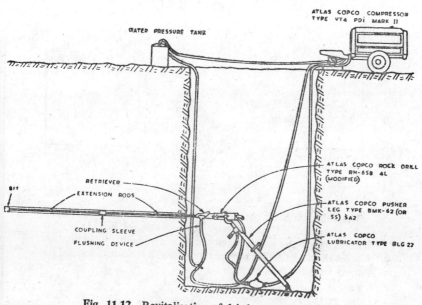

Fig. 11.12 Revitalisation of dried-out open wells.

well. This method of drilling 36-48 mm diameter and 15-30m long holes in rock by using extension drill steel equipment and compressed air, Fig. 11.12 and 11.13, is called extension hole drilling. The extension rods are connected as drilling proceeds by means of coupling sleeves. A pusher leg is used for horizontal hole drilling. A rock drill can also be utilised to drill short holes (1.5m) to enable charging and blasting of the hard rock for deepening the well. Compressed air is the motive power for working the rock drill which is a percussion-cum-rotary machine (about 2,000 blows per minute and 200 rpm). The rock drill imparts the blows on either the extension equipment or the drill steel which has a tungsten carbide insert brazed on it and shatters the rock; the cuttings are flushed out either with compressed air or water and the hole is drilled. A proper cycle of operation has to be

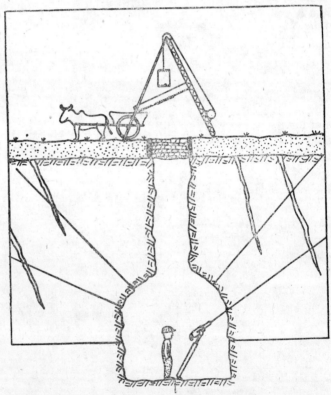

The well should, for economical reasons, be sunk with minimum diameter and only the bottom portion stopped out to provide storage for water which is percolating during night. Revitalisation holes should be drilled to puncture water-filled fissures located far away from the well.

Fig. 11.13 Proposal for sinking of open wells in hard-rock areas with low percolation in the ground.

Water Well Drilling

established in which drilling, blasting and mucking are carried out one after the other.

Blasting Techniques

The various explosives and accessories employed for well sinking are manufactured by Indian Explosives Ltd., Gomia. Explosives used for well sinking must have the following properties.

(i) **High density:** Comparatively smaller holes are drilled in well sinking and it is important that the charges should be concentrated at the bottom of the holes.

(ii) **High water resistance:** Since the wells are sunk in water bearing strata, it is essential that the explosive has good water resistance. Special Gelatine 80% will be usually effective but if the strata is particularly hard, special Gelatine 90% strength may be used. Special Gelatines are supplied in 25 mm dia × 200 mm length cartridges packed in cases containing 25 kg net explosives.

Either electric or safety fuse shot firing can be employed for well sinking. Plain detonators, comprising a metal tube closed at one end, filled with a powerful explosive consolidated under pressure, are used in conjunction with safety fuse, the 'spit' from the fuse causing the detonator to explode. The safety fuse consists of a thin core of black powder wrapped in layers of jute textiles and waterproof coatings. On lighting, it burns at controlled rate from one end to the other. The types of safety fuse recommended are given in Table 11.1.

Table 11.1 Types of fuses

Type of fuse	Application	Burning speed sec/m
Blue sump	Damp conditions	100–120
WCPS	Wet and very rugged conditions	100–120

Other accessories for preparation of charges and stemming the short holes, are crimper, pricker, scraper and stemming rod.

In electric shotfiring, an electric detonator is used. On passage of electric current through the fuse head coupled to a pair of leading wires, a flash is produced which ignites the detonator charge. Instantaneous electric detonators fire instantaneously on application of current. Short delay detonators

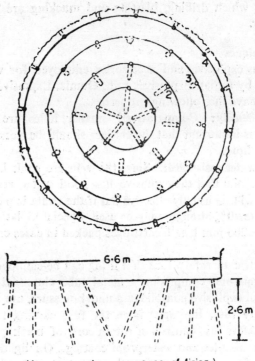

Fig. 11.14 Layout of holes in sinking wells—delay blasting.

Ring	No of holes per ring	Charge per hole (N)	Total Charge per ring (N)
Sumpers	5	14	70
Inner Easers	8	11	88
Outer Easers	12	9.5	114
Trimmers	16	9.5	152
	41		424

Volume of rock — 88 m³
Charge ratio — 0.2 m³ per N

with a nominal delay interval of 25 min and long delay detonators with a nominal delay interval of 300 min are available in delay numbers from 0 to 10. Other accessories in electric shotfiring are exploder (Beethoven Dynamo Condenser exploder capable of firing up to 100 shots is commonly used). Ohm-meter (manufactured by Gyro Laboratories Ltd. Bombay), shot firing cable (a well insulated 2-core cable about 300 m long, each core consisting of a conductor of at least four copper wires of not less than 0.46 mm diameter), stemming rod, scraper and pricker may be used.

Delay blasting: The usual technique in well sinking is to drill concentric rings of holes. The inner ring known as the 'sumper' or the 'cut' is drilled so that the bottoms of the holes almost meet at the centre, thus forming a cone which is blasted first and provides a free face for the subsequent rings. The cavity is enlarged further by the subsequent rings of holes around it spaced at 0.6 to 0.9 m. The last ring holes or 'trimmers' at the side of the well should be drilled with slightly smaller burden and should be more closely spaced than in the preceding rings. By this means the overbreak can be reduced and there will be less likelihood of the wall being damaged. The placement of holes for a well of 6 m diameter is shown in Fig. 11.14. Short delay detonators of 1, 2, 3 and 4 delay numbers are used respectively in the round, with the approximate charges of explosives as shown in Fig. 11.14. The charges are initiated inversely, i.e. the primer cartridge is placed at the bottom of the hole with detonator pointing towards the main charge. This reduces the chances of a misfire due to cutoff. The depth of pull depends upon the diameter of the well, and is restricted to a maximum of half the diameter. The minimum angle of the cut holes cone should be 45°. The blasting ratio generally varies between 0.05 and 0.2 m^3 of solid rock broken per N of explosives used, depending on the nature of the rock and the size of the wells. The smaller the well, the higher is the charge of explosive required per cubic meter of rock blasted. If instantaneous electric detonators are used to initiate the charges, the various rings of holes will have to be blasted separately.

Instantaneous blasting: In narrow wells, say 3 m or less in diameter, short vertical holes (1 m depth) are drilled at a regular spacing of 0.5–0.6 m only and charged to within 0.15 m of the collar. All the holes in a round are fired simultaneously. The spacing between the peripheral holes may be reduced to 0.3–0.4 m for proper breakage.

When well sinking is carried out in built up areas or in the vicinity of buildings, vibrations can be minimised by reducing the amount of charge fired at a time, and also by using delay detonators. Flying debris can be controlled by covering the mouth of the well by strong wire nets weighted down by sand bags.

Thus with a gadget of RB units which comprises air compressors, pneumatic hammer or rock drills, portable magazines and detonators, and a few debris removers, it is possible to deepen the wells quickly and revitalise them for agricultural or drinking water purposes. The diameter of agricultural wells generally varies from 3 to 7.5 m, whereas that of drinking water wells ranges between 1.8 and 6 m only. The wells are deepened to attain a depth of about 12–15 m.

QUIZ 11

I. Match the items in A and B:

Group 1

A

i. Driving
ii. Hand boring set
iii. Well fitted with hand pump
iv. Bailer

v. Loss of mud circulation
vi. Permeability of the formation
vii. Artesian water bearing formation
viii. Avoid bailing up and the bit being stuck up
ix. Caving of the hole in reverse rotary
x. Strata encountered during drilling can be easily and accurately identified
xi. Avoid crooked holes

B

a. Drinking water ≈3,000 lph
b. To remove cuttings in percussion drilling
c. Shallow wells
d. Increase in the volume of mud fluid and the mud fluid rapidly loses its weight and viscosity
e. Well points
f. Examination of drill cuttings
g. Porous formation, drilling fluid invading the formation
h. Cable tool
i. Rig perfectly levelled and checked at regular intervals of 3–10 m
j. Rise and drop drill pipe 1–1.5 m at a time, the pipe being rotated rapidly
k. Sudden drop in mud fluid level from the ground level

Group 2

A

i. 36–48 mm dia horizontal and inclined bores 15–30 m
ii. A gadget of RB units such as air compressors, rock drills, pusher leg extension drill steel, portable magazines and detonators, debris removers
iii. Core drilling in hard rocks and core recovery
iv. Burning speed of 100–120 sec/m
v. Cavity well
vi. Drilling in boulder formations
vii. Dug-cum-borewell (in hard rock areas)
viii. Drilling in limestone areas

B

a. Safety fuse for blasting in open wells
b. Calyx drilling
c. Extension hole drilling
d. Well revitalisation
e. Clayx drill or lighter DTH rigs
f. A good water bearing stratum with a strong roof of hard clay of 5–6 m
g. Cable tool
h. Air rotary

Water Well Drilling

II. State whether 'true' or 'false'; if false give the correct statement.
   1. The capacity of the percussion rig depends upon the weight of the tool which it can handle safely.
   2. Cable tool requires more water for drilling than rotary rig and can drill in a wide range of formations, with faster drilling.
   3. The function of the mud pump is to circulate the drilling fluid at the desired pressure and volume. It is usually a double acting piston type force pump developing a pressure of 8–10 kg/cm^2.
   4. The kelly helps in rotating the string of tools and drill rods.
   5. The mud cake is washed out by placing the drill rods at the top of the bore and forcing clean water often mixed with sodium hexametaphosphate.
   6. The pull down mechanism is used for applying a thrust load on the drill bit when the penetration rate is very high.
   7. Drill pipe sections are hollow seamless tubes, made from high grade steel, usually of 6 m long and of 60–115 mm diameter.
   8. Reamers are roller type cutters used for enlargement of holes.
   9. Size of screen openings and gravel should be selected before making the sieve analysis of the aquifers.
   10. In air rotary (direct) drilling, compressed air is used instead of mud fluid.
   11. Rotary-cum-percussion rig has combined features of direct rotary and percussion rigs and can be used in all types of formations.
   12. In down-the-hole hammer drilling, the hammer is actuated by compressed air which strikes directly on the head of a tungsten carbide bit while the bit itself is not being rotated along with the tool string.
   13. The DTH rig can drill in hard rocks 2–3 times faster than air rotary and many a time faster than the percussion rig.
   14. In reverse rotary, the bore hole is prevented from caving by the hydrostatic pressure of water in the bore hole and also due to the film of the fine-grained material deposited on the wall by the circulating water.
   15. The bits must be sharp and dressed to proper gauge, so that undue eccentricity during drilling is prevented.
   16. An advantage of the trailer mounted rig is that the truck is available for other transport purposes during the drilling operations.

III. Select the correct answer(s):
   1. The primary functions of the drilling or mud fluid are:
      a. to remove cuttings from the bottom of the bore and allow the cuttings to settle in mud pit which depends upon the viscosity and velocity of the drilling mud
      b. to support the wall of the bore to prevent caving by its own hydrostatic pressure
      c. to prevent loss of circulation of mud in porous formations
      d. cool and clean the drill bit
      e. all the above functions
   2. In reverse rotary drilling
      a. a large capacity centrifugal pump is used for lifting the cuttings and mud fluid.
      b. drill pipes of larger size, 15 cm diameter, provided with flanges are used.

c. speed of rotary table is comparatively very high
d. drilling fluid is bentonite mud
e. drills large diameter holes of 45–90 cm in alluvial formations, containing gravel and small boulders
f. well can be drilled much faster because a higher velocity of drilling fluid can be maintained

3. The choice of a particular type of rig for drilling depends on
   a. the formation to be drilled
   b. size of bore
   c. type of well—surface bore, filter points, cavity well, dug-cum-borewell, etc.
   d. type of screen and its installation
   e. speed in drilling operation
   f. approach condition in the area, i.e. permissible loadings on highways and bridges, width and clear height of path for mobility of the rig
   g. availability of water for drilling
   h. costs involved
   i. all the above factors

4. ......bits are used in very hard and compact formations, especially for core drilling. These are very rarely used in water well drilling.
   (drag bits, rock roller bits, diamond bits)

IV. Methods of drilling in different formations and methods of installation of screens are given in columns A, B and C below. Match them properly. (Hint: More than one method may be suitable.)

A Methods of drilling	B Hydrogeological formations	C Methods of installing screens
		Naturally developed wells
1. Driven	a. Unconsolidated	i. Pull-back
2. Manual	b. Consolidated	ii. Open hole
3. Jetting	c. Hard rock	iii. Bail down
4. Percussion/cable tool	d. Boulder	iv. Wash-down
5. Straight rotary/Rotary	e. Sand	v. Driving
		Gravel-packed wells
6. Reverse circulation/Reverse rotary	f. Soft rock	vi. Bail down
7. Rotary cum hammer drilling/DHD	g. Medium hard	vii. Double casing (modified pull-back)
		viii. Open hole
		ix. No screen is required, only casing should be driven for the top loose soil or weathered rock.

12

Water Well Construction

Installation of Well Screens
The method of installing well screens is influenced by the design of the well, the drilling method and problems encountered during drilling. The common methods adopted in the case of naturally developed wells are given in the following.

Pull-back method: In this method, the casing is driven to the full depth of the well. Then the screen is lowered inside the casing and allowed to rest on the bottom. The casing pipe is then pulled up far enough to expose the full length of the screen in the water bearing formation. Using the swedge block, the lead packer provided at the top of the well screen is expanded to make a sand-tight seal between the screen and the inside of the casing, Fig. 12.1. This method is commonly used in cable-tool drilled wells as well as in rotary drilled wells.

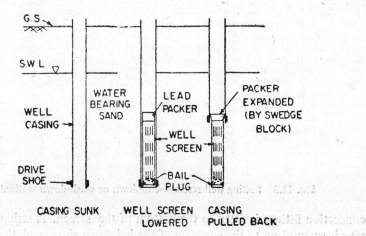

Fig. 12.1 Setting well screen pull-back method.

Open-hole method: In this method the casing is first sunk to a depth a little below the desired position for the top of the well screen. An open hole is then drilled in the water-bearing sand, the casing being filled with the mud fluid. The well screen is then lowered in position and the lead packer is swedged to the casing, Fig. 12.2. This method is applicable to rotary drilled wells.

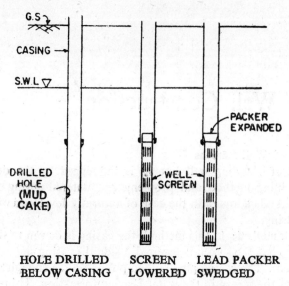

Fig. 12.2 Setting well screen—open-hole method.

Bail-down method: In this method the casing is driven to the intended position of the top of the screen, Fig. 12.3. A bail-down shoe with special

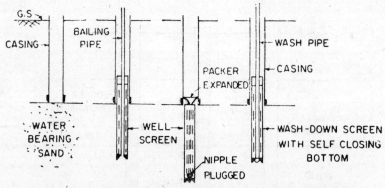

Fig. 12.3 Setting well screen—bail-down or wash-down method.

connection fittings is fitted to the bottom of the screen. A string of bailing pipe is screwed on to the coupling of the baildown shoe and the screen is suspended on this string. The screen is then lowered inside the casing till it

Water Well Construction

reaches the bottom of the bore. Using a bailer or sand pump through the bailing pipe, sand is bailed out from below the screen, when the screen settles down. Before removing the bailing pipe, the special nipple on the bail-down shoe is plugged and after removal, the lead packer is expanded with the swedge block. This method is applicable to rotary drilled well as well as cable-tool drilled well.

Wash-down method: The well casing is first set to the desired depth. A high velocity jet of light-weight drilling mud or fluid issuing from a special wash-down bottom fitted to the end of the screen loosens the sand and allows the screen to sink, Fig. 12.3. The sand is brought up around the screen and into the casing with the return flow of the fluid. After the screen reaches the desired depth, water is circulated through the wash pipe to remove the drilling mud. The lead packer is expanded after removing the wash pipe.

Driving: After setting the casing to the desired depths, the well point with a turned coupling or a packer attached is dropped through the casing and is driven by a driving weight. When driving relatively long well points, a long driving bar attached to a string of pipe or a stem is employed to deliver the driving force directly on the solid bottom of the screen, Fig. 12.4.

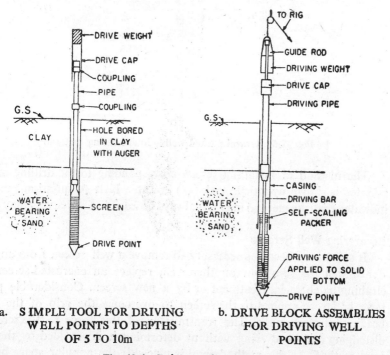

a. SIMPLE TOOL FOR DRIVING WELL POINTS TO DEPTHS OF 5 TO 10m

b. DRIVE BLOCK ASSEMBLIES FOR DRIVING WELL POINTS

Fig. 12.4 Driving well points.

The methods generally adopted in the case of artificially gravel packed wells are the bail-down, open-hole and double-casing methods. The double-casing method, which is most commonly adopted, uses one string of casing corresponding to the outside diameter of the gravel pack (well casing) and a second string of alternate lengths of plain pipe and slotted pipe (well screen to face different aquifer locations). The outer casing is first sunk to the full depth of the well. The inner string consisting of the casing, the well screen and the bail plug at the bottom is then lowered to the bottom. Gravel is put in the annular space around the screen. After filling a certain depth with gravel, the outer casing is pulled back a short distance and the well is developed by compressed air. The steps of placing more gravel, raising the outer casing and developing by compressed air are repeated till the level of gravel envelope is sufficiently above the top of the uppermost screen. The outer casing can be pulled out completely or some length of casing can be left in the top portion, Fig. 12.5.

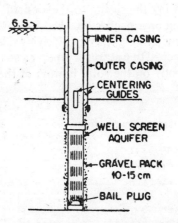

Fig. 12.5 Gravel packed well—double casing method.

The method of installing screen corresponding to the drilling method adopted is shown by a tick mark ($\checkmark$) in Table 12.1. A question mark (?) indicates that the method may not always be feasible since caving may occur.

Recovering Well Screens

It sometimes becomes necessary to remove a well screen from an abandoned well for re-use in another well, replace an encrusted screen after cleaning by chemical treatment or by a new screen. Considerable pulling force has to be applied to the screen to overcome the grip of the water-bearing sand around it. The transmission of this force to the screen for dislodging and recovering without deforming it is provided by the sand-joint method. In this method sand is placed in the annular space between a pulling pipe and the inside of the well screen to form a sand lock or sand joint by tying 5–10 cm strips of sacking to the lower end of the pulling pipe immediately above a coupling ring welded to the pipe, and arranging

Watering Well Construction

Table 12.1 Method of installing screen in drilled wells

Method of installing screen	Driven	Manual	Jetting	Cable tool	Straight rotary	Reverse rotary
Naturally developed						
1. Pull-back	—	✓	—	✓	✓	—
2. Open-hole	—	✓(?)	—	✓(?)	✓	—
3. Bail-down	—	✓	—	✓	✓	—
4. Wash-down	—	—	✓	—	✓	—
5. Driving	✓	—	—	—	—	—
Gravel-packed						
6. Bail-down	—	✓	—	✓	✓	—
7. Double casing (modified pull-back)	—	✓	—	✓	✓	—
8. Open-hole	—	—	—	—	✓	✓

evenly the upper end of the sacking strips around the top of well casing as the pulling pipe is lowered into the well near the bottom of the screen, Fig. 12.6. The pulling pipe is moved up and down and a small stream of water is injected while pouring sand to prevent bridging. About two-thirds of the screen is filled with sand. The pulling pipe is then gradually lifted

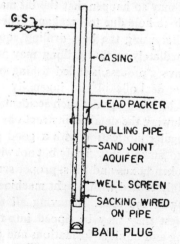

Fig. 12.6 Pulling well screen—sand-joint method.

to compact the sand and develop a firm grip on the inside surface of the screen. Ultimately the screen is pulled out and the sand joint is broken at the surface by washing out the sand with a stream of water. Sizes of pulling pipe and quantity of sand required for sand joints are given in Table 12.2.

Pre-treatment of the screen with said by filling the whole length of the screen with a mixture of equal proportions of hydrochloric or muriatic acid and water, using a string of black pipe or plastic pipe and allowing to stand for several hours or overnight, serves to loosen the rust and incrustation and thus reduces the force required to obtain initial movement of the screen.

Table 12.2 Sizes of pulling pipes for sand joints

Nominal size of screen (cm)	Size of pulling pipe (cm)	Quantity of sand (l/m)
10	3.5	2.9
12.5	5.0	5.2
15.0	7.5	6.5
20	10.0	12.4
25	12.5	23.6
30	15.0	33.5

Fishing Operations

During drilling it may so happen that the bit may be stuck, a tool may be deposited in the bore hole due to breakage. These can happen even to the most capable driller using the best drilling equipment. It is desirable to recover them immediately so that drilling may progress without losing much time. The recovery process is called fishing operation. Fishing operations require a great deal of skill and ingenuity of the driller. The precautions that should be taken against such accidents are proper care and use of drilling tools, following the detailed instructions given in the manufacturer's catalogues, regular lubrication with a good grade of lubricant free from acid or alkali, to make joints firmly but not with excessive pressure as this can result in broken boxes and pins, proper screwing of the tool joints, the threads being thinly coated with a light machine oil and putting screw caps to avoid sticking of the mud, removing all tools immediately away from the bore hole lest they may be tipped into the hole, and exercising utmost care in caving and boulder formations and crooked holes.

In anticipation of the inevitable fishing job, it is necessary to record the exact dimensions of everything used in the well so that information will be at hand for designing or selecting a suitable fishing tool. If the fishing becomes much complicated and difficult, the cost of the fishing tools and the time consumed may greatly exceed the cost af the tools and even the hole, and it may be found economical to move the drill and start a new hole. There are many types of fishing tools suitable for different fishing jobs and special tools have to be manufactured if they are not already there. Many of the tools are rarely used.

Well Development

Well development is the process which causes reversals of flow through the screen openings so as to wash out the fines and rearrange the formation particles in a naturally developed well and form a graded filter (reversed filter) with rings of increasing porosity and permeability towards the well in an artificially gravel packed well, so that ultimately the well will yield clear sand-free water. But causing reversals of flow around the screen, the tendency for several small particles to bridge between large particles is overcome. Figure 12.7 illustrates how the aquifer material has been rearranged after proper development has been made, along with the grain size distribution curves of the aquifer material before and after development. It can be noted that all particles smaller than 1 mm have been removed by the development with an effective size of 1 mm as against 0.33 mm of the

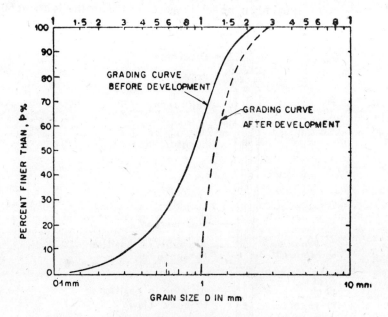

Fig. 12.7 Grading curves of formation material before and after natural development.

original formation, and a uniformity coefficient of 1.4 against 2.9, both of which imply a greatly increased permeability.

Proper development increases the well efficiency and the well loss coefficient C determined by conducting a step drawdown test is indicative of proper development.

Some of the methods of well development in vogue are:

(*i*) *Mechanical surging using solid type or value type surge plunger*: In this method the plunger is operated up and down in the casing like a piston in a cylinder, which produces the required alternate reversals of flow.

(*ii*) *Using compressed air*: This method will prove to be rapid and effective when properly used under favourable conditions. The process involves in combination of surging and pumping. By means of sudden release of large volumes of air, a strong surge is produced by virtue of the resistance of water head, friction and inertia. Pumping is done with an ordinary air lift. The equipment required are:

(a) Air compressor capable of developing a maximum pressure of 700–1000 kN/m^2 and capacity of providing about 6 litres of free air for each litre of water at the anticipated pumping rate.

(b) Pumping (educator or drop) pipe and air line with suitable means of raising and lowering each independently of the other.

(c) Accessories such as flexible high pressure hose, relief valve, quick opening valve, pressure gauge, etc.

The arragement of the pumping pipe and air line is shown in Fig. 12.8. Best results are obtained when the submergence of the air line is about 60%.

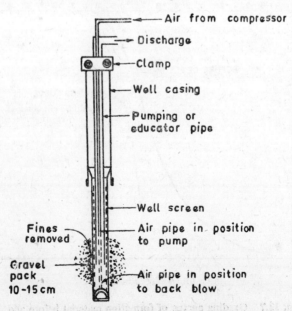

Fig. 12.8 Well development compressed air.

Submergence is calculated by dividing the length of the air line submerged in water by the total length of the air line, i.e. if an air line of length 50 m is submerged in water up to 30 m, then the submergence is (30/50) 100 = 60%. The efficiency of development drops off when the submergence becomes less than 60%. The recommended sizes of pumping pipe and air line for various sizes of well is given in Table 12.3. On wells of 10 cm and smaller size, the well casing itself serves as a delivery pipe, and 18–25 mm airline is used.

Table 12.3 Sizes of pumping pipes and air lines for air lift

Size of well (casing/screen) (cm)	Size of pumping pipe (cm)	Size of air line (cm)
10	5	1.2
12.5	7.5	2.5
15	10	3.2
20	12.5	3.8
25	15	5.0

Development is started by lowering the pumping pipe to 0.6–1.0 m above the bottom of the screen. Then the air line is held about 0.6–1.0 m above the bottom of the educator pipe and water is pumped out by air. As soon as the pumped water appears sand-free, the valve at the air receiver tank is closed and the air line is dropped so that its lower end is about 0.6–1.0 m below the bottom of the educator pipe. The valve is then quickly opened which allows the water to surge outward through the screen openings. Then the air line is pulled up inside the educator pipe and the process is repeated till the well yields clear sand-free water. The position of the entire air-lift assembly then is moved up to a higher portion of the screen (0.6–1.0 m) and such cycles are repeated until the entire length of the screen has been so developed.

(iii) *High velocity jetting of water*: A high velocity jet of water directed horizontally through the screen openings with the tip of the nozzle at about 12–25 mm from the inner wall of the screen is generally the most effective method of well development. A jetting tool coupled to the end of a pipe is lowered into the well. The top of the pipe is connected by a hose to a high pressure pump. As soon as the pump is started the jetting tool is slowly rotated and gradually raised or lowered so that the entire surface of the screen receives the jetting action. The well is pumped by another pump during the jetting operation to maintain the hydraulic gradient so that water and the loosened fine particles will keep entering the well. The jetting tool consists of 2-4 horizontal nozzles having 6-12 mm orifices. The jet velocity shouid be in the range of 30-50 m/sec, which requires a pressure head of 7-14 kg/cm^2. The disadvantage of this method is that considerable supply of water will be required for effective operation. This method is not effective where perforated or slotted pipe is used as a screen.

(*iv*) *Over-pumping and back-washing*: Over-pumping means pumping the well at a higher rate to create excess drawdown, i.e. higher gradients to wash out the fine particles. But since there is no surging effect or reversals of flow, bridging of sand can occur. A higher capacity pump is required, but sometimes the pump intended for regular use in the well is employed for over-pumping.

Back-washing provides a surging effect for well development but it is often found not to be vigorous enough. A deep-well turbine pump without a foot valve is required for this purpose. The pump is started and the water is lifted to the surface or to a large tank above the ground. Then the pump is stopped and the water in the column falls back into the well. This process is repeated, which produces reversals of flow through the well screen.

Another simple method of developing is by back-washing by pouring water into the well as fast as possible and then bailing out with sand pump or bailer. The head built up by pouring water causes reversal of flow through the screen, though it may not be very effective or rapid.

(*v*) *Dispersing agents*: Sometimes it is necessary to add a chemical agent to disperse the clay particles in the mud cake or in the formation to avoid their sticking to sand grains, and to speed up the development process. For this purpose several polyphosphates are used like tetrasodium pyrophosphate, sodium tripolyphosphate, sodium hexametaphosphate and sodium septaphosphate. About 600 g of the chemical is added to every 100 litres of water in the well. The mixture is allowed to stand for about an hour before starting development.

Well Completion

After a well has been constructed, proper sanitary completion is necessary to produce water that is safe by the public health standards. Well completion operations include

(i) Grouting and sealing the casing.
(ii) Completion of the top of the well.
(iii) Disinfection of the well.

Grouting the casing means the filling the annular space between the outside of the casing and the inside of the drilled hole with a cement grout. The grout is a fluid mixture of cement and water of such a consistency that can be forced through the grout pipes and placed as required. Grouting and sealing the casing in water wells serves to prevent the downward seepage of sewage or other polluted surface waters along the outside of the casing, to

Water Well Construction

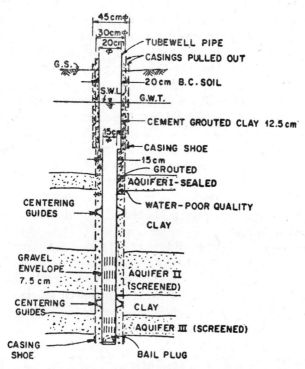

Fig. 12.9 Telescopic well cable tool drilled-(Shallow aquifer of inferior quality sealed; aquifer II & III screened).

seal of aquifers yielding water of poor quality, to make the casing stay tight in the drilled hole and to form a protective sheath around the casing against exterior corrosion, thereby increasing its life, Fig. 12,9.

The top of the casing should normally extend as least 50 cm above the general level of the surrounding surface, well above the maximum flood water levels and isolated from direct contact with accumulating drainage wastes and sudden drainage discharges. The space around the casing should be grouted to a depth of about 6 m to seal of the well from the entrance of surface drainage. A concrete platform should be constructed around the casing at the ground surface. The top of the casing should be provided with a sanitary seal consisting of suitable bushing or packing glands that makes a water-tight seal between the pump column pipe and the well casing.

Abandoned wells should be sealed by filling with puddled clay or cement grout to avoid possible movement of inferior water from one aquifer to another and conserve the water in pumped wells. This is a necessary precaution even if the well casings are perforated in only one aquifer, since casings may eventually deteriorate, permitting interconnection of ground water bodies.

Well Disinfection

After completion of construction, the well and its appurtenances like the casing, pump and pipe systems have to be disinfected or sterilised promptly, for which chlorine solution is the simplest and most effective agent. Highly chlorinated water can be prepared by dissolving dry calcium hypochlorite, liquid sodium hypochlorite or gaseous chlorine in water. A solution containing about 100 ppm of available chlorine should be used. This can be obtained by adding 125 g of dry calcium hypochlorite containing 70% of available chlorine to every 1000 litres of water standing in the well or about 2 litres of liquid bleach containing 5% of available chlorine to every 1000 litres. A solution is made by mixing this total amount in a small quantity of water and is poured into the well through the top of the casing, before it is sealed. The water in the well is thoroughly agitated and allowed to stand for several hours or overnight. The well is then flushed to remove all of the disinfecting agent.

Flowing artesian wells may be disinfected, if found necessary, by lowering a perforated tube, capped at both ends, filled with an adequate quantity of dry calcium hypochlorite to the bottom of the well. The natural upflow of water in the well will distribute the dissolved chlorine throughout the depth of the well. A stuffing box may be provided at the top of the well to restrict the flow and reduce the loss of chlorine.

Well Maintenance

While the expected service life of a well depends upon the design, construction, development and operation of the well, proper maintenance helps to improve the performance and increase the life of the well. Proper records of power consumption, well discharge, drawdown, operating hours, periodical chemical analysis of water and other such observations will help in devising proper maintenance procedures. The sudden pressure drop and increase in the entrance velocity near the screen due to high pumping rates releases carbon dioxide and causes precipitation of calcium carbonate and iron deposits near the screen. The change in entrance velocities results in precipitation of iron and manganese hydroxides. The presence of oxygen in the well can change soluble ferrous iron to insoluble ferric hydroxide. The perforations can be cleaned by adding hydrochloric (muriatic) acid or calgon followed by agitation and surging, which removes the incrusting deposits. Normally the volume of acid required for a single treatment will be about 1.5 to 2 times the volume of water in the screen. Sulphuric acid can also be used instead of hydrochloric acid but its action is a little slower and requires a longer contact time in the well, Fig. 12.10.

The yield of the well may decrease due to the deposition of incrustation of fine particles of silt and clay near the screen. This can be removed by the use of a dispersing agent such as polyphosphates. For effective teaatment, 15 to 30 kg of polyphosphate is added to every 1000 litres of water

Water Well Construction

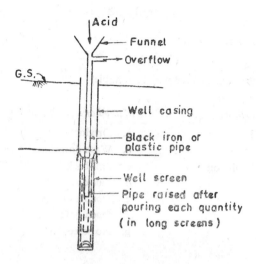

Fig. 12.10 Acid treatment.

in the well. 1 kg of calcium hypochlorite should be added for every 1000 litres of water in the well to facilitate the removal of iron bacteria and their slimes, and also for disinfection purposes. The solution of polyphosphate and hypochlorite is poured into the well and a surge plunger or the jetting technique is used to agitate the water. The well may be treated 2–3 times for better results.

The perforations may become plugged with algae or bacterial growths. Chlorine treatment of wells has been found more effective than acid treatment in loosening bacterial growths and slime deposits which often accompany the deposition of iron oxide. Since a very high concentration of 100 to 200 ppm of available chlorine is required, the process is known as shock treatment with chlorine. Calcium or sodium hypochlorite may be used and the chlorine solution in the well must be agitated by using the high velocity jetting technique or by a surge plunger.

Faulty well construction such as poor casing connections, improper perforations or screens, defective gravel packs and poorly seated valves should be located and set right immediately. Sudden failure of a casing pipe or strainer, resulting in the entry of sand, will require replacement of the well as a whole. An additional precaution against corrosion of the screen is provided by cathodic protection by suspending in the well a rod of a metal low on the electrochemical scale, such as magnesium. This rod will corrode instead of the metal perforations or the screen, and can be replaced when necessary from time to time. Development methods by compressed air or dry ice are sometimes effective in cleaning up corrosion deposits.

Depletion of ground water supply can sometimes be remedied by decreasing pumping drafts, resetting the pump or deepening the well.

QUIZ 12

I. Match the items in A and B:

Group 1
(*Hint*: more than one method may be suitable)

A Method of drilling	B Method of installing screen
i. Cable tool	a. Driving
ii. Straight rotary	b. Wash-down
iii. Reverse circulation	c. Pull-back
iv. Jetting	d. Bail-down
v. Driven	e. Double casing
vi. DTH	f. Open-hole
vii. Calyx	g. No screening

Group 2

A	B
i. Recovering well screen	a. Fishing operation
ii. Recovering lost or broken tools, twisted pipe, parts, etc. from the bore holes	b. Well development
iii. $>60\%$ submergence	c. Air compressor capacity for well development.
iv. Removal of all fine particles near the screen and obtaining sand-free (clear) water from the well	d. Sand joint method
v. Supply of ≈ 6 litres of free air for each litre of water at the anticipated pumping rate and at a pressure of 7–10 kg/cm^2	e. Air pipe for development with compressed air

Group 3

A	B
i. Treatment with polyphosphates	a. Removal of incrustation due to carbonates of Ca and Mg and hydroxides of Fe and Mn
ii. Treatment with calcium hypochlorite	b. Development with surge plunger
iii. Acid treatment	c. Prevent entry of polluted waters
iv. Grouting the case	d. Cathodic protection of the well screen
v. Reversal of flow to avoid bridging of sand grains	e. Disperse clay particles sticking to sand grains and screen
vi. Suspension of a metal rod low on electrochemical scale (magnesium) in the well	f. Removal of bacterial slimes (Fe and Mn bacteria) and well disinfection

Water Well Construction

II. State whether 'true' or 'false'; if false give the correct statement.
1. Acid treatment is more effective than shock treatment with chlorine in removing the bacterial slimes, which often accompany the deposition of iron oxide.
2. The gravel pack is effective in increasing the specific capacity of the well and minimise the rate of incrustation by using a large screen slot.
3. All the screens and plain pipes are numbered and placed serially on the ground surface so that the screens face the water bearing formations when lowered.
4. The surge plunger can be operated by the spudding mechanism of the percussion rig on long strokes; in case of a rotary rig, the movement of the plunger is controlled by the hoisting mechanism of the rig.
5. Types of locations of screens, assembly and size of gravel should be selected with the help of the strata chart before making the sieve analysis of the material.
6. The length of the bail plug should be 3–5 m, with an eye fixed inside.
7. To fish out a tool lost in the bore, it is essential that the diameter of the fishing tool should be at least 5 cm smaller than the bore hole.
8. Stainless steel is not used in the water wells in which the water is corrosive.
9. The corrosion resisting materials commonly used in well screens are yellow brass, red brass, stainless steel, galvanised pipe, PVC pipe and fibre glass.
10. The higher the entrance velocity, lesser will be the incrustation.
11. The cavity well should be developed with compressed air at maximum discharge.
12. Partial clogging of screen is indicated by increased drawdown in the well.

III. Select the correct answer(s):
1. The development of a tubewell should be carried out by compressed air followed by overpumping till
    a. the gravel pack has completely established around the well screen in a graded pattern with increasing permeability towards the screen.
    b. the yield from the well ceases to increase
    c. the drawdown in the well ceases to increase
    d. the drawdown in the well ceases to decrease
    e. sand-free (clear) water is obtained, i.e. sand particles not greater than 10 ppm in the pumped water sample collected in a glass bottle.

13

Yield Test and Selection of Pumpsets

Testing for Yield

After a well is constructed and developed, it should be tested for its yield and the corresponding drawdown. The equipments required for conducting an yield test are:

(i) A test pumping unit consisting of a submersible pump (with 3-phase motor) and a diesel generating set (since power may not be available at the time of test near the well site).

(ii) Measuring devices for time, discharge and water level.

The pump and power unit for an yield test should be capable of operating continuously at constant discharge rate for long periods of time. Preferably the well should be pumped at a slightly higher rate than the anticipated routine pumping rate. Even before the well is completed, some indications as to the probable yield of the completed well is usually obtained. In cable tool drilling such indications are given by the sudden changes in water level, rate of recovery after slush bailing, caving, etc. In hydraulic rotary drilling, mud loss, mud thinning, caving etc. are indicative of the water bearing properties of the formation drilled, while in air rotary drilling, the driller can practically observe how much water is being blown out of the well along with the cuttings. In fact in down-the-hole hammer drilling, the air compressor is continuously operated for 3-5 hours after completion of the total depth of drilling, and the water blown out is measured over a 90° V-notch, which gives an approximate yield. It has been found by experience of drilling in hard rock areas of southern India that the actual yield of the well exceeds the measured yield over a V-notch by about 10%. Once a rough idea of the probable yield is thus obtained, a suitable test pumping unit may be selected depending upon the size of the well, the yield range and the probable heads or drawdowns in the operating area. A quotation for two test pumping units suitable for 10 and 15 cm diameter wells drilled in hard rock areas of the southern India is given in the following.

Yield Test and Selection of Pumpsets

Quotation for test pumping units

From Atlas Copco (India) Private Ltd.   June 10, 1968

Quantity	Description of item	Price each (Swedish currency)
		Sw. Kr.
	One test pumping unit for 10 cm wells consisting of:	
1	Jacuzzi submersible pump type C3 S4D-30 with 3-phase motor, 3, hp, 380, V, 50 Hz	2,000.00
	Electric cable RDVT, 4×1, 5 mm² with cable clamps and quick change coupling	5.40/m
1	Supporting clamp for 40 mm pipes	50.00
1	ASEA circuit breaker type DEF 10-53	125.00
1	Diesel generating set 7.25 kVA driven by Lister diesel engine, 10 hp at 1,500 rpm mounted together on a frame. Electric starting device, instrument panel with voltmeter, 11 litre fuel tank and silencer	8,375.00
	One test pumping unit for 15 cm wells consisting of:	
1	Jacuzzi submersible pump type C5 S6B-8 with 3-phase motor, 5 hp, V, 50 Hz	2,281.25
	Electric cable RDVT, 4×2.5 mm², with cable clamps and quick change coupling	7.95/m
1	Supporting clamp for 65 mm pipes	87.50
1	ASEA circuit breaker type DEF 10-53	125.00
1	Diesel generating set 11.25 kVA driven by Lister diesel engine, 15 hp at 1,500 rpm mounted together on a frame. Electric starting device, instrument panel with voltmeter, 11 litre fuel tank and silencer	10,875.00

The prices for the pumps are given FOB Milan, Italy and for the remaining material FOB Gothenberg, Sweden

The discharge rate should preferably be kept constant throughout the test. The flow rate is measured at regular intervals and adjustment is made to keep it constant. This can be done by adjusting a valve on the delivery pipe rather than by changing the speed of the pump, since a valve gives more accurate control. Small discharges can be accurately measured by noting the time taken to fill a container of known volume. Some of the other methods of measuring the discharge are given below.

(i) *Integrating type of water meter installed in the pipe*: The pipe should run full and so it is given a vertical bend beyond the meter position, Fig. 13.1. The dial of the meter indicates the total volume discharged through

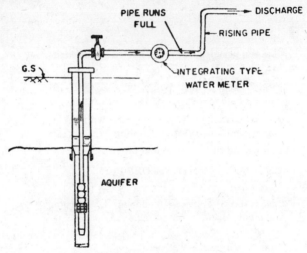

Fig. 13.1 Well discharge measurement.

the meter. By noting the time lapse between two dial readings the flow rate can be calculated.

(ii) *Orifice meter*: An orifice metre is installed in the horizontal length

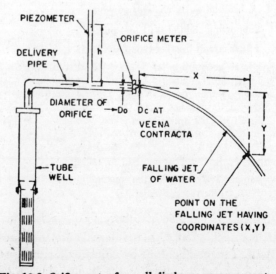

Fig. 13.2 Orifice meter for well discharge measurement.

of the delivery pipe, Fig. 13.2. From a piezometer inserted in the pipe ahead of the meter, the pressure head (h) can be read. By noting the horizontal and vertical distances (x, y) to a point on the falling jet of water (trajectory) and the area of the orifice (A_0), the flow rate (Q) can be calculated as

$$Q = C_d A_0 \sqrt{2gh} \tag{13.1}$$

where the coefficient of discharge (C_d) is given by

$$C_d = \left(\frac{D_c}{D_0}\right)^2 \frac{x}{2\sqrt{yh}} \tag{13.1a}$$

where D_c = the diameter of the jet at the contracted section (veena contracta)

D_0 = the actual diameter of the orifice

(iii) *90° V-notch*: If the flow is led to a field channel in which a 90° V-notch is installed, Fig. 13.3, the flow rate (Q) can be determined by measuring the head (H) over the V-notch as

$$Q = 1.38 \, H^{2.48} \tag{13.2}$$

and can be obtained from tables (Appendix B).

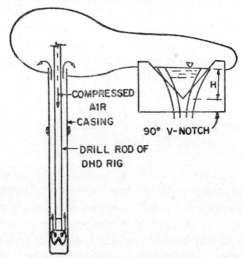

Fig. 13.3 Well discharge measurement—90° V-Notch.

Method of Measuring Water Level

Since the water level drops fast during the first one or two hours of the test, readings should be taken at short intervals which are gradually increased as the pumping continues. Fairly accurate measurements of water level can be made by using an electric sounder, by the wetted tape method (the steel tape is coloured with chalk and lowered inside the well; the reading to the wetted portion is read) or by the air line method, Fig. 13.4. The instant of each reading is noted by a stop watch.

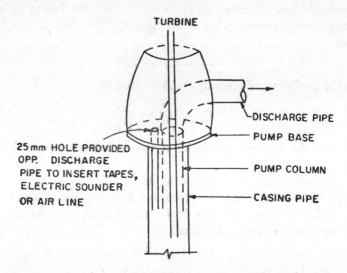

Fig. 13.4 Hole provided on well top for water level measurement.

Sometimes the yield test is conducted by using an air compressor. It involves an air line inside an educator pipe lowered into the well, Fig. 13.11. The compressor should be capable of developing a pressure of 7–10 kg/cm^2 depending on the depth from which water is to be pumped. For determining the compressor capacity the thumb rule is to provide 6 litres of free air for each litre of water at the anticipated pumping rate. The air lift produces best results when the submergence of the air line is about 60%, i.e. when 60% of the total length of the air line is under water.

Slug Tests

Carrying out pump tests becomes problematic in hard rock borewells due to high drawdown and small diameters and the compressor yields not being reliable due to inadequate submergence or capacity. A preliminary evaluation of yield may be made by slug tests. Slug test on a borewell involves sudden injection of a known quantity of water, say a 10 litres slug, and taking a rapid series of water level measurements to define the shape of the head dissipation hydrograph, i.e. residual head versus elapsed time since slugging (Ferris et al. 1962, Cooper et al. 1967). Similarly a known quantity of water can instantaneously be bailed out and the recovery trend of the water level can be analysed by the method outlined by Skibitzke (1958). Similar tests have been described by Houtkamp and Jack (1972).

Slug tests were carried out on 16 borewells tapped in weathered or fractured granitic gneisses. The transmissibility values were obtained by the

Yield Test and Selection of Pumpsets

Cooper method and were correlated with the compressor yields. For hard rocks, T of the order of 100 m³/day/m may be considered good. The dissipation of head is very rapid for T values of over 50 m³/day/m and is very difficult to measure when it is beyond 100 m³/day/m. For ready use by the uninformed well driller, head dissipation hydrographs as shown in Fig. 13.5 can be prepared for the size of the hole his rig is capable of drilling. A constant slug of 10 litres will facilitate comparison of results from different borewells. The slug test transmissibility value is a better indicator of the intrinsic yield capacity of the well rather than the compressor yield. Slug tests may also prove useful in finding out the water yielding capacities of rocks at different depths during drilling. If tests are carried out at different depths, vertical variation in permeability can be determined, which can throw light on the potential water yielding horizon.

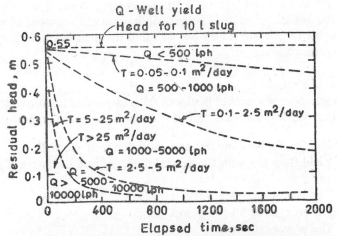

Fig. 13.5 Slug test—head dissipation hydrographs (10.1 slug) for 15 cm borewells in weathered granitic gneisses (after K.R. Karanth, 1974).

Bailing

For small yields, water can be bailed out from the well at a constant rate till the water level in the well stabilises (constant drawdown). The total number of bailings and the total time taken are recorded. Knowing the volume of the bailer and the average number of bailings per unit time, the flow rate can be determined. The observed drawdown may not be equal to the actual drawdown when the well is pumped at the same rate, since the bailing is intermittent. A bailer test, however, gives an idea of the productivity of the well.

The yield test is usually run for 10-15 hours and the sustained yield of the well for a constant drawdown is noted. The specific capacity of the well is its yield per unit drawdown and is true for the given pumping period.

The efficiency of the well may be determined by the step drawdown test. Thus the two characteristics of a well, i.e. the specific capacity and 'efficiency, are determined. Thus the operation of testing for yield consists of:

(i) Measuring the static water level (before pumping begins), i.e. the depth below a measuring point (MP) or ground level.
(ii) Pumping at a maximum until the water level in the well stabilises, i.e. the well yields a constant discharge (sustained yield) after pumping for 10-15 hours and the drawdown becomes constant.
(iii) Noting the drawdown by measuring the depth of the stabilised level of water in the well.
(iv) Measuring the discharge from the well, i.e. the sustained yield under a constant drawdown after a long duration (10-15 hours) pumping.

The results of the yield test, i.e. a sustained yield under a constant drawdown after pumping for 10-15 hours, will help in the selection of a suitable pumping set to be installed on the well and the data has to be furnished to the pump manufacturers.

Selection of Pump Sets

The selection of a proper pumping set is important to ensure continued satisfactory yields from wells and the factors to be considered are:

(i) Finished inside diameter and total depth of the well.
(ii) Yield from the well, the desired pumping rate and hours of pumping per day.
(iii) The lowest pumping water level (in the dry season).
(iv) The total head on the pump.
(v) The power required.
(vi) The quality of water, whether corrosive, clear or sandy.

The characteristic curves of a pump show the relationship between pump discharge, total head, horsepower consumed and pump efficiency at a particular operating speed of the pump, Fig. 13.6a. Pump manufacturers also give the characteristics of their pumps at different speeds within a limited range. The discharge-drawdown relationship of a well is known as the well characteristics and is obtained by pumping tests, Fig. 13.6b. An accurate test of a water well in advance more than pays for itself by the saving that can be made in the selection of the proper pump and in the reduction of power costs. The characteristic curves of the well and the pump enable the selection of a pump which best suits the well. It is usual to draw the well characteristic curve on a tracing paper in the same scale as that of the pump characteristic curves. The pump characteristics are matched with the well characteristics and the pump which gives the maximum efficiency at the desired head and discharge is selected. The pump selected will give a

Yield Test and Selection of Pumpsets

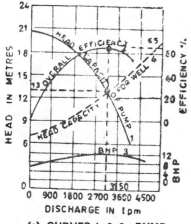

(a) CURVES 1, 2, 3—PUMP CHARACTERISTICS
CURVE 4—WELL CHARACTERISTICS

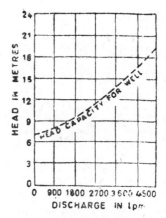

(b) WELL CHARACTERISTIC CURVE

Fig. 13.6 Matching pump with well.

discharge of 3,150 lpm a total head of 13 m with an efficiency of 65% and requires 12 hp. Some pump manufacturers provide information of their products as tabular data. Though precise matching of pump and well characteristics is not possible with tabular data, it is possible to select an efficient pump by checking the data of a large number of pumps. When the well yield exceeds the irrigation requirement, it is economical to utilise in full the available well yield. The excess water may be sold to the neighbouring farmers. Pumps could be of plunger, displacement, deep well turbine, submersible, air lift or jet type. The suitability of each type is given in the following.

Plunger pumps: Hand pumps with the cylinder at the ground surface or the hollow brass cylinder lowered below the ground surface (depending upon the ground water level) can discharge 20–60 lpm. A plunger which is connected to the pump handle rod moves up and down in the cylinder. It can be hand operated or driven by a 1 or 2 hp (V-belt driven) motor, Fig. 13.7. Pump handles often get broken in constant village use; handles should be kept in spare as also extra bolts, nut and bushings needed at the moving joint of the handle. Villagers should be taught to handle the pumps gently. The pump should be set so that the cylinder is within 6.7-7.6 m of the lowest water table, and can pump of a total head of 45-60 m.

Jet pumps: Jet pumps are often practicable for pumping rather small flows (40-90 lpm) under low heads (15-45 m) when the water level is beyond 7.6 m from the ground surface. Their capacity reduces as the lift increases. A jet pump consists of a pump and a jet, Fig. 13.8. Water is recirculated from the delivery side of the pump to the bottom of the suction pipe and is injected through a nozzle to impart additional kinetic energy. This gives

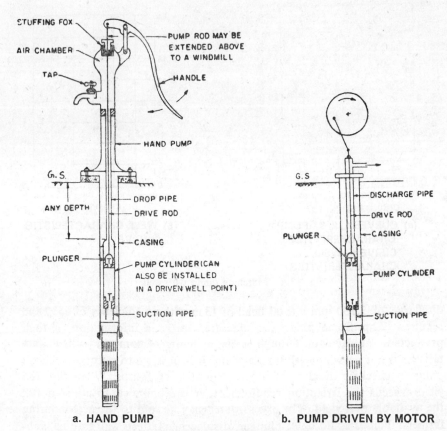

a. HAND PUMP   **b. PUMP DRIVEN BY MOTOR**

Fig. 13.7 Plunger pump-operated by hand or motor.

additional suction lift by creating a partial vacuum at this point. The advantage of the jet pump over most other types of deep-well pumps is that the pump and motor may be set away from the well. Jet pumps are usually used for residential buildings and hotels. There are two types – twin type for borewells 15 cm and above, and packer type (duplex) borewells less than 15 cm.

Deep-well vertical turbine pumps: Deep-well vertical turbine pumps are most widely used for large tubewells. The bowl-assembly (impellers) is kept below the lowest pumping water level, but the driving unit—electric motor or petrol or diesel engine—is on the ground surface and is connected by a long shaft, Fig. 13.9. Usually deep-well turbine pumps are used for fairly high flows under high heads. This type of pump has the advantages of high efficiency, high head pumping capability and excellent serviceability. The impellers can be obtained semi-open or fully enclosed. This pump requires sufficiently straight and plumb well for installation and proper operation and is subject to abrasion from sand. The maintenance problem is severe

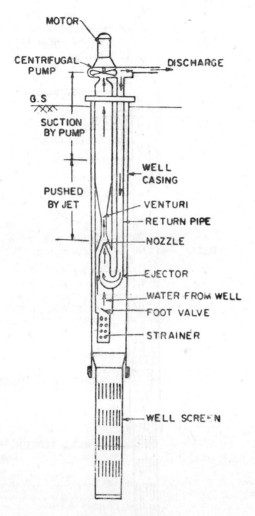

Fig. 13.8 Jet pump.

when pumping corrosive water unless pump, column, shaft, etc. are made of non-corrosive materials. Lubrication and vertical alignment of shaft is critical. The overall efficiencies of turbine pumps range from 50 to 80%.

Submersible pumps: Submersible pumps have the motor and the bowl assembly as a unit submerged below the lowest pumping water level. A water proof cable supplies power to the motor. Submersible pamps to fit inside 10, 15, 20 and 25 cm borewells are available in India, Fig. 13.10. They can be used for flow rates from 40-3,000 lpm and heads from 15-150 m. They can be installed in crooked wells but repair to motor or pump requires removal from well and is subject to abrasion from sand. This type of pump

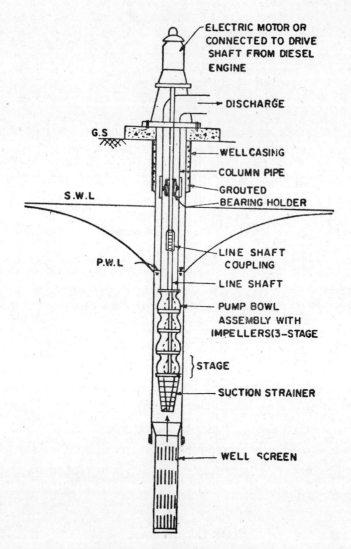

Fig. 13.9 Deep well turbine pump.

has the advantage that it can be installed when there is little or no floor space to install the unit and in locations that require quiet operation. They can be either water or oil lubricated. Their initial costs are lower than those of vertical turbine pumps. Their repair and maintenance costs, however, are high. The new type of voltage regulated starters have solved the problem of over loading.

Air-lift pumps: Air-lift pumps have efficiencies ranging from 20 to 35%; the efficiency greatly depends upon the percentage submergence and is reasonable when the percentage submergence is 50 to 60%, Fig. 13.11. These

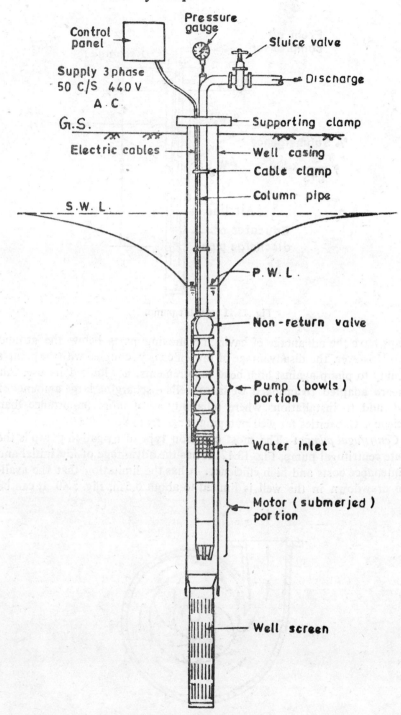

Fig. 13.10 Submersible pump.

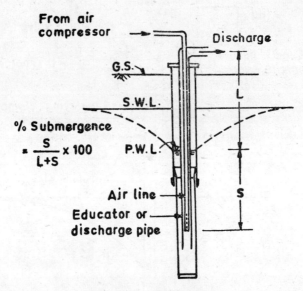

Fig. 13.11 Air-lift pump.

pumps have the advantage of having no moving parts below the ground level. However, the disadvantage of low efficiency, coupled with the pump's inability to pump against high head requirements, has limited its use. Air lifts are adapted to crooked wells, to wells discharging large amounts of sand, and to installations where reliability is of more importance than efficiency. Capacities for well pumping range from 90 to 9000 lpm.

Centrifugal pumps: The most common type of irrigation pump is the volute centrifugal pump, Fig. 13.12. It has the advantage of low initial and maintenance costs and high efficiency. It has the limitation that the available drawdown in the well is limited to about 6.5 m, Fig. 5.40. It can be

Fig. 13.12 Centrifugal pump—volute type.

used on a tubewell if a pit is made to house the pump, so that the pumping water level is within the suction lift of the pump, thereby saving in the initial cost of a deep well pump. It is also not very efficient under low heads, Fig. 13.13.

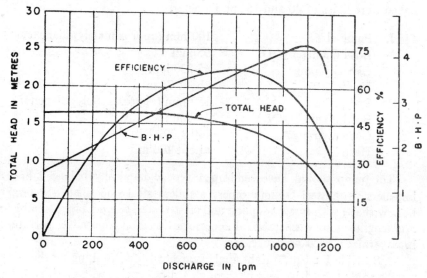

Fig. 13.13 Characteristic curves of centrifugal pump.

The propeller pump is ideally suited for low head high discharge pumping. For pumping from rivers, canals and large tanks, the propeller pump is often the most efficient.

Information to be Supplied to Pump Manufacturers

The following information is usually desired by pump manufacturers so that they can determine the type and size of pump needed to fit the characteristics of the well.

(i) Depth of well.
(ii) Inside diameter of well casing.
(iii) Depth to static water level.
(iv) Drawdown-yield relationship.
(v) Seasonal fluctuation in water table.
(vi) Capacity of pump.
(vii) Depth to end of suction pipe.
(viii) Whether strainer is required.
(ix) Total head on the pump (the height and/or distance to which water has to be delivered).
(x) Type of driver: Electric—voltage, phase, cps
  Gasoline/Diesel
  Natural Gas/LPG Power take off

A typical tender specification for the supply of two 'Turbine pumps' to the Director, Central Ground Water Board, Vedavathi River Basin Project No. 2, Bangalore is given in the following.

A. Water lubricated vertical turbine pump, suitable for installing in naked bore holes of 20 and 15 cm diameter.

I.	Pump size	100 mm (approximately) diameter
	Bowl diameter (OD)	95 mm
	Capacity range	45-230 lpm
	Head	about 75 m
II.	Pump size	150 mm (approximately) diameter
	Bowl diameter (OD)	140 mm
	Capacity range	460-1,090 lpm
	Head	about 100 m

The pump should have semi-open type dynamically balanced bronze impellers, with steel column pipes, stainless steel line shaft (preferably hollow shaft) with screw coupling and class-C steel column pipe assembly of 3 m lengths along with standard accessories such as foot valve, perforated brass strainer, pressure gauge, etc.

B. Suitable 1 : 1 ratio right angle gear head for installing on the above pumps so that the drive may be through a diesel engine with an intermittent spicer shaft coupled with universal joint cross.

Note: These pumps are proposed to be utilised as test pumps in borewells and should be capable of withstanding continuous running at a single stretch of 8 to 10 days.

Detailed specifications and catalogues, etc. should accompany the tenders.

Example 13.1: A borewell was drilled in the hard rock area of Bangalore environs (south India) to a depth of 80 m with a finished inside diameter of the bore 16 cm. The s.w.l. was found to be at a depth of 15 m b.g.l. The approximate yield as measured over the V-notch, when the air compressor of the Ingersoll Rand Rig was operating continuously for about 3 hours in the borewell, was 280 lpm (clear water). Assuming a drawdown of 25 m and that the water has to be lifted to a point 5 m a.g.l. and 60 m away from the borewell, select a suitable pumpset for installation. The actual yield from the borewell may be taken as 300 lpm, i.e. 7% higher than the compressor yield.

Solution: The total head (H) is calculated as follows

Depth to s.w.l.	15 m
Drawdown during pumping	25 m
Delivery head	5 m

Yield Test and Selection of Pumpsets

Friction loss (assuming a 75 mm outlet
and delivery pipe to convey 300 lpm)   2.5 m
Minor losses (due to bends and elbows)   0.5 m

Total head   48 m

Allowing a safe margin, say $H = 50$ m

Yield $Q = 300$ lpm, or 0.005 m^3/sec

Power of the pump required assuming an overall efficiency $\eta = 60\%$ for the pumpset

$$P = \frac{\rho g Q H}{\eta} \text{ watts}$$

where $\gamma = \rho g$, ρ being the mass density of water = 1,000 kg/m^3, and $g = 9.81$ m/sec^2.

Therefore

$$P = \frac{1,000 \times 9.81 \, (0.005) \, 50}{0.6} = 4090 \text{ W, or } 4.09 \text{ kW}$$

From the performance charts or characteristic curves of the available pumpsets, a suitable submersible pump may be selected like 'K—57 type 5-stage 6 kW ATALANTA' submersible pump having maximum outside diameter of 137 mm and outlet size of 75 mm.

QUIZ 13

I. Match the items in A and B:

Group 1

A	B
Method of drilling	*Indications of water yield*
i. Cable tool	a. Mud loss, mud thinning and caving
ii. Hydraulic rotary	b. Amount of water blown out along with cuttings
iii. DTH	c. Sudden changes in water level and rates of recovery after bailing
iv. Air rotary	d. Air compressor continuously operated for 3-5 hours after completion of drilling and the water blown out, measured over a 90° V-notch.

Group 2

	A Type of pump		B Suitability
i.	Centrifugal pump	a.	Pump bowl submerged; head of 30-240 m; requires vertical well and long drive shaft
ii.	Plunger pump	b.	Can pump large quantity of water containing sand and silt; low efficiency of 30-35%; high initial cost
iii.	Jet pump	c.	Pump bowl and motor submerged; head of 16-120 m; alignment of well less critical; repair of motor or pump requires removal from well; noiseless
iv.	Vertical turbine pump	d.	Maximum suction lift of 6.5 m; head of 30-45 m; suitable for open and dug-cum-borewells; low initial and maintenance costs
v.	Submersible pump	e.	Pump cylinder set within 6-7 m of the lowest water table; head of 15-60 m; low-capacity and high lift; can pump water containing sand and silt; operated by hand or motor; suitable for drinking water schemes; high maintenance cost
vi.	Air-lift pump	f.	Head of 6-45 m; capacity reduces as lift increases; high suction lift of 6-30 m; pump can be installed away from the well; simple in operation and maintenance

II. State whether 'true' or 'false'; if false, give the correct statement:

1. The yield test is usually run for 10-15 hours and the sustained yield of the well for a constant drawdown gives the true yield.
2. An air-lift pump gives the accurate yield when the submergence of the air-line is less than 60%.
3. Slug tests are not usually made on wells, since the head dissipation is very rapid and difficult to measure.

III. Select the correct answer(s):

1. In DTH drilling, the compressor yield (water blown out) after completion of drilling
   a. gives exactly the yield of the well
   b. gives approximately the yield of the well, the actual yield being $\approx$ 7-10% more than the compressor yield
   c. gives a rough idea of the yield range and probable drawdown for selecting a test pumping unit
   d. none of the above answers

2. The exact yield of a drilled well can be determined by
   a. a test pumping unit, usually a submersible pump driven by a diesel generating set
   b. bailing, for small yields
   c. slug tests

Yield Test and Selection of Pumpsets

     d. an air compressor
     e. all the above methods
3. The water level(s) during pumping can be measured by
     a. an electric sounder
     b. wetted tape method
     c. air line method
     d. all the above methods
4. The discharge during pumping can be measured by
     a. a water meter
     b. the time required to fill a tub of known volume
     c. an orifice meter
     d. a 90° V-notch
     e. measuring the coordinates of a point on the falling jet of water from the end of the pipe outlet
     f. all the above methods
5. The factors to be considered in the selection of a suitable pumping set are
     a. finished inside diameter of the well
     b. total depth of the well
     c. static water level
     d. the yield from the well
     e. the desired pumping rate
     f. hours of pumping per day
     g. the highest pumping water level
     h. the lowest pumping water level
     i. the total head on the pump
     j. the type of drive—electric motor or diesel engine
     k. the quality of water—corrosive, clear or sandy
     l. all the above factors

PROBLEMS

I. A borewell was drilled in the hard rock area of Bangalore environs to depth of 70 m with a finished inside diameter of the bore of 15 cm. The SWL was found to be at a depth of 12 m b.g.l. The approximate yield as measured over the V-notch when the air compressor of the Ingersoll Rand Rig was operating continuously for about 4 hours in the borewell, was 250 lpm (clear water). Assuming a drawdown of about 18 m and that the water has to be lifted to a point 6 m a.g.l. and 80 m away from the borewell, select a suitable pumpset for installation. The actual yield from the borewell may be taken as 8% more than the compressor yield.   (3.5 kW)

II. A pump has to be installed to lift water from a shallow well to irrigate 2.4 hectares of paddy. The level of the field to be irrigated varies from RL + 34.00 m to + 37.00 m and the field is 360 m away from the well. The lowest water level in the well during pumping may be taken as RL + 20.00 m. Assuming an overall efficiency of 60%, determine the hp of the pumpset reqired, size of suction and delivery pipe, and the monthly electricity bill for running the pump for 12 hours a day, on an average. The rate per unit of electric power (kWhr) may be taken as 15 ps.

(*Hint*: Peak consumptive use of paddy = 7 mm/day, Irrigation efficiency = 70%)

(2kW, 6×6 cm size pump, 15 cm concrete pipe, Rs. 99.50 p.m.)

14

Ground Water Pollution and Legislation

Ground water is an economic resource and more than 85% of the public water supplies are obtained from wells. Ground water supplies for rural areas have certain advantages over surface water. The supply is invariably close at hand, the water is of more uniform character and relatively free from harmful bacteria, and can be developed at a small capital cost in a short time. Ground water may become contaminated due to improper disposal of liquid wastes, defective well construction and failure to seal the abandoned wells. These provide possible openings for the downward movement of water into subsurface formations without the process of natural filtration. Contamination may also take place through movement of waste water through large openings such as animal burrows, fissures in rocks, coarse gravel formations or manmade excavations. Contaminated ground water may appear clear and yet contain pathogenic organisms. Bacteria from the liquid effluents from septic tanks, cess pools, pit prives, etc., are likely to contaminate shallow ground water aquifers. Sewage effluents discharged directly into water-bearing formations through abandoned wells or soil absorption systems contaminate the ground water.

The depth of the water table below ground level is a governing factor in determining pollution since, as the water table approaches nearer the ground surface, the greater is the risk of contamination. The area around the well should be protected—either uncultivated or simply laid down to pasture. The number of harmful organisms is generally reduced to tolerable levels by percolation of water through 2–3 m of fine grained soil.

Bacterial and viral contamination are among the most significant health hazards that must be considered in protecting ground water quality. Bacteria also affect ground water quality indirectly in beneficial ways. Studies of bacteria and virus in ground water systems have suggested that these organisms may travel only short distances in sand aquifers but long distances in short periods of time in more porous and permeable aquifers.

The source of pollution of ground water can be traced by chemical and bacteriological methods. The most common chemicals used for tracing pollution are the dyes which may be added as concentrated solutions, to give colourations of water even when diluted several million times. A strong alkaline solution of Fluoresine is the commonly used dye and should be used with care. Where turbid water is to be investigated, the red dye Rhodamine B may be employed for the detection of leaks from drains. Either of these liquids may be poured, in some quantity, on the ground surface around the well and pump pipe. If the inside of the well is illuminated, one can observe the indicator slowly trickling into the well. The quicker the advent of the test liquid, the greater is the opportunity for pollution. Common salt and sodium phosphate can also be used as tracers. Several different bacteria have been successfully used as tracers. Serratia marcescens is easily identified by the red colour of its colonies when it is cultured. Strains of aerogenes-like bacteria which ferment lactose at 44°C may also be used under certain selected conditions as test organisms. Whichever chemical or organism is used, it must first be established that it does not appear naturally in water which is under investigation.

When ground water wells are constructed, proper sanitary protection should be provided against surface contamination. A well should be located at a safe distance from all possible sources of contamination. Some recommended distances are given in Table 14.1, as a guide.

Table 14.1 Recommended minimum distances between a ground water well and sources of contamination.

Contamination sources	Recommended distance (m)
Building sewer	15
Septic tank	15
Disposal field	30
Seepage pit	30
Cess pool	45

The well site should be prevented from being flooded and should be graded so as to facilitate the rapid drainage of surface water away from the well. The area should be filled if necessary and maintained to prevent the accumulation or retention of surface water within a radius of 15 m from the well. For a well on a hill side, adequate intercepting ditches should be constructed on the uphill side of the well in order to keep the runoff at least 15 m away in all direction. Pump platforms, pump floors or well covers should be located at least 60 cm above the maximum flood level. The annular space around the casing should be filled with neat cement grout. In addition to protecting the supply against surface pollution grouting also serves to provide a protective sheath around the casing against corrosion, to seal off water of unsuitable chemical quality in strata above the desirable water bearing formation and to stabilise soil or rock formations which are of a caving nature. The ratio of water to cement for a suitable cement grout is 22 to 27 litres per 50 kg bag of cement, which will keep shrinkage to a minimum. Bentonite (3% by weight of cement) may be added to reduce shrinkage and improve fluidity of the mixture. Bentonite and water should be mixed first and cement added to the clay-water suspension. The water used in mixing the slurry should meet reasonable standards for drinking water. The top of the casing should be extended at least 60 cm above the general level of the surrounding surface. Whereever possible, a concrete platform should be constructed around the casing at the ground surface. The top of the casing should then be provided with a sanitary well seal to fill the annular space between the pump column pipe and well casing. The sanitary well seal consists of suitable bushing or packing glands making a water-tight seal at the top of the casing. If the pump is not installed immediately after the construction of the well, the top of the casing should be securely closed with a metal cap either screwed or tackwelded in place. After installing the pump, the well and its appurtenances including casing, pump and pipe systems should be disinfected thoroughly in order to kill any pathogenic organisms that may be present. A solution containing about 100 ppm of available chlorine is the simplest and most effective agent for disinfecting the well and its appurtenances. Pump room floors should drain outside. The pump house should be well lighted and ventilated. It is desirable to provide a water sampling tap on the discharge line from the pump which can also be used for releasing any trapped air in the system.

Abandoned wells should be properly sealed to prevent the ingress of surface waters and the sealing materials include concrete, cement grout, neat cement, clay, sand or a combination of these materials.

The main factors affecting quality of ground water are:

(i) Salt water intrusion.
(ii) Organic, and inorganic and heat pollution by sewage and industrial wastes.
(iii) Pollution of good quality aquifers by bad quality aquifers because of faulty construction methods.

Sea water intrusion can be halted or prevented by maintenance of ground water levels well above sea level. This can be accomplished by reducing ground water extractions, modifying the pumping patterns and also by augmenting the natural replenishment by artificial recharge of local or imported water supplies. Another approach to control sea water intrusion is to form a subsurface dam by constructing a cutoff wall of sheet pile, concrete or puddled clay. An impervious zone, to prevent ground water movement in the aquifer, might also be created by injection of cement grout, emulsified asphalt, bentonite, silica gel, calcium acrylate or suitable plastics. A series of spreading grounds or injection wells, or a combination of both, could be utilised, along the coast as dictated by the geological conditions encountered to create a ground water mound or ridge. The fresh water utilised to maintain the ridge augments recharge to the ground water basin which is available for re-use. A pumping trough can be developed by a line of pumping wells, properly spaced along the coast. These wells produce a mixture of saline and fresh water resulting in waste of considerable quantities of fresh water. The pumping costs involved and the waste of otherwise usable waters are major factors to be considered in evaluating the practicability of protecting a ground water basin by maintaining a pumping trough.

In locations downstream from heavily irrigated areas, the water may be too saline for satisfactory crop production. The removal of salinity is exceedingly expensive. A possible solution is to dilute with waters of lower salt concentration so that the resulting water after mixing in suitable for use.

There is need for understanding pollution as a first step for its evaluation and control. The hydrologic environments that cause pollution include the interdependence of factors such as permeability, sorption, hydraulic gradient, position of water table and distance from contamination source. Waste disposal and salinity are the two major sources of pollution. For extensive literature on man-caused ground water pollution including causes, occurrences, procedures for control and methods for monitoring reference may be made to the classical work: POLLUTED GROUND WATER by D.K. Todd and D.E. Orren McNulty (Ref. 36). However an outline of ground water pollution is given in chart 14.1 to arouse curiosity and indicate lines for detailed study on ground water pollution, evaluation and control.

Ground Water Legislation

Large scale ground water exploitation is planned for:

(i) Carrying out 'green revolution' schemes for increasing agricultural production in the country.
(ii) Providing drinking water supply in all the villages in the country within a specified period, i.e. national rural water supply schemes.
(iii) Providing water for municipal and industrial uses and promote rapid growth of industries.

Ground Water Pollution and Legislation

Haphazard development will create a lot of problems requiring heavy expenditure and time for setting things right at a later stage. For proper regulation, control and development of ground water and efficient ground water management, it is necessary that the subject of ground water is brought under legislation. The bill of legislation should cover the following aspects.

(a) Notification of area for control and regulation of ground water development.

(b) Registration of existing user of ground water in the notified area with details as follows.

  (i) Source of water such as a well, tubewell or spring.
  (ii) Water lifting appliance.
  (iii) Date since water was first used.
  (iv) Purpose (s) water being used.
  (v) Quantity of water used.
  (vi) Period of use in each year.
  (vii) If the well is used for irrigation, the location and extent of the area irrigated.
  (viii) In the case of state, municipal or community run water supply schemes, the area of land involved, the number of people utilising such water, the quantity of water used, diversion or pumping points and their locations.

(c) Grant of permit for extraction and use of ground water for persons other than the existing users in the notified area after obtaining full details of the proposal(s) as listed in Clause (b) above.

(d) Regulation of drilling activity in the notified area.

(e) Grant of licence for sinking wells or tubewells and for drillers.

(f) Powers for changes in the permit and cancellation of permit or licence for valid reasons.

(g) Right of access and other powers to enter, investigate, inspect and obtain useful relevant information.

(h) Powers of imposing penalty for contravening the provision to the bill.

(i) Indicate methods of appeal.

(j) Indicate jurisdiction of courts.

(k) Power to amend or make further rules.

QUIZ 14

I. Select the correct answer(s):
  1. Ground water can be protected against pollution by
      a. locating a well at a safe distance from all possible sources of contamination

b. providing drainage of surface water away from the well
  c. grouting and sealing the well casing
  d. sealing of the poor quality aquifers by grouting
  e. taking measures to halt and abate sea water intrusion
  f. all the above methods
2. Ground water pollution occurs due to
  a. improper disposal of liquid wastes, sewage and industrial effluents
  b. failure to seal the abandoned wells
  c. bacterial and viral contamination
  d. sealing of the poor quality aquifers by grouting
  e. sea water intrusion
  f. all the above factors
3. Ground water legislation is required to
  a. minimise the use of ground water
  b. protect the ground water against pollution
  c. raise the irrigation efficiency
  d. check indiscriminate draining of ground water
  e. minimise energy consumption
  f. all the above reasons
4. ....is the commonly used dye to trace the ground water pollution.
  (Rhodamine, Serratia Marcescens, Fluoresine)

II. State whether true of false:
1. The nearer the water table to the ground surface, greater is the risk of contamination.

15

Ground Water Recharge

With the increasing use of ground water for agricultural, municipal and industrial needs, the annual extractions of ground water are far in excess of net average recharge from natural sources. Consequently, ground water is being withdrawn from storage and water levels are declining, resulting in crop failures, adverse salt balance, sea water intrusion in coastal aquifers and land subsidence in areas where drafts result in compaction of sediments.

In many instances, the overdraft is due to the diminishing opportunity for natural recharge of ground water basins due to such factors as:

(i) Lining of stream channels and concentration of surface runoff by flood control projects.
(ii) Discharge of sewage and industrial wastes through closed sewage disposal systems.
(iii) Sealing of natural recharge areas with impervious side walls, streets, air ports, parking lots and buildings.
(iv) Diversion and export of waters which might otherwise percolate naturally in the stream channels.

Artificial recharge (or replenishment) is one method of modifying the hydrological cycle and thereby providing ground water in excess of that available by natural processes. It is accomplished by augmenting the natural infiltration of precipitation or surface water into underground formations by some method of construction, by ponding or spreading of water, or by artificially changing the natural conditions.

The following are the favourable conditions for natural or artificial recharge.

(i) Formation of sand, gravel, or highly fractured rocks either underground or exposed over a large area or in stream channels.
(ii) The presence of caverns, fractured or faulted zones or numerous

small cavities in rock formations (limestone areas) either underground or exposed on the land surface or stream channels.
(iii) Karsts or sinkhole topography.
(iv) The absence of barriers for horizontal or vertical movement of ground water.
(v) Feasible locations for installation of recharge wells, dams, diversions or other recharge structures.
(vi) Wide braided streams, broad alluvial fans and glaciofluvial deposits may present excellent opportunities for water spreading.

The methods of artificial recharge are by direct flooding, water spreading in basins, ditches and furrows, Fig. 15.1, irrigation, modified stream bed or natural channel, percolation dams, pits and shafts, injection wells, and

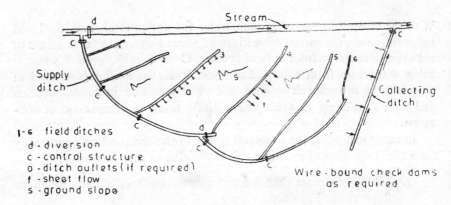

Fig. 15.1 Ditch and flooding-type recharge project.

induced recharge by lowering the water table by pumping water from wells, collectors, galleries located near surface water sources like lakes or streams, Fig. 15.2a, b. For confined aquifers or shallow beds where ponding is not practicable, recharging is effected by pumping water down the wells at rates rather less than the corresponding withdrawal rates. Mostly recharge water is excess surface water, but industrial waste water, sewage and uncontaminated cooling water from industrial and airconditioning plants are used in some countries.

In areas where the previous formations are at a shallow depth, recharging is done by digging pits or shafts. Abandoned gravel pits have been utilised occasionally. If storm waters are to be recharged through shafts, consideration should be given to removal of silt. Injection rates of 20 m/day was noticed in the first 1/17 ha of Peoria pit at Illinois, USA, Fig. 15.3.

Injection wells may be drilled downstream of a dam and the water released from the spillway is conveyed into the wells. In areas of cavernous limestones and gypsum, recharge wells may be placed upstream of a flood

Ground Water Recharge

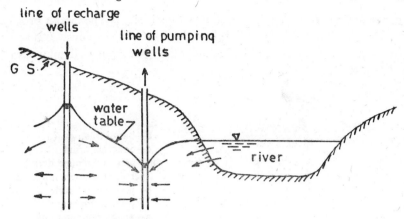

a. Recharge induced by wells

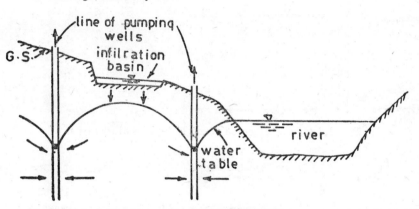

b. Recharge induced by wells and infiltration basin

Fig. 15.2 Induced recharge.

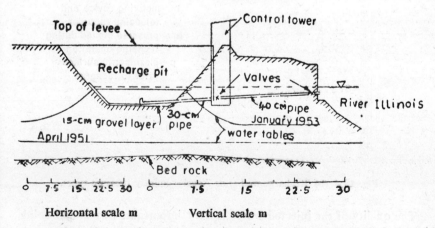

Fig. 15.3 Recharge pit at Peoria, Ill. U.S.A. (after Suter, 1956).

water retarding basin or other structure, Fig. 15.4a, b. The intake, provided with trashrack, should be well below the crest of the spillway but several metres above the bottom to aid in desilting. A pumping well may also be used as a recharge well.

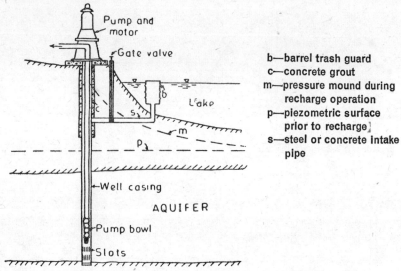

Fig. 15.4 (a) Recharge well (Texas high plains district type).

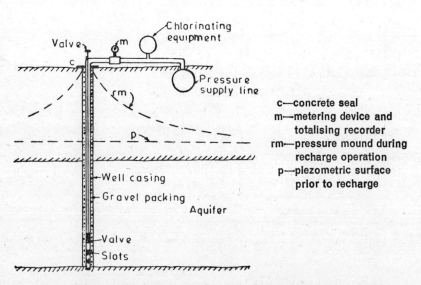

Fig. 15.4 (b) Recharge well (Los Angeles country pressure type).

The quality of the injected water is very important. Suspended solids, biological and chemical impurities, dissolved air and gases, turbulence and

Ground Water Recharge

temperature of both the aquifer and injected water will have an effect on the life and efficiency of a well by clogging or corrosion of the well screen. Large amounts of dissolved air in the recharge water tend to reduce the permeability of the aquifer by 'air binding'. The injected water should have temperature only slightly higher than the temperature of the aquifer.

The use of fresh water barriers by ground water recharge to prevent sea water intrusion, is practised extensively on the sea coast in southern California, Fig. 15.5. The fresh water barrier should be far enough inland to force all the wedge back seaward. Otherwise, the fresh water will separate the wedge and force the landward edge still farther inland, creating a saline wave. A series of spreading grounds or injection wells or a combination of both, could be utilised as dictated by the geologic conditions encountered.

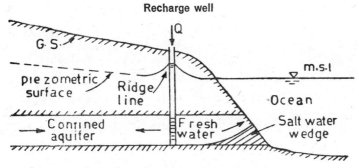

Fig. 15.5 Pressure ridge to control sea water intrusion.

Injection rates are maintained in wells along the coast of Manhattan Beach in Southern California by using chlorinated water, free from suspended solids. The high cost of chlorinated water is justified in this case, since the system of injection wells protects an inland ground water basin from sea water intrusion.

In Netherlands, the fine-sand beds are recharged with treated water from the Rhine, to provide water storage for supply as well to act as a barrier against the inward seepage of sea water from the North Sea.

Spreading in natural stream channels that are not subject to year round flow is an effective method. No additional land is required and the stream beds tend to be self cleaning. Inexpensive small levee systems can be constructed between storm periods to maximise coverage, or permanent drop type structures can be incorporated.

Water meandering in canals over a part of 32 ha Rohrer Island, Dayton, Ohio has recharged ground water at the rate of 13.2 ha-m/day, (Fig. 15.6).

A typical plan of a basin type recharge project is shown in Fig. 15.7. In projects designed principally for the purpose of spreading storm waters, multiple basins are advantageous since the first of a series of basins can be utilised for settlement of silt. The desilting or detention basin should be

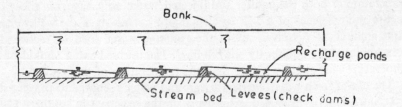

Fig. 15.6 Spreading in stream channels.

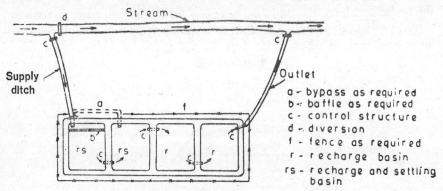

a - bypass as required
b - baffle as required
c - control structure
d - diversion
f - fence as required
r - recharge basin
rs - recharge and settling basin

Fig. 15.7 Basin type recharge project.

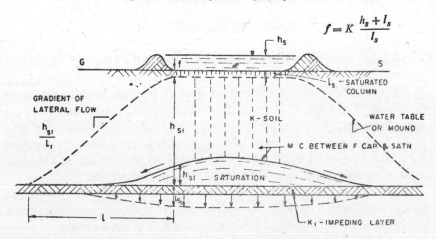

$$\text{Percolation Rate} = K_1 \times \frac{h_{s1} + l_{s1}}{l_{s1}}$$

When mound meets saturated soil column, infiltration is controlled by area through which lateral flow moves $\times K \times \dfrac{h_s}{l}$ + area through which water moves into impeding layer $\times K_1 \times \dfrac{h_s + l_{s1}}{l_{s1}}$

Fig. 15.8 Mechanics of spreading (after unknown, 19- —).

Ground Water Recharge

large enough to reduce the velocity of flow substantially, and its inlet and outlet facilities should be so located that short circuiting is prevented.

The mechanics of recharge by spreading is illustrated in Fig. 15.8. The infiltration rates were determined with infiltrometers—open ended cylindrical or square units driven into the soil and constant heads maintained, prior to the construction of ELRIO spreading ground, Fig. 15.9. In the figure, if $h_s = 30$ cm, $l_s = 7.5$ cm, $K = 30$ cm/day, the infiltration rate is

$$f = Ki = 30 \left(\frac{30 + 7.5}{7.5} \right) = 150 \text{ cm/day} \quad \text{or} \quad 1.5 \text{ m/day}$$

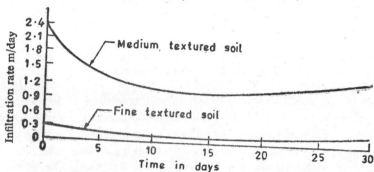

Fig. 15.9 Infiltration rates (from infiltrometers) at proposed Elrio water spreading ground (after unknown, 19- —).

Beneath the saturated soil column the soil moisture content trends to be between field capacity and field saturation, Fig. 15.10. Water reaching an impeding layer creates a water table or mound which moves upward and outward or laterally. The amount of water moving through the impeding layer largely depends on its permeability. If for clay soil $K_1 = 0.06$ cm/day, average height of the water table on the impeding layer is 3 m and the length of the saturated column beneath the top of the impeding layer is 7.5 cm. The percolation gradient is very high

$$i_p = \frac{3 + 0.075}{0.075} = 41$$

but the percolation rate is only $V = K_1 i_p = 0.06 \times 41 = 2.46$ cm/day compared to f.

If for a mound height of 6 m, the water has moved out 30 m, the lateral flow gradient $= 6/30 = 0.2$. If the lateral flow is not great enough, the mound will build up high enough to contact the saturated column at the surface, i.e. there is hydraulic continuity between the impeding layer and soil surface. When this occurs the infiltration rate declines and is controlled largely by the lateral flow. The same situation occurs when percolating water builds up on a water table. This is an important consideration in the design as the type, size and shape of the basin affect the rate of lateral flow.

A multipurpose spreading operation which emphasises the disposal of

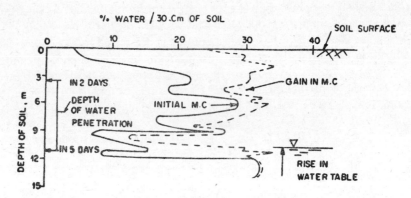

Fig. 15.10 Change in soil moisture after recharging 12 days—wood ville spreading basin (after unknown, 19- —).

suitably treated agricultural waste waters and spreading of local storm or flood waters and imported water for additional ground water replenishment, provides a promising method. Rotational spreading basins may be used to allow significant drying periods for recovery in infiltration rate. The infiltration rates can be increased by certain soil treatments like vegetation, organic residues (bacteria), chemicals (ferric sulphate), grits and sand materials, and certain physical approaches like spreading when the infiltration rate is high (rates decline with time due to microbial sealing) and using relatively high depths of water to increase the gradient. Cotton trash consisting of boll hulls, leaves, stems, a few seeds and a small amount of lint, when mixed with soil and given a moist incubation period, is effective in increasing infiltration rates. Various grasses, particularly the Bermuda, improve the intake rate of the underlying soil. Alternate wetting and drying periods of 7-14 days with cultivation during the dry cycle gives maximum spreading rate. Drying kills microbial growths and this, along with scarification of soil, reopens soil pores.

Strips or portions of an area may be used where subsurface layers limit the flow. Water will accumulate and spread laterally on a subsurface layer. Schiff (1954) suggested spacing strips on the infiltration—percolation rate ratio. If the ratio is 10, about the same amount of recharge could be obtained by using 1/10 of the land as by using the entire area. Strips treated to increase the infiltration rates, or strips in the form of channels or shafts in the bottom of channels may further reduce the areas required.

Multiple rectangular basins have more lateral flow opportunity for recharge. This is illustrated in the following example:

Assume a 400 ha basin with uniform medium textured and ($f = 0.6$ m/day) up to 60 m and little or no permeability below 60 m. Recharge at $f = 0.6$ m/day for 3 months = 21,600 ha-m. Assuming a specific yield of 24%, water stored between depths 3 to 60 m = $57 \times 400 (0.24)$ = 5472 ha-m.

Lateral flow to adjacent area = 21,600 − 5472 = 16,128 ha-m. The greater the perimeter for a given area the greater the lateral flow opportunity:

a. A 400 ha square basin of size 2,000×2,000 m has a perimeter of 8000 m.
b. 10 square basis of 40 ha each of size 630×632 m have a total perimeter of 25,280 m.
c. 10 rectangular basins ($L = 4B$) of 40 ha each of size 1264×316 m have a total perimeter of 31,600 m.

The ratio of perimeters or lateral flow opportunity

$$a : b = 1 : 3.2$$
$$a : c = 1 : 3.9$$
$$b : c = 1 : 1.25$$

Hence ten rectangular basins of 40 ha each may be adopted for spreading. Thus more lateral flow will occur if a number of spreading areas are used.

Water spreading ranges from 0.3 to 3 m/day in USA. Estimates of the rates of recharge in full scale basins conducted in USA indicate an average discharge rate of 1,125 lpd/m^2, recharge through wells of 180 to 3900 lpm, recharge through wells of storm drainage 360 to 5000 lpm or more and sewage and waste water 0.06 to 0.36 m/day. Treated sewage can also be recharged through wells. Continuous operation is possible with regular chlorine injections and redevelopment by pumping about 4% of the recharged water. Pit recharge rates of 0.3, 0.4 and 3 m/day are used for the underlying aquifers of sandstone, fine sand, and sand and gravel, respectively.

In Maharashtra a number of percolation dams are built and the artificial lake created will improve the ground water conditions in the areas in the vicinity, where a number of wells may be sunk.

Recharge wells. Deep, confined aquifers can be recharged by a recharge well. Its flow is the reverse of a pumping well but its construction may or may not be the same.

If water is passed into a recharge well, a cone of recharge will be formed which is reverse of a cone of depression for a pumping well, Fig. 15.11. The approximate steady-state equations for recharge rate Q_r into a completely penetrating well are:

for confined aquifer : $Q_r = \dfrac{2\pi (Kb)(h_w - h_0)}{\ln \dfrac{R}{r_w}}$

for an unconfined aquifer : $Q_r = \dfrac{\pi K (h_w^2 - H^2)}{\ln \dfrac{R}{r_w}}$

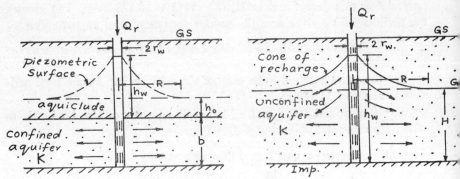

Fig. 15.11 Recharge wells.

Though the above equations are similar to discharge equations from a pumping well, the recharge rates are seldom equal to pumping rates.

Well recharge rates in USA vary from 500 to 5000 m³/day. Initially the intake rates are high and gradually decrease or become constant. High intake rates are found in porous formations like limestones and lavas.

QUIZ 15

I. Match the items in A and B

A

i. Gravel formations, fractured or faulted zones, karsts, caverns, wide braided streams, and absence of barriers for horizontal and vertical movement of ground water

ii. Percolation dams

iii. Confined aquifers

iv. Excess surface water, industrial waste water, sewage and uncontaminated cooling water from industrial and air-conditioning plants

v. Pervious formations at shallow beds

vi. Excess dissolved air in recharge water

vii. A series of spreading grounds or injection wells

viii. Bermuda grass

ix. Spacing strips according to infiltration–percolation ratio

x. Augment the natural replenishment of ground water storage by some artificial measures

B

a. Injection wells

b. Recharge water

c. Recharging by digging pits or shafts

d. Reduce permeability of aquifer by 'air binding'

e. Prevent sea water intrusion

f. Favourable sites for recharge

g. Artificial recharge

h. Schiff, 1954

i. Improves infiltration rate

j. Large scale well sinking programme in the vicinity

Ground Water Recharge

II. State whether 'true' or 'false'; if false, give the correct statement:
   1. A pumping well cannot be used as a recharge well.
   2. Injected water should have temperature slightly lower than the temperature of the aquifer.
   3. The first of a series of spreading basins should be large enough to reduce the velocity of flow.
   4. Rotational spreading basins are preferred to obtain higher infiltration rates.
   5. Multiple rectangular basins for spreading storm water have more lateral flow opportunity.
   6. Recharge rates through injection wells are higher than the corresponding withdrawal rates.

16

Ground Water Basin Management and Conjunctive Use

For optimum development of water resources of any basin and their management, the step by step studies to be made and data to be collected are given in the following:

    a. Identify the basin boundary, the main river and its tributaries and other physiographic features.

    b. Divide into sub-basins of controllable size depending on factors like steep hill slopes, forest areas, irrigated and unirrigated lands, fallow areas, etc.

    c. Establish a hydrometeorological set-up for each sub-basin.

    d. Select a convenient base period for the hydrologic equation.

The hydrologic equation simply states that all water entering a river basin or sub-basin during any period of time should either go into storage within its boundaries or leave the basin during the same period and a water balance is obtained. The different components of inflow to and outflow from the basin and the method of estimation of each item is given in Table 16.1. A base period is selected so as to

(i) Allow direct determination of as many items in the equation as possible.

(ii) Cover a length of time during which the investigator has reasonable confidence that the items used reflect truly average conditions.

(iii) Eliminate some items that are negligibly small during the period selected.

If the base period does not represent long time mean climatological conditions or the proper stage, present or future, of water use for which an answer is being sought, adjustments have to be made.

    e. Apply the water balance equation for the basin for the period selected as in item (d) above. The components of water balance equation and methods of their estimation are given in Table 16.1.

Table 16.1 Components of water balance equation

S.No.	Supply into the basin	Disposal from the basin	Method of estimation
1.	Surface inflow	Surface outflow	Stream gauging—current meter gaugings, bridge openings, weirs, Chezy formula and AWLR installations.
2.	Sub-surface inflow	Sub-surface outflow	(i) Pumping test and flow net analysis $$T = \frac{Q}{n_f \Delta h}$$ (ii) Darcy's law ($Q=TiW$) if the aquifer cross-section, permeability and hydraulic gradient (as shown by the wells piercing the aquifer) are known; otherwise treated as unknown.
3.	Precipitation on the basin		Rain gauge network—Thiessen polygon and isohyetal methods
		Evaporation from soil and water surfaces	(i) Tanks and lysimeters (ii) Open-pan evaporation data and formulae
		Evapotranspiration (consumptive use) from irrigated and unirrigated fields	(i) Climatological data $U = \sum \frac{ktp}{100}$ (ii) Open-pan evaporation data $E_t = kEp$ (iii) Field experimental plots (iv) Colarado sunken tanks
		Domestic, municipal and industrial uses	Empirical and semi-empirical methods

(*Contd.*)

Table 16.1 (Contd.)

S.No.	Supply into the basin	Disposal from the basin	Method of estimation
4.	Imported water and sewage	Exported water and sewage	Hydraulic methods
5.	Decrease in surface storage	Increase in surface storage	Water levels in reservoirs, lakes and ponds (AWLR)
6.	Decrease in soil moisture storage	Increase in soil moisture storage	Soil sampling and soil moisture meters
7.	Decrease in groundwater storage	Increase in groundwater storage	(i) Specific yield—water table aquifer (ii) Storage coefficient—confined aquifer (iii) Water levels in observation wells (AWLR)

f. Determine average depth of rainfall over the basin or sub-basin by Thiessen polygon or isohyetal methods.

g. Draw stage and discharge hydrographs at control points.

h. Draw cropping pattern maps for different seasons and estimate evapotranspiration. Evapotranspiration figures for different crops may be obtained either by field experiments or by climatological data. Even open-pan evaporation data are helpful in this regard.

i. Determine evaporation from soil and water surfaces.

j. Draw soil map of the basin or sub-basin and conduct infiltration studies in different soils and irrigated and unirrigated lands.

k. Determine isobath and ground water level contours monthly or at least for two different seasons, i.e. driest (LWL) and monsoon (HWL) seasons.

l. Correlate hydrographs of monthly rainfall, river stage and discharge and ground water levels.

m. Change in ground water storage = change in GWL × involved area of the aquifer × specific yield in the case of water table aquifer or storage coefficient for confined aquifer.

n. Determine specific yield in the laboratory and storage coefficient by pumping tests in the field.

o. Directions of flow can be determined from water table (or piezometric surface in the case of confined aquifers) contours.

p. Arrive by water balance the monthly ground water accretion. Water balance studies for the Noyyil river basin, Coimbatore District, Tamil Nadu (South India) conducted by the author is given in Table 16.2.

q. Construct plots of monthly rainfall, ground water levels and cumulative change in ground water storage for period selected as in item (d) above, Fig. 16.1 and Fig. 16.2.

r. The usable capacity of the ground water reservoir can be developed by planned extractions of ground water during periods of low precipitation while subsequent replenishment can be made during periods of surplus surface supply.

s. The ground water investigation team is mainly concerned with location of sites of discharge (borewells) and recharge and design of suitable recharge facilities; and also to prevent sea water intrusion in coastal aquifers.

t. Aerial and infra-red photography, electrical resistivity surveys and well logging techniques can provide valuable information in regard to item(s) above. The US Geological Survey, by the use of an infra-red scanner, has published an atlas of Hawaii's coastal areas, pinpointing the location of underground fresh water flows.

Thus with the hydrometeorological data combined with geophysical and hydrogeological investigations, test drilling and pumping tests, it is possible to develop and manage the ground water resources of a basin. For example

Table 16.2 Water balance studies for the Noyyil river basin (3260 km^2), Coimbatore district, 1970-72

Year	Month	Inflow into the Noyyil basin (I)		Rainfall over the Noyyil basin (P)		Average evapotranspiration from the basin (E_t) M.m^3	Outflow from the Noyyil basin (O) in the Noyyil river at Muttur causeway (D/s end of the basin), M.m^3	Balance for ground water recharge (+ ve) or discharge (− ve) $I+P-E_t-0$ M.m^3	Cumulative ground water storage (datum as on Jan 1, 1970)* M.m^3
		By LBP irrigation canal at mileage 74/2 M.m^3	By the PAP Irrigation canals M.m^3	Mean depth (mm)	M.m^3				
1970	Jan	4.09	—	4.67	15.25	81	1.92	−63.58	−63.58
	Feb	14.70	—	17.80	58.10	81	0.82	−9.02	−72.60
	Mar	13.20	—	8.96	29.20	81	0.75	−39.35	−111.95
	April	9.80	—	24.01	78.50	81	0.69	+6.61	−105.34
	May	3.33	—	57.30	187.00	81	2.52	+106.78	+1.44
	June	—	—	7.34	23.90	162	0.20	−138.30	−136.86
	July	—	—	7.72	25.20	162	—	−136.80	−273.66
	Aug	—	—	21.08	68.80	162	—	−93.20	−366.86
	Sept	21.40	—	48.90	159.50	162	0.13	+18.77	−347.09
	Oct	21.70	2.66	204.60	668.00	162	15.10	+515.26	+168.17
	Nov	21.86	4.31	92.40	301.00	81	13.45	+232.72	+400.89
	Dec	11.47	4.28	5.42	17.67	81	4.35	−51.93	+348.96
	Total	121.52	11.25	500.20	1632.12	1377	39.93		

(Contd.)

1971	Jan	—	4.75	6.76	22.04	81	2.51	−56.72	+292.28
	Feb	—	2.85	1.96	6.39	81	0.09	−71.85	+220.43
	Mar	—	0.48	27.76	90.40	81	1.04	+8.84	+229.27
	April	—	—	35.20	114.90	81	0.12	+33.78	+263.05
	May	—	—	54.16	176.70	81	0.57	+95.13	+358.18
	June	—	—	48.90	159.50	162	0.07	−2.57	+355.61
	July	—	0.88	20.45	66.75	162	—	−94.37	+261.24
	Aug	—	0.34	81.68	266.20	162	2.69	+101.85	+363.09
	Sept	—	0.73	85.99	280.50	162	4.22	+115.01	+478.10
	Oct	—	0.48	150.47	491.00	162	17.20	+312.28	+790.38
	Nov	—	0.42	28.28	92.30	81	7.45	+4.27	+794.65
	Dec	—	0.25	70.04	228.60	81	20.45	+127.40	+922.00
	Total	—	11.18	611.65	1995.28	1377	56.41		
1972	Jan	17.8	0.10	15.20	49.60	81	2.92	−16.42	+905.58
	Feb	14.3	—	21.17	69.00	81	2.22	+0.08	+905.66
	Mar	13.4	—	3.50	11.42	81	2.40	−58.58	+847.08
	Total	45.5	0.10	39.87	130.02	243	7.54		

The Eastern boundary of the Noyyil river basin has been terminated about 19.3 km upstream of the actual boundary for all calculation purposes.

*Subsurface inflow and outflow have not been considered.

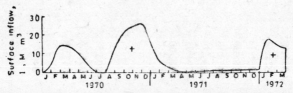

a. INFLOW FROM IRRIGATION CANALS INTO NOYYIL BASIN, I

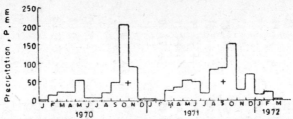

b. MEAN DEPTH OF PRECIPITATION OVER NOYYIL BASIN, P

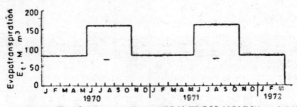

c. AVERAGE EVAPOTRANSPIRATION FROM NOYYIL BASIN, E_t

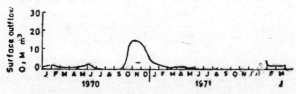

d. OUTFLOW FROM THE NOYYIL BASIN, O

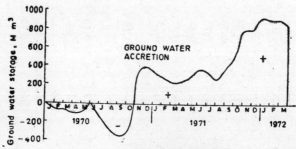

e. CUMULATIVE GROUND WATER STORAGE IN NOYYIL BASIN
(Subsurface inflow and outflow not considered)

Fig. 16.1 Water balance studies of Noyyil river basin, Coimbatore district, Tamil Nadu (South India).

in Israel the success ratio in ground water drilling has been raised from 35 to 85% with the help of geophysical surveys.

Conjunctive utilisation of river-aquifer systems, mathematical and economic models of alternative well field configurations, pumping patterns, cropping patterns and irrigation sequences must be studied in detail before arriving at a final design.

Methods of systems analysis techniques and computer applications provide a better insight for the problems of ground water management. An excellent example of the application of the computer is the study of ground water conditions in southern California (Ref. 13).

Conjunctive Use

Optimum development of water resources can be achieved by the conjunctive use of surface and ground waters. Ground water recharge occurs in nature by seepage from canals and reservoirs and return flow from irrigation. It can be augmented by artificial methods such as spreading of storm water in ponds or basins, recharge wells, pits and shafts. The usable capacity of the ground water reservoir can be developed by planned extractions of ground water during periods of low precipitation while subsequent replenishment can be made during periods of surplus surface supply. Such a coordinated operation of surface and ground water supplies is possible if there is sufficient ground water storage to meet the requirements for regulation of local and imported water supplies and if the aquifers possess sufficient transmissibility to permit the movement of recharged water to the area of extraction. Also is underground storage is devoid of losses due to evaporation, quality deterioration due to pollution, etc. Their reduction from danger of destruction of reservoir structures and wide dispersion of outlet facilities in earthquake areas, in places liable for atomic attacks, make ground water basins of inestimable value as an emergency supply. Large ground water reservoirs thus developed not only meet the deficiencies of the surface supplies in seasons of drought but also supplement them to a large extent. These conjunctive operations result in a more economic yield as they provide more water at a lower average cost. Tubewell schemes can be integrated with the canal irrigation scheme by suitably spacing them along the drainage lines in the distribution area. The benefits accruing from the conjunctive use of waters are:

(i) A large sub-surface storage at a relatively lower cost and safe against any risk of dam failures.

(ii) Provides water supplies during a series of drought years while a surface storage can at the most tide over one such year.

(iii) Efficient water use from well spaced wells due to smaller surface distribution system than a canal irrigation scheme.

(iv) Water table can be controlled by pumping from wells and prevent water logging in canal irrigated areas and reduce land subsidence due to reduced ground water levels particularly in confined aquifers.

(v) Both water conservation and flood protection can be achieved simultaneously.

(vi) A sub-surface scheme can be developed in a shorter period while it takes 10-15 years for the completion of a big surface water project.

(vii) No evaporation and percolation losses, thus obviating the construction of expensive storm and seepage drains.

(viii) In project under conjunctive use of waters, tubewell loads can be reduced by releasing surface water for irrigation during periods of peak power demand thus resulting in lower power costs.

(ix) Crop water requirements can be ensured right through the year using surface water during the monsoons and ground water supplies when the surface water is not available.

(x) Ground water and surface water can be mixed in proper proportions to obtain a desired water quality for irrigating certain crop types (particularly when the ground water has a higher salt concentration); when the ground water has a higher salt concentration only certain salt tolerant crops can be grown.

(xi) Integration of the two types of schemes can be obtained with the existing water resources without loss of earlier investment.

Figure 16.3. broadly depicts various aspects of system approach for planning conjunctive use of surface and ground water resources.

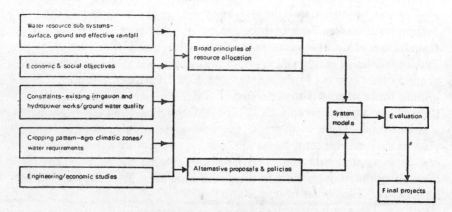

Fig. 16.3 Aspects of system approach.

Mathematical Modelling of a Dual Aquifer System

A typical grid network used in mathematical modelling of a dual aquifer system (Williams, 1973) concept in alluvial valleys is shown in Fig. 16.4. A multi-layered mathematical model of the Gujarat area was constructed and tested by UNDP. Results have provided guidelines for optimum development of ground water resources with due considerations and artificial recharge.

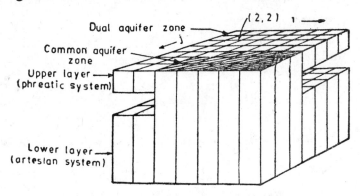

Fig. 16.4 Typical grid of dual aquifer system
(after Dennis E. Williams, 1973).

Digital computer models are currently being used as a management tool in the water resources field whereby the actual field situation is recreated mathematically and proposed changes studies on computer outputs before implementation in the field.

MATHEMATICAL MODEL FOR A BASIN

If Q is the net inflow-outflow to the system per unit area (draft, subsurface flow, recharge due to precipitation, etc.), Eq. 4.65 can be written in the form (de Ridder, 1972).

$$\nabla . T \nabla a - S \frac{\partial h}{\partial t} - Q = 0 \tag{16.1}$$

for two dimensional flow*, h being the height of water table above a datum.

In the Tyson-Weber scheme (Tyson and Weber, 1964), Eq. (16.1) is replaced by an equivalent system of difference-differential equations, the simultaneous solution of which gives the wanted function 'h' at a finite number of node points lying within the boundaries of the aquifer. The basin is subdivided into a number of polygonal areas, each having a node point which is the control point for the polygon, and all inputs and outputs to the polygon are assumed concentrated at the node. For a polygon associated with a typical node B in Fig. 16.5, the difference-differential equation can be written as

$$\left[\frac{h_i^{j+1} - h_B^{j+1}}{L_{iB}} \cdot T_{iB} \cdot W_{iB} \right] = \frac{A_B S_B}{\Delta t} (h_B^{j+1} - h_B^j) + A_B Q_B^{j+1} \tag{16.2}$$

and
$$Y_{iB} = \frac{T_{iB} \cdot W_{iB}}{L_{iB}}$$

---

* $\frac{\partial}{\partial x}\left(T \frac{\partial h}{\partial x}\right) + \frac{\partial}{\partial y}\left(T \frac{\partial h}{\partial y}\right) = S \frac{\partial h}{\partial t} + Q$

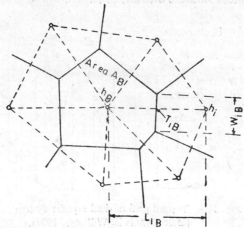

Fig. 16.5 Typical element of polygonal network.

where i, B = adjacent continuous nodes and node in question
A_B = area of polygon associated with node B
W_{IB} = length of perpendicular bisector associated with nodes i and B
T_{IB} = transmissibility at midpoint between nodes i and B
L_{IB} = distance between nodes i and B
Y_{IB} = conductance of path between nodes i and B
S_B = storage coefficient of polygonal zone associated with node B
Q_B = net volumetric flow rate per unit area at node B
h_I, h_B = water table elevations at nodes i and B

Superscripts $j, j+1$ represent continuous points along the time axis, i.e. $t^{j+1} = t^j + \Delta t$.

Equation (16.2) states that,

| Summation of subsurface flows between a given area and its surrounding areas | = | Rate of change of storage in the given area | + | Surface inflow or outflow rate, to or from the given area |

If the polygon borders the basin boundaries, any subsurface flow crossing the boundary is usually included in $A_B Q_B$.

The system of Eq. (16.2) can be solved on a digital computer. Initial value of water table elevations is impressed at each node. The flows, subsurface storage and extraction are balanced at each node by setting their sum equal to the residual term for any time step. Water table elevation at the node is then adjusted by the magnitude of the residual attenuated by a relaxation coefficient given by

$$\text{RELAX}_B = \frac{1}{\Sigma Y_{IB} + \frac{A_B S_B}{\Delta t}} \qquad (16.3)$$

Ground Water Basin Management and Conjunctive Use

when all the water table elevations have been adjusted, the sum of nodel residuals is formed and compared with an error criteria. The calculations are repeated till the sum of these residuals become $\leq \varepsilon$ the permissible error. At this stage the calculations for that time step are complete. These values then become the initial water level elevations for the next step in time. By comparison of the water table elevations thus computed with the historical records, the model is verified. Then the model can be used to predict the future response of the basin for varying inputs, extractions, etc.

The case history of a mathematical model for the Varuna Basin, U.P. (Satish Chandra and Pande, 1975) to determine the average annual recharge and water balance available for the year 1972-73 for which water table elevation records were available, is given in the following:

GROUND WATER BALANCE STUDY OF THE VARUNA BASIN DURING 1972-73, (Fig. 16.6).

The hydrologic balance equation for ground water, considering long term averages can be written as

$$R + I_Q + S_I = O_Q + S_E + E_t + D_{GW} + \Delta GWS \qquad (16.4)$$

where R = Recharge into ground water due to rainfall; irrigation water percolating down and seepage from canals

I_Q = Inflow into the basin from other basins

S_I = Influent seepage from streams

O_Q = Outflow from basin to other basins

S_E = Effluent seepage from streams

E_t = Evapotranspiration from the region in direct contact with the aquifer

D_{GW} = Draft from ground water

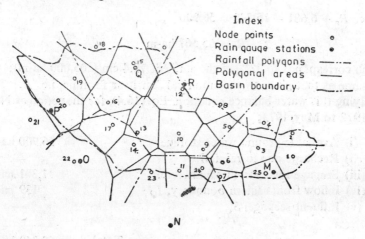

Fig. 16.6 Mathematical model for Varuna basin, U.P.

CWS = Change in ground water storage of the aquifer

Area of basin = 2,58,250 ha; a.a.r. = 79.4 cm. Applying Eq. (16.4) to the Varuna basin, U.P. for the period of June to October 1972:

(i) Recharge due to rainfall	R_r
(ii) Seepage from canals, tubewells and return flow from irrigated fields	6,581 ha-m
(iii) Inflow from eastern boundary, I_Q	100 ha-m
(iv) Influent seepage, S_I	—
Total	R_r + 6,681 ha-m

(v) Outflow from western boundary O_Q	118 ha-m
(vi) Effluent seepage, S_E	—
(vii) Evapotranspiration from waterlogged and forest areas (assuming 1.25% of the total area), E_t	2,170 ha-m
(viii) Ground water draft, D_{GW}	13,060 ha-m
Total	15,348 ha-m

During this period the rise in water table is 2.286 m

$$\therefore \Delta GWS = \text{area of the aquifer} \times \text{Rise in GWT} \times \text{Storage coefficient}$$

$$= 258{,}250 \text{ ha} \times 2.286 \text{ m} \times 0.0963$$

$$= 56{,}900 \text{ ha-m}$$

Hence R_r + 6,681 − 15,348 = 56,900

$$\therefore R_r = 65{,}567 \text{ ha-m}$$

which corresponds to 32% of the a.a.r. of 79.4 cm (205,000 ha-m); this is the same as that given by the empirical Amritsar formula, i.e. R_r=31.2%. Applying this water balance equation, Eq. (16.4), for the period of November 1972 to May 1973:

(i) Available ground water storage	56,900 ha-m
(ii) Recharge due to rainfall	—
(iii) Seepage losses from canals, etc.	11,301 ha-m
(iv) Inflow from eastern boundary, I_Q	139 ha-m
(v) Influent seepage, S_1	—
Total	68,340 ha-m

(vi) Outflow at western boundary, O_Q 163 ha-m
(vii) Effluent seepage, S_E 36,403 ha-m
(viii) Evapotranpiration, E_t 2,730 ha-m
(ix) Ground water draft, D_{GW} 24,344 ha-m

Total   63,640 ha-m

Ground water balance at the end of May 1973 = 68,340 − 63,640

= 4700 ha-m

available for future development.

Development of the mathematical model: (i) The formation constants S and T were determined by the pump test data using the Boulton's method (for an aquifer with delayed yield): $S = 0.0963$; $T = 1240$ m^2/day, these values were used for the entire basin.

(ii) The entire basin was divided into 25 Thiessen polygons, Fig. 16.6, such that the water levels at the nodes were known for some period. The six rain gauge stations in the area are shown in Fig. 16.6 and Thiessen polygons were drawn for these stations also.

(iii) The canal command area was delineated.

(iv) For each polygon A_B, W_{tB}, L_{tB} were measured and Y_{tB} determined using $T = 1,240$ m^2/day.

(v) The inputs to various nodes were due to precipitation and canal input. The precipitation at each rain gauge station being known for different months, the average value for each polygon associated with any node was worked out on the basis of proportion of the area of that node under the influence of a given rain guage station (with the help of Thiessen polygons for rainfall). The canal input was assumed zero for polygons other than those falling in the canal command. The input was distributed uniformly over the total area of those polygons. The irrigation efficiency was assumed as 60%.

(vi) The tubewell draft (output) was distributed uniformly over the area of all the polygons except those falling in the canal command. The irrigation efficiency was assumed as 65% and conveyance losses as 20%.

(vii) The forest land and waterlogged area was assumed as 1.25% of the total for computation of evapotranspiration.

(viii) The water levels at node points were known for the years 1972 and 1973.

(ix) The water levels in January 1972 were taken as the initial values. The time step taken was one month. For each node the input was determined for the month of January. This included the recharge due to rainfall and canal input, etc. The recharge due to rainfall (R_r) was taken $R \times C$ where C is a fraction and R the rainfall. The extractions from the node were the tubewell draft, evapotranspiration, etc. The algebraic sum of these replenishments and extractions gave the net recharge $A_B Q_B$ for the polygonal area. The storage factor $A_B S_B$ for the polygon was calculated. The water levels at the beginning of February 1972 were calculated adopting the scheme outlined in Eq. (16.2) and (16.3).

(x) The computations of water levels were carried out and compared with the available historic data. If they did not agree within a permissible error (ε_1), a new value of C was chosen and the computations repeated. In this manner the computations were carried out on a digital computer for all nodes for the entire period of 2 years. Figure 16.7 shows the flow diagram used for the computations. The value of C differed from node to node and from month to month (deviations = $\pm 30\%$ of the average value of 32%).

(xi) Thus a mathematical model, based on Tyson-Weber scheme of polygonal areas was prepared for the basin. The recharge for each polygon was worked out using the water table elevation data for two years. The model, with the recharge values for individual polygons as worked out can be used for predictions of future response of the basin under different hydrologic conditions being imposed.

(xii) The model can be refined with some more data on the variation of S and T values and verification with historical data of a longer duration. The refined model can be used for optimum utilisation of the water resources of the basin.

Finite Element Method

In this method the solution to flow system through a basin is obtained through an equivalent variational functional, rather than through a finite difference solution of the differential flow equation. With the finite element technique, the solution of the differential flow equation, Eq. (4.34), with a source or sink term Q added to its left side, is obtained by finding a solution for the head h that minimises an equivalent variational functional of the form (Guymon, 1974; Pricket, 1975)

$$F = \iint_D \left[\frac{K}{2}\left(\frac{\partial h}{\partial x}\right)^2 + \frac{K}{2}\left(\frac{\partial h}{\partial y}\right)^2 + \left(S\frac{\partial h}{\partial t} - Q\right) h \right] dx.dy \qquad (16.5)$$

To find the solution, the flow domain (D) is divided into a number of small areas or finite elements which are of triangular and quadrilateral pattern for two-dimensional systems, Fig. 16.8, and tetrahedral or parallelepiped for three-dimensional systems. The elements are of irregular pattern to facilitate representation of irregular boundaries and are smallest where the flow is concentrated. For the solution of each step, the parameters K, S and Q are kept

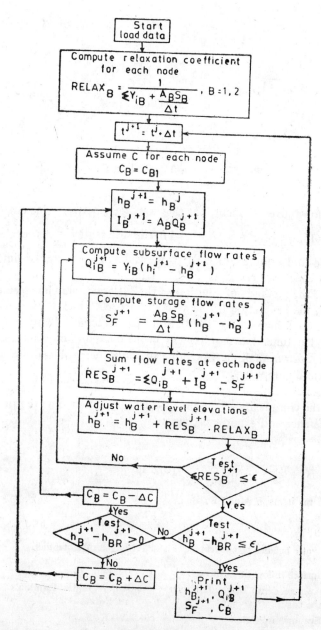

Fig. 16.7 Flow diagram for digital computer solution
(Abridged from Satish Chandra & P.K. Pande).

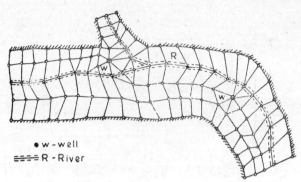

Fig. 16.8 Finite element method for the solution of flow systems.

constant for a given element, but they may vary between elements. To minimise Eq. (16.5) with respect to head h, the differential $\frac{\partial F}{\partial h}$ is evaluated for each node and equated to zero; this results in a number of simultaneous equations which are readily solved by computer.

One of the methods of solution of the system of simultaneous equations is through the generation of the system matrix of block tridiagonal form by a partitioning scheme (Das, 1975). Since the matrices of only one block are handled at a time, the requirement of the computer core storage is considerably cut down, making it possible to solve the problems on small and medium sized computers.

The choice between finite-element and finite-difference methods will depend on the complexity of the flow system, computer time required, individual preference and experience, etc.

QUIZ 16

I. Match the items in A and B:

A	B
Water balance component	Method of estimation
1. Surface inflow or outflow	a. Thiessen polygon
2. Average depth of rainfall over the basin	b. Blaney-Criddle, formula or Pan evaporation data
3. Evapotranspiration	c. AWLR, S_y, S

4. Change in surface storage	d. Pumping tests:
	   i. Darcy's law: $Q = T\,i\,W$
	   ii. Flow net analysis; $Q = T\,n_f\,\Delta h$

5. Change in ground water storage   e. Stream gauging
6. Subsurface inflow and outflow   f. Water levels in reservoirs

II. Obtain a water balance equation from the components:

> Ground water recharge, G_{wr}
> Precipitation, P
> Evapotranspiration, E_t
> Runoff coefficient, C
> Area of basin, A (in ha)
> Change in soil moisture, S_m

Take a base period of rainfall within the rainfall cycle, so that changes in soil moisture can be neglected.

III. Calculate the approximate ground water potential of Bihar from the following data:

Formation	Region	Area M. ha	Precipitation cm	Evapotranspiration cm	Runoff coefficient
Hard rock	Chhota-Nagpur including Santhal-Parganas	9.70	125	55	0.5
Alluvial	Northern Bihar	4.18	112.5	51	0.3
Alluvial	Southern Bihar	4.05	102	54	0.3

Hint: $G_w = (P - P \times C - E_t) A$ (2.62 M.ha-m)

IV. Write down the storage equation for the ground water reservoir in Haryana state with the following components:

Recharge due to rainfall, R_r
Recharge from streams, drains and canals, R_{sdc}
Recharge from canal irrigation, R_{ci}
Return flow from well irrigation, R_{wi}
Subsurface inflow, R_{Gi}
Recharge by leakage through aquiclude, R_{la}
Evapotranspiration, E_t
Draft by effluent streams, drains and canals, D_{sdc}
Draft by deep and shallow tubewells, and open wells, D_w
Subsurface outflow, D_{Go}
Draft by leakage through aquiclude D_{la}
Increase or decrease in ground water storage, Δ GWS

V. Select the correct answer (s):
1. The benefits of conjunctive use are
   a. efficient water use with high irrigation efficiency
   b. prevent water logging
   c. water requirements ensured right through the year

d. desired water quality can be obtained particularly when the ground water has a higher salt concentration
   e. effective flood control
   f. water needs during period of draught or years of subnormal precipitation be met
   g. optimum utilisation of water resources
   h. all the above benefits
VI. State whether 'true' or 'false':
   1. Leakage through an aquiclude depends on the permeability of the confining bed and the available difference in head between the aquifer and the source bed for water.
   2. It is the subsurface inflow and outflow component that does not pose problem in water balance studies.

17

Irrigation Systems

The cropping pattern should be such as to ensure all the year round use of the available water resources and maintain the fertility of the soil. The canal supply, wherever possible, may be supplemented by open wells and tubewells in the area (conjunctive use). Ground water resources can be used to provide supplemental irrigation for starting the crops on the right time or for proper ripening of crops when there is scanty rainfall at late stages. By adopting this approach it may be possible to take two to three light irrigated crops in place of one paddy crop. The rainfall pattern and the soil type have to be taken into account and cropping patterns along with water management schedules should be worked out and implemented for different agro-climatic regions, to ensure maximum water use efficiency. It has been shown that due to irregular distribution, deep percolations and uneven application, the losses are as much as 25 to 30%. The remedy lies in scientific planning of the project, proper land development (land levelling and shaping), efficient distribution of water as needed, growing the right type of crops in the appropriate manner and providing for good drainage of agricultural lands. Lining of field channels (water courses), underground pipe conveyance systems and latest water harvest techniques are some of the water and soil conservation measures.

Depending upon the type of soil, the slope of land, the kind of crop and the size of the stream (water supply available) there are many methods of irrigating crops such as flooding, flat bed, strip and border strip, ridge bed, furrow, drip, contour ditch and basin.

Border Irrigation

In this method, a thin sheet of water advances down the narrow strip between line ridges and water infilters into the soil as the sheet advances, Fig. 17.1. The border is usually adopted where topography permits precise land levelling at a reasonable cost and where relatively large irrigation

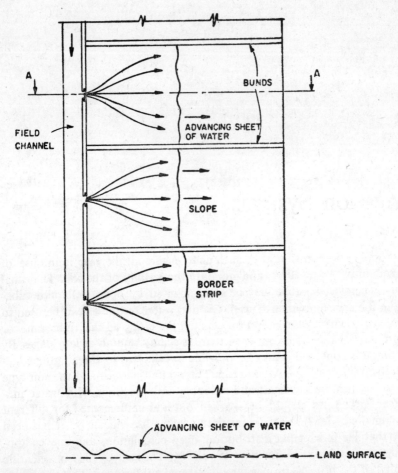

Fig. 17.1 Border strip method of irrigation.

streams are available. The width and length of each border strip varies from 4 m × 30 m for sandy soils to 15 m × 300 m for clayey soils. The strips are separated by small bunds 15 cm high. The land is graded smoothly along the natural slope (0.1 to 0.6%) in the direction of irrigation with no cross slope. Border irrigation is good for gentle slopes, close growing crops like wheat, grain (barley, bajra and berseem) and row crops like cotton, maize, jower and sugarcane (furrow irrigated in borders).

The advancing sheet of water should be adjusted so that there is enough time for the water to soak into the soil to replenish the soil moisture reservoir. When the required depth of water has been applied, the stream is shut off and diverted to another strip. The time of irrigation is given by the formula

$$t = 2.303 \frac{y}{I} \log_{10} \frac{q}{q-IA} \tag{17.1}$$

Irrigation Systems

where A = area covered with water in time t
I = infiltration rate
q = size of stream
t = time necessary to cover the strip
y = average depth of water as it flows over the land.

[All the above should be in consistent units (see Example 17.1).]
Maximum area that could be covered by the stream

$$A = \frac{q}{I} \tag{17.1a}$$

Field trials with borders have proved that uniform water applications may be obtained by proper control, with high water use efficiencies and less time required to cover the area. The time of irrigation is also given by the empirical formulae:

$$t = \frac{15d}{I}, \text{ for level borders with } I \leqslant 2.5 \text{ cm/hr} \tag{17.2}$$

$$t = \frac{60d}{I}, \text{ for graded borders with } I > 2.5 \text{ cm/hr} \tag{17.2a}$$

where i = time of irrigation in minutes
d = depth of watering (irrigation) in cm
I = final intake rate in cm/hr.

A unit stream, frequently used in border irrigation, is the size of the stream required to irrigate 100 m² of border area, i.e. an area 1 m wide and 100 m long.

The border method of irrigation has been introduced in India in some of the new irrigation project areas, such as the Tungabhadra project in Karnataka. Since the land is sloping, wet land rice cannot be grown on borders. The Government hopes to save water that way, so that more farmers can make use of the water from the large dam projects, by growing crops other than rice.

Suitable dimensions for border strips on different soils are given in Table 17.1.

Example 17.1: How long will it take for a stream of 60 lps to irrigate a border strip of 8×250 m if the infiltration capacity of the soil is 2 cm/hr assumed to be constant throughout the period of irrigation. The depth of flow over land is 7.5 cm. If the moisture content before irrigation and field capacity of the soil are 20% and 40% respectively, and the soil has an apparent specific gravity of 1.64, what is the average depth of penetration of water. What is the maximum area that should be covered by the stream?

Table 17.1 Recommended dimensions for border strips

Soil type	Optimum grade of border (% slope)		Size of stream q lps		
			30	30 to 50	50 to 100
Sand	0.4–05	W m L m	6–10 60–100	10–12 100–120	10–12 140
Loam	0.2–0.3	W m L m	8–10 80–120	12 140–220	12 140–220
Clay	0.05–0.1	W m L m	10 140–220	12 220	15 220–300

Solution:

Time of irrigation

$$t = 2.303 \frac{y}{I} \log_{10} \frac{q}{q-IA}$$

$$= 2.303 \frac{7.5}{2} \log_{10} \frac{0.06 \times 60 \times 60}{0.06 \times 60 \times 60 - \frac{2}{100}(8 \times 250)}$$

$$= 0.765 \text{ hr, or } \mathbf{46 \text{ min}}$$

If the depth of penetration is D cm, depth of water applied (depth of irrigation d)

$$d = \frac{w_f - w_i}{100} G_m D$$

$$= \frac{40-20}{100} \times 1.64 \times D$$

$$= 0.328 \, D$$

Equating the volume of water applied to qt, $(qt = Ad)$,

$$0.328 \, D(8 \times 250) = 0.06 \times 46 \times 60$$

$$D = \mathbf{0.25 \text{ m}}$$

The maximum area that could be covered by the stream

$$A = \frac{q}{I} = \frac{0.6 \times 60 \times 60 \text{ m}^3/\text{hr}}{2/100 \text{ m/hr}} = 10{,}800 \text{ m}^2$$

Irrigation Systems

Example 17.2: The available field capacity of soil having an apparent specific gravity of 1.5 is 6%. What depth of water should be appiled if the depth of effective root zone for the growing crop of wheat is 1.3 m? If the period of irrigation is 18 days, what is the peak consumptive use of water for wheat?

Solution:

Depth of application, $d = wG_mD = 0.06 \times 1.5 \times 1.3$

$$= 0.117 \text{ m, or } 11.7 \text{ cm}$$

Peak consumptive use for wheat $= \dfrac{\text{depth of water applied}}{\text{period of irrigation}}$

$$= \frac{11.7}{18} = 0.65 \text{ cm/day}$$

Furrow Irrigation

Furrow irrigation is suitable for cultivation of row crops like maize, sorghum (*jowar*), sugarcane, cotton, tobacco, groundnut, potatoes, chillies and also for orchards and vegetables. The water is applied in furrows between crop rows and water soaks into the root zone of crops, Fig. 17.2a.

The spacing of furrows is usually decided by the spacing of the crop rows and varies from 60 to 120 cm. Furrows should be close enough so that wet areas meet. The furrows may run down the slope when the slope is within reasonable limits (0.7 to 1%) and across the slope or on the approximate contour, to reduce grade and prevent erosion, on steeper slopes. The size of the stream depends on the type of soil, slope and the length of run. The length of run depends upon the infiltration capacity of the soil, the size of the stream, the root depth of crop and furrow gradient. The water should run in the furrow till the desired penetration is reached. Furrow grades of 0.2 to 0.5% are found satisfactory. Furrow streams of 5 to 30 lps are common. The land must be graded so that the water moves down the entire length of the furrow without ponding. Short tubes (2 to 5 cm diameter) of metal, rubber, wood or tile, and rubber or plastic syphon tubes may be used for letting in water into the furrows from the channels, Fig. 17.2b. Furrow sizes (depth × width) vary from 15 × 25 cm to 30 × 60 cm. A suitable combination of discharge and length of run should be made for a given soil of known infiltration rate and slope. Maximum efficiency can be achieved by selecting a suitable combination of hydraulic variables, Table 17.2.

For furrow irrigation

$$t = \frac{Lwd}{10\, q}$$

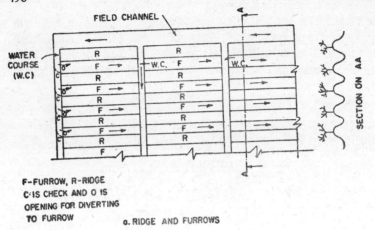

F-FURROW, R-RIDGE
C IS CHECK AND O IS
OPENING FOR DIVERTING
TO FURROW

a. RIDGE AND FURROWS

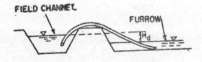

b SYPHON TUBES FOR DRAWING WATER
TO FURROWS FROM THE FIELD
CHANNEL OR WATER COURSE
Hd = HEAD ACTING ON THE SYPHON

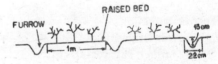

d. WIDE RIDGES BETWEEN FURROWS FOR
GROWING GINGER, TURMERIC AND SURAN

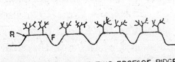

c PLANTING NEAR THE TWO EDGES OF RIDGE

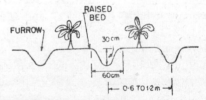

e. RAISED BEDS FOR GROWING BANANA

Fig. 17.2 Furrow irrigation.

Table 17.2 Furrow streams in lpm* (after water has reached the end of the furrow)

	Soil infiltration rate								
	High > 4 cm/hr			Medium 1.5–4 cm/hr			Low 0.25–1.5 cm/hr		
Length of run (m) →	100	200	400	100	200	400	100	200	400
Land slope ↓									
0–0.2	40	90	200	18	45	112	9	18	40
0.2–0.5	18	40	90	14	30	70	7	16	35
0.5–1.0	14	30	70	7	15	35	4	9	18

*With furrows, initial flow rate will be 2–3 times the indicated rate to fill the run as quickly as possible and then the flow cut back to the indicated rate. The furrow stream should reach the lower end of the field within one-fourth of the total time required for irrigation.

Irrigation Systems

where t = time of irrigation (min)
d = net irrigation application (cm)
w = furrow spacing (cm)
q = average furrow intake rate (lpm) or furrow stream
L = Length of furrow (run), m

The maximum nonerosive flow rate in furrows (lpm) is estimated by the empirical formula

$$q_{max} = \frac{36}{S}$$

where S = slope of furrow, expressed as %

With level furrows, the initial stream is continued till the end of irrigation.

Example 17.3

Furrows 80 m long and spaced at 75 cm apart are irrigated by an initial furrow stream of 100 lpm. The initial furrow stream reached the lower end of the field in 40 min. The size of the stream was then reduced to 30 lpm. The cutback stream continued for 1 hr. What is the average depth of irrigation?

Solution

Initial stream: $qt = Ad_1$

$$d_1 = \frac{qt}{A} = \frac{0.100 \times 40}{80 \times 0.75} = 0.067 \text{ m or } 6.7 \text{ cm.}$$

Cutback stream: $d_2 = \frac{qt}{A} = \frac{0.030 \times 60}{80 \times 0.75} = 0.03$ m or 3 cm

Average depth of irrigation $d = d_1 + d_2 = 6.7 + 3 = 9.7$ **cm.**

Corrugation Method

The corrugation method is an adoption of furrow irrigation for heavy soils, small streams and close growing crops. Corrugations are shallow furrows and are close enough so that the moisture is got both by capillary action and gravity, i.e. the soil between them will be wetted by lateral movement of water and, at the same time, moisture will reach the bottom of the roots. The spacing vary from 40 cm for sandy soils to 60 cm for heavy soils. Corrugations are 10 cm deep and 12–15 cm wide. A stream of 5–20 lpm is applied to each corrugations. The length of corrugation varies from 100 to 200 m.

In this method small ridges are made on the bed with openings at alternate ends, Fig. 17.3 and the crops are planted on the sides of the ridges. Water is admitted at the upper end and and passes through the furrows in a zigzag way. After it reaches the lower end of the plot, the supply is cut off. Thus enough time is provided for the water to percolate into the soil.

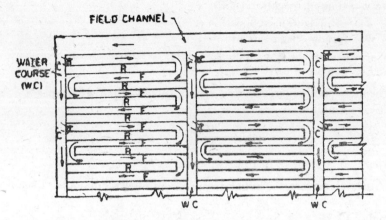

C-CHECK, O-OPENING FOR DIVERTING WATER TO CORRUGATIONS
R-RIDGE, F-FURROW

Fig. 17.3 Corrugation method of irrigation.

The field efficiency is 70% for slopes up to 0.5% and the efficiency is reduced for increased slopes.

This method is most suitable for almost level plots and raising crops like sweet potato and turmeric.

Basin Method

In this method small basins are made round the plant to receive water from the channel which gradually percolates to the root zone, Fig. 17.4. A hill is formed around the stem of the plant by putting up earth to protect the

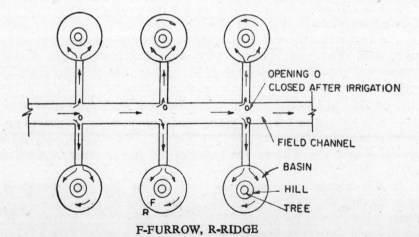

F-FURROW, R-RIDGE

Fig. 17.4 Basin method of irrigation.

Irrigation Systems

tree from direct contact with water. Basins can be made larger as the plants grow. This method is good for orchards and steep lands, especially if the soil is heavy.

Sprinkler Irrigation

Sprinkler irrigation is suitable for uneven topography, steep slopes, easily erodible or shallow soils, soils too porous or heavy, black retentive soils, and for irrigation stream too small for efficient distribution by surface irrigation. Much land levelling and laying out of the field, etc. are not needed, which means a great saving in cost. Liquid fertilizers, fungicides and insecticides, etc. can also be sprayed. Water can be used economically with high water use efficiency in places of inadequate water supply. It is suitable for all crops (except rice) on any irrigable area and highly permeable soils except in hot windy areas because of high evaporation loss. The initial investment, annual depreciation of equipment and labour cost on transportations are the main disadvantages. It is therefore usually recommended for cash crops which can pay for the investment. Irrigation by sprinkling of commercial crops helps in crop cooling, frost protection and application of soluble fertilisers. When water is already being pumped to the point of use, the additional horse power needed for sprinkling is to provide an extra head of about 40 to 45 m with a minimum of additional capital investment. The details of a sprinkler head are shown in Fig. 17.5a.

The layout of the sprinkler equipment is most important. It requires rather highly technical know-how. The source of water supply, amount available, water table, evaporation and natural precipitation rate, wind, velocity and direction, contour map of the area, soil types and crops and discharge-drawdown relationship of the well, etc. will determine the size and type of the equipment.

The salient points in the design of a sprinkler irrigation system are:
(i) The depth of application will depend on the available moisture-holding capacity of the soil.
(ii) The precipitation rate of the sprinklers should not be greater than the infiltration capacity of the soil.
(iii) The whole area should be covered within the irrigation interval (= depth of application/daily consumptive use).
(iv) Natural precipitation should be taken into account and sprinklers have to be designed only for providing supplemental irrigation.
(v) Sprinkler heads generally operate at pressures of 2.1 to 3.5 kg/cm^2; the precipitation rates vary from 0.5 to 1.2 cm/hr for 12 to 16 hours of operation of sprinklers per day. Sprinkler spacing on the lateral should be 0.3 to 0.5 of the wetted diameter of the sprinklers, and the spacing of the laterals should 0.5 to 0.7 of the wetted diameter to get the necessary spray overlaps. The wetted diameter for most sprinklers vary from 15 to 45 m for moderate pressures, Fig. 17.5b.

Ground Water

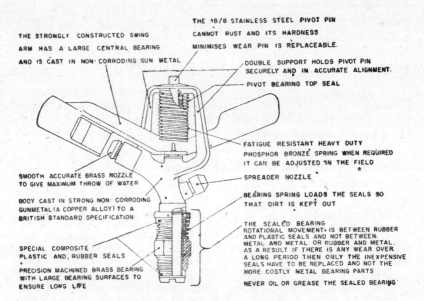

Fig. 17.5a Sprinkler construction.

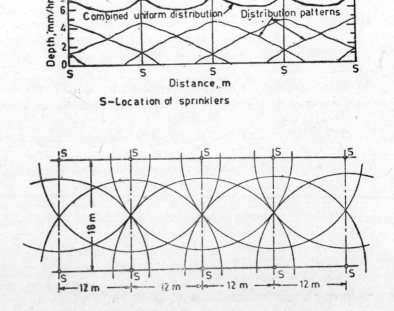

Fig. 17.5b Overlapping sprinkler position to give uniform depth.

Irrigation Systems

The economics of sprinkler irrigation from borewells drilled in hard rock formations of Karnataka is given in the following example.

Example 17.4

Project	Nuggehalli village, Hassan district, Karnataka,
Economic status of farmers	Poor with individual holdings 1/4 to 4 ha
Average annual rainfall (a.a.r.)	59 cm with 30 rainy days; 58 cm during the months of April to November
Soil	Sandy loam with water holding capacity of 3 cm per 30 cm depth
Crops recommended	Maize, jowar, ragi, groundnut, chillies, sunflower, soyabeans, etc.
Location of borewell sites	Hydrogeological survey with electrical resistivity method is conducted on 91 ha of project area and borewell sites are located in the low resistivity areas, Fig. 17.6 and the vertical resistivity curves for six borewell sites are shown in Fig. 17.7. 15 cm borewells are drilled by the Ingersoll Rand (DHD) Rig and their details are given in Table 17.3.

Since the farmers have small or marginal holdings, the community irrigation project under a cooperative farming registered society is the only solution for the farmers to avail the loans advanced from the cooperative land development banks.

Design of sprinkler irrigation system: For the types of crops grown in the area, a peak consumptive use of 4 mm/day and effective depth of root zone of 90 cm are assumed. Since the moisture holding capacity for the soil is 3 cm per 30 cm depth, the depth of irrigation required to bring the soil moisture from 50% depletion level

$$= 3 \times \frac{90}{30} \times 0.5 = 4.5 \text{ cm}$$

Irrigation interval in days $= \dfrac{\text{depth of irrigation}}{\text{daily consumption use}}$

$$= \frac{4.5}{0.4} = 11 \text{ days}$$

Since the rainfall during the 8 months of April to November (i.e. 244 days) is 58 cm, the average rainfall available per cycle of irrigation (of 11 days)

$$= 58 \times \frac{11}{244} = 2.62 \text{ cm}$$

Assuming 50% supplemental irrigation from natural rainfall, depth of irrigation to be applied $= 4.5 \times \frac{1}{2} = 2.25$ cm.

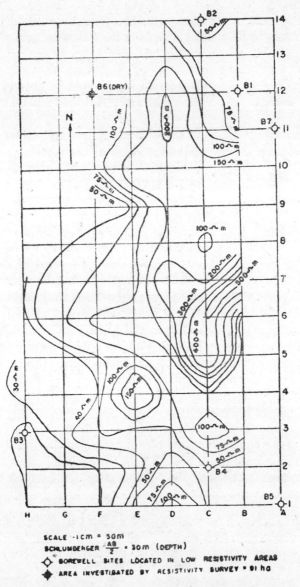

Fig. 17.6 Isoresistivity map of Nuggehalli, Hassan district, Karnataka (south India).

Assuming a spacing of sprinklers as 12×18 m, i.e. spacing of sprinklers on each lateral as 12 m and the spacing of laterals as 18 m, hours of operation of sprinklers as 12 hours per day in 3 shifts, i.e. 4 hours per shift and an application efficiency of 70%, the precipitation rate of the sprinklers is

$$r = \frac{2.25 \text{ cm}}{4 \text{ hr} \times 0.7} = 0.8 \text{ cm, or 8 mm/hr.}$$

Irrigation Systems

Table 17.3 Details of borewells drilled at Nuggehalli, Hassan district, Karnataka for sprinkler irrigation of 39.6 ha (50% supplementary to natural rainfall)

Borewell No. (Fig. 17.6)	Total depth drilled m	Depth of casing m	Yield m³/hr	Cost of borewell Rs.	Power of submersible pump hp	Power of submersible pump kw	Cost of submersible pump installation including piping Rs.	Hard rock formation drilled	Electrical resistivity curve obtained (Fig. 17.7)
B 1	52.7	28.2	19.1	8910	8	5.9	15,000	Gneisses	C 1
B 2	37.5	26.2	36.4	6700	10	7.4	17,000	Gneisses	C 2
B 3	60.3	23.4	5.9	8655	5	3.7	14,000	Amphibolites	C 3
B 4	51.8	15.5	19.1	7070	10	7.4	17,000	Schist	C 4
B 5	52.1	21.6	11.4	7709	10	7.4	17,000	Pegmatite	—*
B 6	45.1	15.9	Dry	6301	—	—	—	Gneisses	C 5
B 7	60.0	26.2	4.5	8999	2.5	1.9	12,000	Gneisses	C 6
Total	359.5	157.0	96.4	54344	45.5	33.7	92,000		

*Due to undulating topography site selected on promising geological features.

(i) Average depth of drilling $= 359.5/7 = 51.3$ m
(ii) Average depth of casing $= 157/7 = 22.4$ m
(iii) Average yield from borewells $= 96.4/6 \approx 16$ m³/hr
(iv) Cost of sprinkler irrigation in hard rock areas (from borewells alone) $=$ Rs. 28.50 per ha-cm
(v) Success rate of borewells—sites selected from Ground Water Survey $= (6/7) \times 100 = 86\%$

Data for 50% supplementary sprinkler irrigation

(i) Area irrigated per borewell (with 12 hr pumping in a day) $= 39.6/6 = 6.6$ ha (for the crop types and soil)
(ii) Power for lifting water (to the common sump) $= 45.5/39.6 = 1.15$ hp/ha or 0.85 kw/ha
(iii) Power for operating the sprinklers $= (2 \times 15)/39.6 = 0.76$ hp/ha or 0.56 kw/ha
(iv) Total power for sprinkler irrigation from borewells $= 1.91$ hp/ha or 1.41 kw/ha
(v) Electrical consumption per day $= 17$ kwh/ha; power cost $=$ Rs. 220 per crop-ha
(vi) Precipitation rate of sprinklers $= 8$ mm/hr (3 shifts per day, 4 hr per shift)

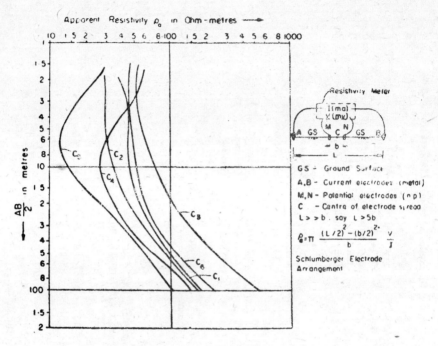

Fig. 17.7 Schlumberger resistivity curves for six borewell sites selected at Nuggehalli, Hassan district, Karnataka.

which is well within the intake rate for the soil type, i.e. < 3 cm/hr. The capacity of the sprinkler

$$Q = (12 \times 18)\frac{0.8}{100} = 1.73 \text{ m}^3/\text{hr}$$

A suitable model of the sprinkler to have a pressure of around 3 kg/cm² at the base of the nozzle and to give a discharge of 1.73 m³/hr for the given spacing and precipitation rate is selected from the models available from the prospective sprinkler manufacturers in the country like, IAEC, Voltas, Premier and Jindal Sprinklers.

Assuming the hours of pumping from six borewells (one borewell located is dry) to be the same as the hours of operation of the sprinklers, the area covered per shift

$$= \frac{\text{available volume of flow rate from six borewells}}{\text{precipitation rate}}$$

$$= \frac{96.4 \text{ m}^3/\text{hr}}{(0.8/100)\text{m/hr}} = 12{,}050 \text{ m}^2, \text{ or } 1.2 \text{ ha}$$

The area irrigated per day

$$= 1.2 \text{ ha} \times 3 \text{ shifts} = 3.6 \text{ ha}$$

Irrigation Systems

The total area brought under sprinkler irrigation
$$= \text{area covered per day} \times \text{irrigation interval in days}$$
$$= 3.6 \text{ ha} \times 11 \text{ days} = 39.6 \text{ ha}$$

Number of sprinklers operating per shift
$$= \frac{\text{available flow rate from six borewells}}{\text{capacity of each sprinkler}}$$
$$= \frac{96.4 \text{ m}^3/\text{hr}}{1.73 \text{ m}^3/\text{hr}} = 56 \text{ sprinklers}$$

With 14 sprinklers on each lateral, 4 laterals are used as shown in Fig. 17.8. By shifting the laterals thrice in a day, the whole area of 39.6 ha is covered in 11 days.

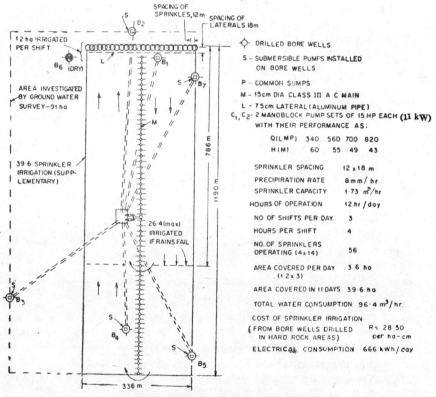

Fig. 17.8 Sprinkler irrigation from borewells.

In designing the lateral, the frictional loss should not exceed 20% of the average pressure at the base of the nozzle. In the main pipe the pressure loss should be limited to 10%, but 3/4 of this loss is considered in the actual design practice. Therefore, if a pressure of 3 kg/cm² is required at the sprinkler head, the pressure at the junction between the laterals and mains should be 3.6 kg/cm² and at the pump outlet 3.9 kg/cm². For reasons of lightness an ease of handling, aluminium pipes are common, but now PVC pipes are getting popular due to flexibility in laying in horizontal and vertical curves.

The total head (H) acting on the pump operating the spriklers=difference in level from the lowest water level in the sump to the highest patch of land+ the height of the riser + the frictional loss in the main pipe and lateral + the pressure head at the base of the nozzle + losses in the suction pipe and foot valve, bends and elbows. If the pumping rate is Q m³/sec, the power of the pump required for operating the sprinklers in kW (excluding the power of the pumps installed on the borewells for lifting water to the sump)

$$P = \frac{\rho g Q H}{1000 \, \eta_0} = \frac{g Q H}{\eta_0} \qquad (17.3)$$

where ρ = density of water (1,000 kg/m³)
$\quad \eta_0$ = overall efficiency of the pump (60% for electrical pumpsets and 40% for pumps fitted with diesel or petrol engine)
See Example 17.6
In the present scheme, two monoblock pumpsets of 11 kW (15 hp) each and 15 cm diameter portable aluminium pipes as laterals are proposed.

Economical viability of the scheme:
I. Capital costs:

(i) Ground water survey for locating borewell sites @ Rs. 300 per site for 7 sites   Rs   2,100
(ii) Cost of 7 borewells drilled   Rs   54,344
(iii) Cost of conducting yield test @ Rs. 500 per borewell for 6 borewells (one being dry)   Rs   3,000
(iv) Cost of installation of 6 submersible pumps and piping   Rs   92,000
(v) Cost of constructing one common sump   Rs   8,000
(vi) Cost of 2 monoblock pumpsets of 11 kW (15 hp) each @ Rs. 5,520 per set   Rs   11,040
(vii) Cost of 2 suction side fitments, common manifold on the delivery side, panel board and switch and constructing a pump house   Rs   10,000
(viii) Cost of constructing 1,190 m of a.c. pressure pipeline (main) 15 cm diameter with 66 outlets   Rs   60,000
(ix) Cost of portable sprinkler line 672 m of aluminium pipes 7.5 cm diameter with quick couplers, 56 sprinklers, risers, etc.   Rs   36,400

Total capital cost   Rs   276,884
say, Rs 2.8 lakhs.

Capital cost per hectare = $\dfrac{2.8 \text{ lakhs}}{39.6 \text{ ha}}$ = Rs 7075

II. Operating Costs:
  (i) Depreciation @ an average of 8% of capital cost ... Rs 22,400
  (ii) Repairs and maintenance of equipment @ 3% of capital cost ... Rs 8,400
  (iii) Interest on capital investment @ 10% flat ... Rs 28,000
  (iv) Electric power charges—
    $(34 + 11 \times 2)$ kW $\times$ 12 hr
    = 672 kW hr $\times$ 0.15 Re
    = Rs 100 per day $\times$ 300 days for triple cropping ... Rs 30,000
  (v) Staff for operation and maintenance @ 5% of capital cost ... Rs 14,000
  (vi) Cost of cultivation @ Rs 750 per ha for the types of crops grown in 8 rainy months (2 crops) and Rs 500 per ha for the third irrigated-dry crop (Dec–March)
    $(2 \times 39.6 \times 750) + (1 \times 39.6 \times 500) =$ ... Rs 79,300

  Total operating cost ... Rs 1,82,100
  say, Rs 1.8 lakhs

Operating cost per hectare under triple cropping
$$= \frac{1.8 \text{ lakhs}}{39.6 \text{ ha}} = \text{Rs } 4500$$

Volume of water supplied by sprinkler irrigation per year under triple cropping
$$= 96.4 \text{ m}^3/\text{hr} \times 12 \text{ hr/day} \times 300 \text{ days/yr} = 347,000 \text{ m}^3$$
Cost of sprinkler irrigation—items II, i to v,
   = Rs 98,840
Cost of sprinkler irrigation per ha-cm, i.e. for 100 m^3 (from borewells drilled in hard rock areas)
$$= \frac{\text{Rs } 98,840}{3,470 \text{ ha-cm}} = \text{Rs } 28.50$$
while for tubewell irrigation without sprinklers in unconsolidated formation, the cost of irrigation per ha-cm is Rs. 10.60.

III. Gross Revenue per Hectare:
  (i) First crop of maize yielding 60 qn/ha @ Rs. 120/qn ... Rs 7,200
  (ii) Second crop of ragi yielding 35 qn/ha @ Rs 120/qn ... Rs 4,200
  (iii) Third crop of millets or pulses yielding 25 qn/ha @ Rs 100/qn ... Rs 2,500

  Total annual revenue per ha ... Rs 13,900

IV. Net Annual Income per Hectare:
Total annual revenue per ha   Rs   13,900
Annual operating cost per ha   Rs   4,500

Net annual income per ha   Rs   9,400

V. Repayment of Loan and Economic Status of the Farmers:
The capital cost has to be repaid in 8 years. The annual repayment of loan per hectare

$$= \frac{\text{Rs 2.8 lakhs}}{8 \text{ yr} \times 39.6 \text{ ha}} = \text{Rs 885}$$

The net annual savings of the farmers per ha

$$= \text{Rs 9400} - \text{Rs 885} = \text{Rs 8615, or Rs 718 per month}$$

Thus the poor farmers in the area holding about 1/2 ha each can have a net savings of about Rs 360 p.m. apart from the labour provided for them from cultivation of their lands. They also have some money (64% of the capital investment) put aside towards the depreciation of the pumpsets, sprinkler equipment, etc. The scheme is envisaged as economically viable and helps to improve the economic status of the poor farmers in Nuggehalli village. They can hope for prosperity after the complete repayment of loan, i.e. from the ninth year onwards, when they can have a net savings of Rs 783 p.m. per ha which can further be increased to Rs. 1000 p.m. by including crops like sunflower, soyabeans, chillies, etc. Crops like ragi and soyabeans are drought resistant and can withstand failure of rain for some time. Thus optimum utilisation of water resources can be achieved by applying the required depth by sprinkler irrigation to supplement natural rainfall, so as to bring maximum area under production and improve the economic status of the poor farmers in the area.

During any crop season if the rains completely fail, the full depth of 4.5 cm has to be sprinkled at a precipitation rate of 8 mm/hr, which takes 8 hours to irrigate 1.2 ha in one shift. If the hours of operation of the sprinklers and pumping from the borewells could be increased to 16 hr/day, 2.4 ha can be irrigated per day in two shifts and the total area irrigated in an irrigation interval of 11 days $= 2.4 \times 11 = 26.4$ ha, which is the maximum area that can be covered by assured irrigation from sprinklers with the available yield from borewells, i.e. 96.4 m^3/hr.

The limitation of this cooperative sprinkler irrigation scheme from borewells is that due to the limited yield in the hard rock formations, the whole area covered by the borewells, which are mostly drilled in the sites selected by the Ground Water Investigation Team after detailed hydrogeological surveys, cannot come under irrigation. This may create misunderstanding among the farmers who own different portions of the land.

Irrigation Systems

Example 17.5: Determine the pumping capacity, number and capacity of sprinklers, and the precipitation rate for a sprinkler irrigation system for 16 ha of maize crop given the following data:

Spacing of sprinklers	12 × 18 m
Sprinkler head	250 k N/m²
Height of riser	1 m
Moisture replaced in soil at each irrigation	60 mm
Irrigation efficiency	70%
Irrigation period	10 days in a 12 day interval
Hours of operation	16 hr/day

If the lowest water level in the well is 15 m bgl at well site and the level of ground at the highest riser is 2 m agl at well site, what should be the monthly electricity bill assuming an efficiency of 60% for the pump and 85% for the motor, the losses in the main-lateral-riser system due to friction, etc. as 3 m, and the power tariff as 3 m, and power tariff as 25 ps per kW hr.

Solution:

i. Depth of application $d = \dfrac{d_i}{\eta_i} = \dfrac{60}{0.70} \approx 86$ mm

ii. Capacity of sprinklers $q = (12 \times 18) \dfrac{86}{1000 \times 16} = 1.16$ m³/hr,

or $\approx$ **20 lpm.**

iii. Precipitation rate of sprinklers $r = \dfrac{86}{16} = 5.4$ mm/hr.

iv. Area irrigated per day $A_d = \dfrac{16 \text{ ha}}{10 \text{ days}} = 1.6$ ha/day.

v. Pumping capacity $Q = (1.6 \times 10^4) \dfrac{86}{1000 \times 16} = 86$ m³/hr.

vi. Number of sprinklers $n_{sp} = \dfrac{Q}{q} = \dfrac{86}{1.16} = 74$

vii. Power of the pump-motor, $P = \dfrac{\rho_g QH}{1000 \, \eta_p \, \eta_m}$ kW

Total head $H = h_s + h_{er} + h_r + h_r + h_f + h_n$

$= 15 + 2 + 1 + 3 + \dfrac{250}{10} = 46$ m

$P = \dfrac{1000 \times 9.81 \, (86/3600) \, 46}{1000 \times 0.60 \times 0.85} = 21.2$ kW

viii. Monthly electricity bill

$= 21.2 \, (10/12 \times 30 \times 16) \, 0.25 =$ **Rs. 2120.**

Example 17.6 For a square pattern of sprinklers with spacing 20 m × 20 m, the depth of water collected (mm) in open cans placed at regular intervals within the sprinkled area bounded by four sprinklers (S_p) during 1 hr are given below. Determine the uniformity coefficient.

Size of sprinklers	5×3 mm
Pressure at nozzles	280 k N/m²
Wind	4 km/hr S—W
Humidity	41%

S_p	9	8	7	S_p
8	8	10	10	9
9	9	9	9	9
10	8	9	9	9
S_p	8	7	7	S_p

Solution:

Uniformity coefficient (C_u) gives an idea of the uniformity of application under given conditions like the nozzle size, spacing of sprinklers, nozzle pressure, humidity, wind, etc.

$$C_u = \left[1 - \frac{\Sigma fD}{\Sigma fx}\right] 100$$

$C_u \geqslant 85\%$ is considered as satisfactory. Data on uniformity coefficient are useful for selecting the combinations of spacings, sprinkling rates, nozzle size and operating pressure, etc. to obtain high values of irrigation efficiency at specific operating conditions. For the given data

Observation	Frequency		Deviation from mean	
x	f	fx	|D|	f|D|
10	3	30	0.95	2.85
9	10	90	0.05	0.50
8	5	40	1.05	5.25
7	3	30	2.05	6.15
	$n = \Sigma f = 21$	$\Sigma fx = 190$		$\Sigma fD = 14.75$

$$\text{Mean} = \frac{\Sigma fx}{n} = \frac{190}{21} = 9.05$$

Irrigation Systems

$$C_u = \left[1 - \frac{\Sigma fD}{\Sigma fx}\right] 100 = \left[1 - \frac{14.75}{190}\right] 100 \approx 92\%$$

which is very satisfactory.

Drip Irrigation

Drip irrigation is a new system of irrigation developed in the deserts of southern Israel. The method had given great hopes of possibilities of agriculture under arid conditions with poor sand soils, high evapotranspiration rates and water supply that is both limited and high in salts. From the water source, a booster pump pressurises about 25 to 50 m³/hr (per ½–1 ha to be irrigated) and delivers it to the control head, Fig. 17.9, where metering,

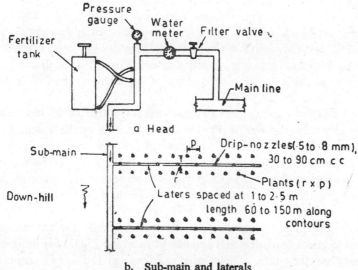

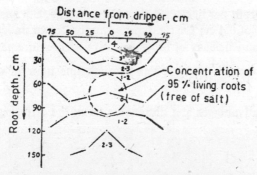

c. Salinity profile after some months of drip irrigation (numbers represent E.C. in milli-mhos/cm in a 1 : 1 soil water extract)

Fig. 17.9 Drip irrigation (after Goldberg et al., 1972).

filtering, pressure regulation and fertiliser injection take place. From the control head the water travels through the conducting pipe (4–5 cm main or feeder line) and then through the distribution tubes (12–16 mm plastic tubes, i.e. laterals laid along the crop rows spaced at 90 cm covering the entire field) on which the drip-nozzles are inserted at intervals of about 50 cm and from the drippers, the steady drops of water are discharged at zero pressure. One dripper supplies water and fertiliser for each plant and the drip-nozzle discharge ranges of 2–10 lph. All the components of the system except the head and fertiliser apparatus are generally of plastic construction. The laterals and drip-nozzles are laid on the soil surface or buried not deeper than 5–10 cm. Water trickles to crop roots and low rates of water are applied frequently. The crop is irrigated daily in many cases. Soil samplings have shown accumulation of salts on the soil surface and at the edges of the wetted areas due to saline irrigation water and high evapotranspiration, but the region in which the plant roots are concentrated has actually the minimum possible salinity content, Fig. 17.9c, with considerable increase in the number and density of feeder roots than when other irrigation methods are employed.

Evaporation losses are eliminated and use of water of higher salinity than normally tolerated by the plant is made possible by the nature of the wetting pattern. Drip irrigation is best suited for row crops and orchards such as tomatoes, corn, grapes, citrus fruits, melons, etc. It has been proved that drip irrigation gives better crop yields than other irrigation methods like sprinkler and furrow, under desert conditions of high evapotranspiration and coarse soil of poor water holding capacity.

The capital cost per hectare works out to Rs. 35,000 for the imported equipment (control head, drippers and plastic) and Rs. 25,000 for indigenous equipment (booster pump, delivery pipe and fencing). India has a very highly developed plastic industry and this item forms the bulk of the investment. The net annual profit per hectare works out to Rs. 25,000. Assuming a daily consumptive use of 0.4 cm for the types of crops grown under this system, and an application efficiency of 75%, the daily water requirement per hectare will be $10,000 \times \frac{0.4}{100 \times 0.75} = 53 m^3$ through one hour of operating the system daily.

Some of the advantages and disadvantages of the drip irrigation are as follows:

Advantages

(i) Does not wet the foliage and aisler.
(ii) 30–50% saving in water.
(iii) Salts accumulate at the outer edge of the wetted pattern.
(iv) Enables spraying, dusting, picking, cultivating, etc.

Irrigation Systems

(v) The dry aisler prevent weed growth.
(vi) No compaction (due to floor irrigation) and erosion. (on hill slopes).
(vii) Increase in yield ranging from 20-30%.
(viii) Roots stay within the moist zone.
(ix) Better grade of tomatoes, firmer strawberries, cotton, sugarcane and potatoes have been reported.
(x) Can be used on hill sides (ideal system) and with all types of mulches.

Disadvantages

(i) Does not offer frost protection as sprinklers do.
(ii) Plastic drip-lines and submains may be attacked by rodents and small animals. However, this is not such a serious problem.
(iii) Requires regular flushing (to clear off the dirt collected near the ends of the drip-lines) and supervision.

Example 17.7: The extract of saturated soil solution has an EC of 10 m mhos/cm and the EC of irrigation water is 1.5 m mho/cm. What is the leaching requirement? If the consumptive use of crop is 6 cm, what depth of water has to be applied?

Solution

$$LR = \frac{EC_i}{EC_d} \times 100\%$$

Assuming EC of drainage water as twice that of the soil extract,

$$LR = \frac{1.5}{2 \times 10} \times 100 = 7.5\%$$

$$LR = \frac{EC_i}{EC_d} = \frac{D_d}{D_i}, D_i = D_c + D_d = D_c + LR\, D_i$$

$$D_i = \frac{D_c}{1-LR} = \left(\frac{EC_d}{EC_d - EC_i}\right) D_c$$

$$D_i = \frac{6}{1 - \frac{7.5}{100}} = 6.5 \text{ cm.}$$

Example 17.8: Design a subsurface tile drainage given the following data: Soil—Sandy loam with $K = 2.3 \times 10^{-4}$ cm/sec. Impermeable barrier at 4 m below land surface as indicated by subsurface borings. Tile drains to be installed at a depth of 1.8 m below land surface. Water lable should be held at a minimum depth of 1.2 m below land surface (from farm experience). Depth of irrigation = 10 cm every two weeks. Deep percolation loss from irrigation = 20% (from consumptive use studies). Field ditch loss = 8% of water applied (from ditch flow measurements). There is no significant natural subsurface drainage from the area.

Solution: Using the Donnan's formula (developed for relief drains based upon certain barrier conditions) the spacing of the tile drains (S) is given by

$$S = \frac{4K(b^2 - a^2)}{Q_d} \tag{17.4}$$

where a and b are the depths as shown in Fig. 17.10

Q_d = drainage coefficient which is the discharge of an underdrainage system expressed as the depth of water to be removed from the area in 24 hr ($\approx 1\%$ of a.a.r. in 24 hr).

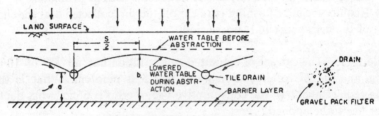

Fig. 17.10 Spacing of tile drains.

Water lost through percolation from each irrigation
$$= 10(20\% + 8\%) = 2.8 \text{ cm in 14 days}$$

Drainage coefficient $Q_d = \frac{2.8}{14} = 0.2$ cm/day

$$a = 4 - 1.8 = 2.2 \text{ m}, \ b = 4 - 1.2 = 2.8 \text{ m}$$

Spacing of the tile drains

$$S = \sqrt{\frac{4K(b^2 - a^2)}{Q_d}}$$

$$= \sqrt{\frac{4 \times 2.3 \times 10^{-4}(2.8^2 - 2.2^2)}{0.2/(24 \times 60 \times 60)}} = 34.5 \text{ m}$$

The flow in the tile drains and required tile sizes may be computed from the irrigation practice, rainfall intensity, soil type and tile drain spacing. The tile drains are usually of 10-15 cm diameter, laid with open joints with graded filters provided around each joint and the tile drain grades varying from 0.05 to 0.10%. Tile drains should be spaced so as to lower the ground water enough for good plant growth within 24 hours after rain. Some desired limits are given in Table 17.4.

For the reclamation of water logged salt affected lands, tile drainage with earthen bell and spigot type tiles with holes or slits are recommended. The cost of tile drainage is around Rs. 2,500 per hectare.

Irrigation Systems

Table 17.4 Desired limits for tile drains

Soil type	Spacing (m)	Depth (m)	Dia. of tile drain (cm)	min. grade %	max. length (for min. grade) (m)
Clay and clay loam	12–21	0.75–0.9	10	0.10	400
Silt loam	18–30	0.9–1.2	12.5	0.07	600
Sandy loam	30–90	1.0–1.3	15	0.05	900

The grain size of filter materials in relation to openings in pipes is given by United States Bureau of Reclamation USBR (1965) as

$$\frac{D_{85} \text{ of the filter material nearest the pipe}}{\text{maximum opening of the pipe drain}} \geq 2 \quad (17.5)$$

Principles of Soil and Water Conservation

1. Contour farming should be adopted on all the sloping lands.
2. Crops are grown in strips or bands at right angles to the slope of the land (strip cropping), Fig. 17.11.

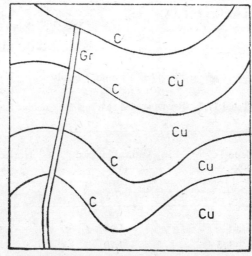

C-Contours, Cu-Cultivated, Gr-Grass waterway
Fig. 17.11 Contour farming (buffer strips).

3. On rolling lands, strips of row and cover crops are sown alternatively to control wind and water erosion.
4. On eroded lands mixed cropping is practised.

5. Leguminous crops and fodder grasses are included in the crop rotations. They add nitrogen and organic matter to the soil.

6. Contour bunds are constructed on sloping lands (<6%) where the rainfall is less than 60 cm for storing surface runoff to improve soil moisture status for increased crop production, Fig. 17.12. The bunds should be so

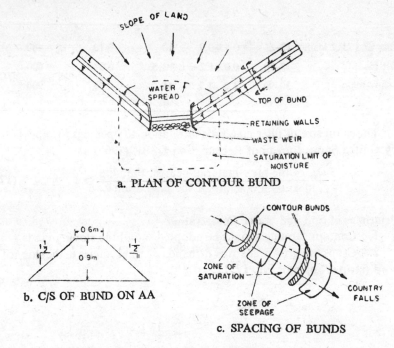

Fig. 17.12 Contour bunds.

Table 17.5 Recommended spacings for Contour bunds

Slope S%	Vertical interval (VI) m	Horizontal interval (approx.) m
0 –1	1.05	105
1 –1.5	1.20	96
1.5–2	1.35	75
2 –3	1.50	60
3 –4	1.65	50
4 –5	1.80	40
5 –6	1.95	36

spaced that the seepage zone of the upper bund should extend up to the zone of saturation of the next lower bund. The spacings shown in Table 17.5 are recommended in Maharashtra for contour bunds on cultivated lands.

7. Graded bunds are constructed in areas where the rainfall is over 60 cm and the soils are heavy. Bunds are aligned with a gentle gradient of 0.1 to 0.5% towards the outlet.

8. When the land slope is steeper than 6% and the rainfall is over 100 cm, terracing can be practised by having narrow strips with 'cut and fill' method, Fig. 17.13 a, b. For graded terraces the spacing is given by

$$VI = \left(2 + \frac{S}{4}\right) 0.305 \tag{17.6}$$

where
VI = vertical interval in metres
S = per cent slope of the land

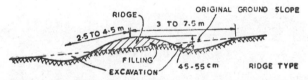

a. INTERCEPTION AND RETENTION TERRACE (CONSTRUCTED FROM BOTH SIDES)

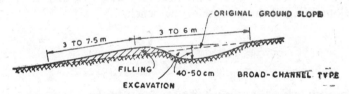

b. INTERCEPTION AND DIVERSION TERRACE (CONSTRUCTED FROM UPPER SIDE)

Fig. 17.13 Terrace cross-sections.

Similarly staggered trenching can be usually adopted for plantations for horticultural crops such as cashew-nuts.

9. Bench terracing and paddy terracing are practised on mountainous and submountainous areas. Bench terraces may be level, sloping outward or slopping inward depending upon local conditions. Terrace spacing is given by Fig. 17.14,

$$VI = \frac{WS}{100-S} \tag{17.7}$$

where
W = width of the terrace
S = percent slope of the land
VI = vertical interval

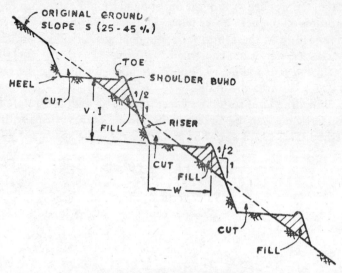

Fig. 17.14 Bench terracing.

Conditions of the depth of soil, slope, rainfall, climate, and farming practices influence the terrace design. A vertical interval of 1.8 m is considered satisfactory. The back slope of the riser may be $\frac{1}{2}$ to 1 or 1 to 1, a flatter back slope being at the expense of the bench.

QUIZ 17

I. State whether 'true' or 'false'; if false, give the correct statement:
  1. The rainfall pattern and the soil type have to be taken into account in determining cropping patterns for different agro-climatic regions.
  2. The precipitation rate of the sprinklers should be greater than the infiltration capacity of the soil.
  3. The furrow stream suould reach the lower end of the field within one-fourth of the total time required for irrigation.
  4. Leguminous crops and fodder grasses are included in the crop rotations.
  5. Tile drains are usually 10–15 cm diameter laid flat with open joints covered with the excavated earth.

II. Select the correct answer(s):
  1. The method of irrigation depends on
     a. the type of soil
     b. the slope of soil
     c. the kind of crop
     d. water supply available
     e. all the above factors

2. The size of tile drains depends upon
   a. the rainfall intensity
   b. the soil type
   c. the tile drain spacing
   d. the irrigation practice
   e. the slope of the land
   f. all the above factors

III. Match the items in A and B

A	B
1. Commercial crops on heavy black retentive soils as well as highly permeable soils with high irrigation efficiency	a. Drip irrigation
2. Orchards and steep lands with heavy soils	b. Corrugation method of irrigation
3. Growing sweet potato and turmeric on almost level plots	c. Sprinkler irrigation
4. Cultivation of row crops like maize, sugarcane, tobacco, chillies, vegetables, etc.	d. Basin method of irrigation
5. Large irrigation streams on almost level lands and close growing crops like wheat	e. Furrow irrigation
6. Irrigation under arid conditions with water supply limited and high in salt concentration	f. Border irrigation
7. Contour bunds	g. 2.1–3.5 kg/cm^2
8. Terracing	h. $VI = \left(2 + \dfrac{S}{4}\right) 0.305$
9. Spacing for graded terraces	i. On sloping lands—slope<6%, rainfall $\leqslant$ 60 cm
10. Bench terracing	j. On sloping lands—slope>6%, rainfall > 100 cm
11. Pressure at sprinkler heads	k. Mountainous areas— $VI = \dfrac{WS}{100-S}$

PROBLEMS

1. What depth of water is needed for a crop having an effective root depth of 1.3 m. The soil has an apparent specific gravity of 1.5 and available field capacity of 6%. (11.7 cm)

2. Design an artificial gravel pack for a drain in a sand formation with

   the finest material : $D_{15} = 0.15$ mm, $D_{85} = 0.45$ mm
   the coarsest material : $D_{15} = 0.30$ mm, $D_{85} = 0.90$ mm

3. A sprinkler irrigation system has to be designed to irrigate 16 ha of corn on a deep silt loam soil for which maximum permissible water application rate is 1.25 cm/hr. Depth of root zone is 90 cm. Peak rate of moisture use by the crop is 4 mm/day. Water application efficiency is 70%. Moisture holding capacity of the soil is 4.5 cm for 30 cm depth. Determine the following:
   (i) Depth of water pumped per application.
   (ii) Number of hectares irrigated per day.
   (iii) Capacity of the irrigation system, number of sprinklers required and the capacity of each sprinkler assuming a spacing of 12×18 m (sprinklers × laterals). (9.64 cm, 1 ha/day, $Q = 96.4$ m^3/hr with 10 hr of operation/day, 46 sprinklers, $q = 2.1$ m^3/hr.)

4. A farmer uses PVC syphon tubes of 5 cm I.D. and 1.20 m long to apply water into furrows from a head ditch. If the furrows are triangular with side slope 1:1 and longitudinal slopes of 0.001 and the water level in the head ditch is 50 cm above the bottom of the furrow, what should be the discharge of a single syphon? Assume uniform flow in the furrow and $n = 0.030$.
$$(C_d = 0.9, d = 17 \text{ cm}, 4.5 \text{ lps})$$

5. Tile drains have to be installed in an agricultural land which has a soil permeability of 25,000 lpd/m^2. An impermeable stratum exists at about 3.2 m below the land surface and it is desired to keep the water level at least 1 m below the land surface. What drain spacing would you recommend if the average annual rainfall is 100 cm? (140 m to remove 1% of a.a.r. in 24 hr)

6. Irrigation water for growing wheat has an electrical conductivity of 490 μmhos/cm. If the conductivity tolerance of wheat is 7,000 μmhos/cm, determine the leaching requirement for wheat. ($LR = 7\%$)

7. Furrows 80 m long, spaced at 70 cm, were irrigated by an initial stream of 2 lps which reached the lower end of the field in 30 min. The stream was then reduced to 0.5 lps and allowed to continue for 1 hr. Estimate the average depth of irrigation. (9.65 cm)

8. Write notes on the following:
   (i) Spacing of production wells
   (ii) Ground water provinces of India
   (iii) Geohydrological properties of laterites of west coast of India
   (iv) Well loss
   (v) Non-Darcy flow regime
   (vi) Ground water rights
   (vii) Dug-cum-bore well and its advantages
   (viii) Saline water intrusion in coastal aquifers
   (ix) Aquifer tests and their purpose
   (x) National policy for ground water use

Appendix A
Crop Details

Table A.1 Data for principal Indian crops

Sl. No.	Crop	Important high yielding varieties	Sowing time	Harvesting time	Duration of crop days	Seed required per ha kg	Average yield per ha qn/t	Spacing row to row × plant to plant, cm	Manures FYM/ Compost t	Fertilisers kg/ha			Seasonal consumptive use cm	Peak consumptive use mm	Cost of cultivation Rs/ha	Total income Rs/ha	Net income Rs/ha	Well yield required to irrigate 1 ha with 12 hr pumping per day & 70% irrigation efficiency lpm
										N	P_2O_5	K_2O						
1	2	3	4	5	6	7	8	9	10	11			12	13	14	15	16	17
1.	Paddy	IR-8, IR-20, Jaya, Madhu, Taichung Native-1, Sona (IET 1991)	June– July	Oct– Nov	105–145	60–80	4–8 t	15×10	5–10 Green manure	100	50	50	90	7	1,500	5,000	3,500	140
2.	Sugarcane	CO-419, CO-449, CO-740, IC-225	Feb– March	Dec– March	300–420	2¾ t (cut pieces)	80–150 t	90×90 in furrows	12.5–25	250	100	125	150	6–8	3,000	6,600	3,600	140

(*Contd.*)

1	2	3	4	5	6	7	8	9	10	11	12	13	14	15	16	17
3.	Cotton	Varalaxmi, Hybrid-4 Jayalakshmi DCH-32	May-June	Nov-Jan	180–210 180–200	2.5	30–40 qn seed 50 qn white lint-80 counts :18 qn	120 × 60–90	10–12.5	150	75	75	4–5	4,000	12,000 seed Rs 60-100/qn	8,000	85
4.	Maize	Deccan Hybrid, Ganga-5, Ganga white, Ambar, Vijai (yellow), composite Jowahar (orange)	June	Sept-Oct	105–120	15–20	5-t	75 × 25	7	150	75	37.5	5	1,500	3,000	1,500	100
5.	Ground-nut	Spanish improved TMV-2, S-206, AK-12-24, Ganga-pur	July	Nov-Dec	100–120	100–125	2–4 t	25 × 15	—	25	75	25	4	1,250	3,000	1,750	80
6.	Jowar (Sor-ghum)	CSH-1, CSH-2, CSH-5 Savarna	July	Nov	100–120	10–13	5 t-grain 10 t-fodder	40 × 10	7	125	75	37.5	4	1,250	2,500	1,250	80

	Crop	Varieties	Sowing	Harvesting	Duration (days)	Seed rate	Spacing (cm)	N	P	K	FYM	Yield (kg/ha)	Irrigation
7.	Wheat	Mexican varieties—Safed lerma, Choti lerma, UP-301, Kalyan sona, S-308 (Sonalike)	Oct-Nov	Feb-March	105-115	100-150	3-5 t	15-20×10	7	100	75	50	40	5	1,300	3,500	2,200	100
8.	Ragi	Purna, Shakti, CO-10	Feb	May	80-100	5	3-5 t	22.5×15	7	100	50	50	40	4	2,300	3,800	1,500	80
9.	Potato*	Kufri varieties-Chandramukhi, Sinduri-120, Shakti CI, Alankar (3649) Sheetman	Oct	Feb	90	1,000 tubers	15-30 t	45-60×15-25 in furrows	40	250	150	45	45	5	4,000	10,000	6,000	100
10.	Tobacco	Virginia gold, NC-95, VFC special (promising)	Feb	May	100-120	10-15 g	8-10 qn cured leaf	90×60 in furrows	12.5	25	75	75	50	5	1,500	4,000	2,500	100
11.	Chillies	CO-1	Throughout the year		120	1.25	7.5-10 t	60×30 in furrows	40	150	75	75	40	4	3,000	15,000	12,000	80

*The tuberous commercial crop of potato comes up well in almost all types of soils (except the saline and alkaline), in the cool hills and the plains. Simla has developed 10 improved varieties—Jyoti, Muthu, Lavkar, Bahar, Alankar and Badsha.

(Contd.)

1	2	3	4-5	6	7	8	9	10	11	12	13	14	15	16	17
12.	Bajra (spiked millet)	HB-1, HB-4, CO-4	July-Aug Oct Jan (summer crop)	80–95	5	3–4 t	45 × 15	6.25	100 60	30 25	4	1,000	2,000	1,000	80
13.	Barley, peas, oats, grams, mustard, linseed, etc.		Sep-Oct Feb-March	88		7–15 qn				30	3	700	1,900	1,200	60
14.	Tomato	CO-1, SIOUS, Pusa Rubi, S-12, best of all, best	May Sept	120	0.5	25–35 t	120–150 × 30 90 × 45	50–100	125 125 100 30	125 30	3	3,000	15,000	12,000	60
15.	Vegetables		Throughout the year	120		7.5–10 t		50–100	125 62.5 75 37.5	62.5 37.5	4–5	1,500	4,000	2,500	85
16.	Banana		Perennial	300–360		26–55 t			275 90	825 120	5	4,500	8,500	4,000	100
17.	Fodders		Kharif	100		250 t per year			75 —	45	5	800	10,800 per year	10,000	100

No.	Crop	Variety	Season	Duration	Seed rate	Spacing									
18.	Sunflower	EC. 68415 (Armavitsky) EC. 68414 (Peredovik) EC. 69874 (Armavertis)	Throughout the year	90–100	10–12.5	20–25 qn 45×22 in furrows	7	100	50	40	4	2,000	4,000	2,000	80
19.	Soyabeans	Hardie, Improved pelican, Brag, Clark-63	May-June	100–115	75–100	25–30 qn 30–45×3	—	20 (40) zinc sulphate—15	40	40	4	1,300	4,500	3,200	80
20.	Mulberry Bivoltine silk		Perennial	90–120 per crop	30 t (leaves) 1,7 t (cacoons) per year	45×15	20	250	100	40 per crop	4	5,000 (cost of cultivation) 3,500 (Rearing cost of silk worms)	12,000 17,000	7,000 (if leaves are sold) 8,500 (if leaves are fed to silkworms)	80

For 1 kg of silk: 7–8 kg of cacoons of bivoltine race
: 10–12 kg of cacoons of multivoltine race

Earning of Rs 1000 for every 100 desease free lavas (dfls)

Table A.2 Size and BHP of the pump and input hp of motor (kW) for paddy cultivation in India

Area cultivated (ha)	Size (suction × delivery) (cm)	Discharge (lpm)	Total static lift (from the lowest pumping water level to the highest patch of irrigated land) in meters							
			6		7.5		10		15	
			BHP	kW	BHP	kW	BHP	kW	BHP	kW
0.4	4×4	90	1/2	0.44	1/2	0.44	1/2	0.44	3/4	0.66
0.8	5×5	180	3/4	0.66	1	0.66	1	0.88	$1\frac{1}{4}$	1.10
1.2	6×5	270	1	0.88	1	0.88	$1\frac{1}{4}$	1.10	$1\frac{3}{4}$	1.54
1.6	6×6	365	$1\frac{1}{4}$	1.10	$1\frac{1}{2}$	1.32	2	1.76	$2\frac{3}{4}$	2.42
2.0	7.5×6	460	$1\frac{3}{4}$	1.54	2	1.76	$2\frac{1}{2}$	2.20	$3\frac{1}{2}$	3.07
3.0	7.5×7.5	685	$2\frac{1}{2}$	2.20	$2\frac{3}{4}$	2.42	$3\frac{1}{2}$	3.07	5	4.00
4.0	10×7.5	910	3	2.64	$3\frac{1}{2}$	3.07	$4\frac{1}{2}$	3.96	$6\frac{1}{4}$	5.50
6.0	10×10	1365	5	4.40	$5\frac{1}{2}$	4.83	$7\frac{1}{4}$	6.36	10	8.8
8.0	12.5×12.5	1820	$6\frac{1}{4}$	5.50	$7\frac{1}{2}$	6.60	10	8.8	13	11.4
10.0	12.5×12.5	2275							16	14.0

Appendix

Assumptions:

1. Peak consumptive use for paddy = 0.75 cm/day
2. Irrigation efficiency = 70%
3. Pump efficiency = 50%
4. Pumping hours = 8 hours per day (If the pumping hours per day is increased, proportionately more area can be cultivated at 12 hr/day 1½ times the area can be cultivated)
5. Efficiency of electric motor and drive = 85%
6. Conveyance losses due to a long field channel is not considered, i.e. the well is assumed to be in the centre of the irrigated land itself.
7. Suction and delivery sizes smaller than given above involve more friction losses and call for a higher hp pump.
8. For light crops (low water requirement) like ragi, maize, jowar, chillies, groundnut, cotton, fodder, etc. the area cultivated can be almost double than indicated.

Proforma A.1 Details of tubewells drilled in _____ district

Well no.................Owner's name.................Village.................Block.................Survey no.................
Location.................Bearing.................Toposheet no.................Quadrant.................Location of MP.................
RL of MP.................Height of MP a g.l.................Water supply/Agricultural/Industrial use.................
Well dimensions; at the surface.................at.................depth reduces to.................
Total depth of well b.g.l.................SWL: Monsoon.................Dry season.................
Bore log: Description   Depth b.g.l.   Water bearing formations:
           ....
           ....

Direction of flow into well: ....
Details of assembly: Slots.................blind pipe.................pump bowl.................
Pump details: Type.................hp/kW.................Head.................Tested yield/pumping rate.................Drawdown.................
Quality of water: Good/Brackish.................pH.................TDS (ppm).................Chloride (ppm).................
No. of days tubewell is run and hours of pumping per day: Kharif.................Rabi.................Summer.................Other season.................
Crops grown and their water requirement: Kharif.................Rabi.................Summer.................Other crops.................

Appendix

Consumption of electricity/diesel: Per hour (average)............Per year............
Per year.............

Annual draft from the tubewell:
(i) Discharge of tubewell (lph) × Total no. of hours pump is run in a year*
(ii) Total water applied for irrigation of kharif, rabi, summer and other crops.

MP—Measuring point, a.g.l.—above ground level, b.g.l.—below ground level, RL—Reduced level

* No. of hours pump is run in a year: (i) Hours of pumping per day (average) × No. of days tubewell is run in a year (ave.) (average of i & ii)
(ii) Total consumption of electricity or diesel in one year ÷ consumption of electricity or diesel per hour (average)

Maps showing lithological logs of all the tubewells (about 10 tubewells in a place) should be prepared and geological cross-sections in longitudinal and transverse directions should be drawn so as to depict the hydrogeology of the underground strata.

Appendix B

Hydraulic Details

Table B.1 Loss of head due to friction in steel pipes in metres per 100 metres of pipe length

Discharge lpm	Diameter of pipe (mm)									
	25	30	40	50	60	75	100	125	150	200
40	11.7	3.1	1.4							
60		6.6	3.0	1.1						
80		11.2	5.2	1.8						
100			7.9	2.7						
120			11.0	3.8	1.3					
140			14.7	5.1	1.7					
160			18.8	6.6	2.2					
200				10.0	3.3	1.4				
240				13.9	4.7	1.9				
280				18.4	6.2	2.6				
320				23.7	7.9	3.3				
360					9.8	4.1	1.0			
400						5.0	1.2			
480					16.8	7.0	1.7			
560					22.3	9.2	2.3			
640						11.8	2.9			
720						14.8	3.6			
800						17.8	4.4			
880						21.3	5.2			
960						25.1	6.2			
1,040							7.2			
1,120							8.2			
1,200							9.3			
1,300							10.7			
1,400							12.4	3.20		
2,000								7.80	3.56	
3,000									8.20	1.32
4,000									13.50	3.21

Appendix

Table B.2 Loss of head due to friction in concrete pipes (with mortar joints) in metres per 100 m length

Discharge lpm	Diameter of pipe (cm)						
	10	15	20	25	30	35	40
170		0.03	0.01				
340		0.11	0.02				
510		0.26	0.06				
680		0.46	0.10	0.03			
850		0.72	0.16	0.05			
1,020		1.04	0.23	0.07			
1,190		1.40	0.32	0.10	0.04		
1,360		1.84	0.41	0.13	0.05	0.02	
1,530		2.34	0.52	0.16	0.06	0.03	
1,700		2.88	0.64	0.20	0.08	0.04	
2,040		4.20	0.92	0.28	0.11	0.05	0.02
2,380		5.60	1.25	0.39	0.15	0.07	0.03
2,720		7.40	1.63	0.51	0.20	0.08	0.04
3,060		9.30	2.07	0.65	0.24	0.11	0.05
3,400		11.50	2.54	0.80	0.30	0.14	0.07
3,740		14.00	3.08	0·95	0.37	0.16	0.08
4,080		16.50	3.65	1.14	0.44	0.19	0.10

Manning's formula (for mean velocity)

$$v = \frac{1.486}{n} R^{2/3} \ S^{1/2} \text{ in feet units}$$

$$v = \frac{1}{n} R^{2/3} \ S^{1/2} \text{ in metric units}$$

Flow rate (discharge): $Q = Av$

where
A = area of cross-section
v = mean velocity of flow

Manning Roughness Coefficients for Various Boundaries

Boundary	Manning roughness $n\,(m^{1/6})$
Very smooth surfaces—glass, plastic, brass, pure cement, plaster	0.01
Very smooth concrete and planed timber	0.011
Smooth concrete	0.012
Ordinary concrete lining	0.013
Good wood, G.I.	0.014
Vitrified clay	0.015
Shot crete, untrowelled and earth channels in best condition	0.017
Straight unlined earth canals in good condition	0.02
Rivers and earth canals in fair condition	
—some growth	0.025
—rough condition	0.03

Hazen-Williams Formula (for water mains)

$$v = 0.848\, C\, R^{0.63}\, S^{0.54}$$

$$Q = Av, \quad A = \frac{\pi}{4} d^2, \quad R = \frac{d}{4}$$

$$Q = 0.2785\, C\, d^{2.63}\, S^{0.54}$$

where
    v = mean velocity of flow in m/sec
    d = diameter of the main in metres
    $S = (h_f/L)$ = loss of head per unit length of pipe
    C = constant depending on pipe roughness

Pipe Roughness Factors

Pipe material	Hazen-Williams C
Coated cast-iron, new	130
30 years old	95
Coated mild steel, new	130
Asbestos cement	140
Concrete	130
Plastic	140

Appendix

Discharge through a 90° V-notch (complete contraction)

$Q = 1.38 H^{2.5}$, where H = head in m and Q = discharge in m²/sec

Head H (cm)	Discharge Q lpm
2	4.7
4	26.5
6	73
8	149
10	255
12	414
15	725
20	1490
25	2580
30	4060

Appendix C

Equipment Dealers

Dealers in Water Well Drilling Rigs
1. LPM Precision Engineering Co. Pvt. Ltd., P.B. No. 7, Bilimora, District Bulsar, Bombay.
   Distributors: William Jacks & Co. (I), Pvt. Ltd.
2. Greaves Cotton & Co. Ltd., 1 Forbes Street, Bombay—400 001.
3. Ingersoll Rand (India) Pvt. Ltd. (a subsidiary of Ingersoll-Rand Company, New Jersey, USA), Khanna Construction House, 44 Maulana Abdul Gaffer Road, Worli Bombay—400 018.
4. Atlas Copco (India) Ltd., Mahatma Gandhi Memorial Building, Netaji Subhas Road, Bombay—400 002.
5. Killick Halco Ltd. (In Technical collaboration with Halifax Tool Co. Ltd., Southowram, Halifax, Yorks, England, UK), 31 Murzban Road, Bombay—400 001.
6. RMT Drill Pvt. Ltd., Industrial Estate Post, Coimbatore—641 021
7. Voltas Ltd., 19 Graham Road, Ballard Estate, Bombay—400 001.
8. Vak Engineering Pvt. Ltd. (Successors to Garlick Engineering, Madras) 'Queen's Court' Ist Floor, 15 Montieth Road, Egmore, Madras—600 008.
9. Longyear Diamond Core Drill, Christensen—Longyear (India) Ltd., 113 Koliwada Road, Sion East, Bombay—400 022.
10. Alvin Core Drills Pvt. Ltd., Plot No. 108, Kandivli Industrial Estate, Bombay—400 067.
11. Radha Core Drill, 100 Feet Road, Tata Bad, Coimbatore—641 012.
12. Hansa Electric & Engineering Corporation, 8-B/94, Mettupalayam Road, Coimbatore—641 011.
13. Qualitex Machinery (Pvt.) Ltd., 11/63 Pusa Road, New Delhi—110 005.
14. Water Development Society, Industrial Estate, Moula Ali, Hyderabad—500 040, A.P.

Pump Dealers
1. Worthington-Simpson Centrifugal Pumps, Best & Co Pvt. Ltd., 13/15 North Beach Road, Madras—600 001.
2. KSB Pumps Ltd., Sales and Service by Protos Engineering Co. Pvt. Ltd.. 173 Jamshedji Tata Road, Church Gate, Bombay—400 020.

Appendix 537

3. Kirloskar Brothers Ltd., Udyog Bhavan, Tilak Road, Poona—411 002.
4. Jyoti Limited, Baroda—390 003.
5. IAEC (India) Ltd., 196, L.B. Shastri Marg, Bhandup, Bombay—400 078.
6. Calama Pleuger Submersible Pumps, Calama Industries Pvt. Ltd., 7 Gaiwadi Industrial Estate, Swami Vivekananda Road, Goregaon, Bombay—400 062.
7. Atalanta Submersible Pumps, Protecto Engineering Pvt. Ltd., 26, Govt. Industrial Estate, Kandivili (West), Bombay—400 067.
8. Rohit Deep Well Turbine Pumps, Shree Viswakarma Engineering Works, Mahadeo Nagar, Bilimora.
9. Wasp-Ejecto and Submersible Pumps, Water Supply Specialists Pvt. Ltd., 151 Thambu Chetty Street, P.B. No. 1707, Madras—600 001.
10. Setco Deepwell Pump (plunger type), 316 Thambu Chetty Street, Madras—600 001.
11. Johnston Pumps Ltd., Mcneil & Magor Ltd., 2 Fairlie Street, Calcutta—700 001.
12. Eusun Engineering Co. Ltd., 5-7 Second Line Beach, Madras—600 001.
13. Batliboi & Co. Pvt. Ltd., Fortes Street, Fort, Bombay—400 001.
14. Larsen & Toubro Ltd., P.O. Box 278, Bombay 400 001.
15. Farm Maschinen, 47 Jayachamarajendra Road, Bangalore—500 002.
16. TEXMO Submersible pumpsets, Texmo Industries, Coimbatore—641 029.
17. Udhya Solar Submersible pumps, Udhayakrishna Engg. works, Coimbatore—641 004.

Dealers in Air Compressors and Blasting Units

1. Atlas Copco (India) Ltd., Mahatma Gandhi Memorial Building, Netaji Subhas Road, Bombay—400 002.
2. Ingersoll-Rand (India) Pvt. Ltd., Khanna Construction House, 44 Maulana Abdul Gaffar Road, Worli, Bombay—400 018.
3. Kirloskar Pneumatic Co. Ltd., Hadapsar Industrial Estate Ltd., Poona—411 013.
4. Compressor Air Equipment, Holman—Climax Manufacturing Ltd., Dolphin Court (2nd floor), 7A Middleton Street, Calcutta.
5. Consolidated Pneumatic Tool Co. Ltd., 301/302 Agra Road, Mulund, Bombay—400 080.
6. Industrial Explosives Pvt. Ltd.. Itwari, Nagpur—440 002.

Dealers in Irrigation Equipment

1. Premier Irrigation, 3 Netaji Subhas Road, Calcutta—700 001.
2. Voltas Ltd., 19 Graham Road, Ballard Estate, Bombay—400 001.
3. Farm Implements, Cossul & Co. Pvt. Ltd., Industrial Area, Fazalgunj, The Mall, Kanpur, U.P.
4. International Tube Trading Corporation, 98 Najdevi Street, Bombay—400 003.
5. The Indian Tube Co. Ltd., 18 Forbery Road, Tank Road PO, Bombay—400 033.
6. The Indian Hume Pipe Co. Ltd., Construction House, Ballard Estate, Bombay—400 001.
7. Hard PVC Pipes, M/s Wavin India Ltd., Ambattur Industrial Estate, Madras—600 058.
   Distributors: Tubes & Malleables Ltd., PO Box No. 60, 99 Armenian Street, Madras—600 001.

8. Jindal Sprinkler Irrigation Systems, Jindal Aluminium Limited, 16th km, Tumkur Road, Bangalore—560 073.

Water Diviners
1. S.P. Premachandra Rao, B.Sc., Member of British Society of Dowsers, Water Diviner, 3-6-476, Himayatnagar, Hyderabad—500 029, A.P. Tel: 61322.
2. Colonel D.M. Hennessey, Technician—Oil, Water & Mineral Radiesthetist, C/o National & Grindlays Bank Ltd., Post Box No. 6747, Bangalore.
3. For 'Ground Water Investigation' and 'Water Well Drilling' contact:
   Action for Food Production (AFPRO)
   Community Centre, C-17, Safdarjung Development Area
   New Delhi—110 016.

Dealers in Miscellaneous Equipment
1. Hindusthan Clock Works, 264-2 Shaniwar Peth, Poona—411 030 (Hydrometeorological equipment)
2. Purandare Electric Laboratories, 572-A, Shaniwar Peth, Poona—411 030. (Hydrometeorological equipment)
3. IAEC (Bombay) Pvt. Ltd, Forbes Street, Bombay—400 001 (Water Treatment plants, sprinklers and pumps).
4. V's Instruments, 1107/5, Shivajinagar, Poona—411 005 (V's Water Level Indicator)
5. Sparkonix (India), 1103/A-15 Shivajinagar, Poona-411 016 (Aquameter—a.c. type frequency resistivity meter similar to the terrameter manufactured by ABEM, Sweden, Rs. 20,000 approx.)
6. National Geophysical Research Institute, Hyderabad—500 007 (Resistivity meter—d.c. type, Rs 5,000 approx.)
7. Khoday Electronics, Bangalore (Terrameter components of ABEM, Sweden, assembled: Rs 30,000 approx.)
8. ABEM Terrameter SAS 300, A.C. low frequency type, 1—Hz range Ni Cd Batteries—rechargeable, 5.6 kg, Also the Terrameter SAS LOG 200 for Borehole logging—200 m depth, N,L, SP, Fluid Rty., Temp. logging tools, dia 40 mm; 15 kg. Atlas Copco ABEM AB Box 20086, S—161 20 Bromma, Sweden. TELEX 13079 ABEM S; Rs. 2 lakhs (approx.)
9. Logmaster, Oklahoma, USA (Electrical well logging—resistivity short & long normal, SP and radioactive logs sondes, US $ 18,000 approx.)
10. Widco Division, Gear Hart—Owen Industries Inc., 1100 Everman Road, PO Box 1936/Fort Worth, Texas 76101 (Widco Porta—Logger, records SP, single point resistivity & gamma ray, logging depth 300 m, US $ 7,000 approx.
11. Suji Engineering Works, 42, IV 'N' Block, Rajajinagar, Bangalore—560 010 (Protectron—electronic auto-dual water level controller.)
12. Lawrence and Mayo (India) Pvt. Ltd., 274 Dr. D. Naoroji Road, Bombay—400 001.
13. P. Orr and Sons Pvt. Ltd., 200 Mount Road, Madras—600 002.
14. Elico Pvt. Ltd., B-17 Industrial Estate, Sanatnagar, Hyderabad—500 018.
15. IGIS—ACR_1: A.C. Resistivity meter, 2.5 Hz, 100 W, 500 mA, 24 V rechargeable battery (int./ext.) Integrated Geoinstruments & Services Pvt. Ltd., 12-13-120 St. No. 3, Tarnäka, Hyderabad—500 017, A.P. Ph. 71632; Rs. 27000.

Appendix D

Tubewell Estimates

Performance of Bits and Drilling Rigs in Hard Rock Areas of Southern India

Table D.1 Bit performance

Type of bit	Rock type	Drilling rate m/hr	Bit life hr
30 cm Ingersoll Rand Button bit	Basalt	4.5–7.5	90–95
	Granite	7.5–6.0	90–95
	Weathered granite	7.5–12.0	—
Ingersoll Rand Carset	Basalt	3.6–6.0	90
	Granite	3.0–4.5	85–90
	Weathered rock	7.5–12.0	375–400
Atlas Copco	Basalt	4.5–4.8	—
	Granite	1.0–1.5	80–110
	Limestone	5.6	—
	Laterite	3.3	—
	Quartzite	3.3	—
30 cm RMT Button	Granite	1.9	90–150
30 cm X. RMT	Granite	1.5	55–130
Roller Cone	Shale/Schist	9–12	50

Table D.2. Performance of Drilling rigs

Drilling rig	Size cm	No. of days per well (average)	Depth of drilling per day (average) m/day
Ingersoll Rand			
TRUCM—3	15	2–3	25–32
RMT, DTH	15	2–3	9–15 (24–30)
Atlas Copco, DTH	COP 4–10 COP 6–15	3–5	18
Larsen Toubro, DTH	15	2–3	9–15
Vollum, DTH	15	2–3	9–15
Killick Nixon, DTH	15	2–3	9–15
Sanderson Cyclone, DTH	20	2.69	16.5
Percussion		10–15	3–4.5
Calyx	10–15	25–35	1–1.8

Appendix

Table D.3 Detailed estimate of a filter point tubewell 30 m deep, 10 cm diameter

Sl. No.	Item	Quantity	Rate Rs.	Cost Rs.
1.	Drilling 10 cm dia bore hole and reaming up to 7.5 cm, 31.5 m deep			
	0–15 m	15 m	18/m	270
	15–30 m	15 m	20/m	300
	30–31.5 m	1.5 m	23/m	35 (say)
2.	Washing the bore hole, lowering the pipe assembly and fitting of the pump		LS	250
3.	10 cm dia black pipe	20 m	55/m	1,100
4.	10 cm dia coir strainer	7 m	55/m	350
5.	10 cm dia bail plug	3 m	LS	250
6.	3-5 hp, monoblock centrifugal pump, with starter, switch board, etc.		LS	5,000
7.	Wiring, earthing, etc.		LS	600
8.	10 cm dia blank pipe with bends, sockets, sluice valve, reducer, etc. for delivery side	3.6 m	LS	600
9.	40 mm dia hand pump (to serve as check valve on suction side and facilitate priming)		LS	150
10.	Pump foundation and delivery cistern		LS	300
11.	Precast pump chamber		LS	1,000
12.	Labour charges		LS	200
			Total	10,405
	Add 10% of the total cost towards transportation costs, contingencies and increase in prices			1,040
				11,445
			Grand total, say, Rs	12,000

Table D.4 Detailed estimate for a tubewell drilled with rotary rig-45 cm dia bore and 20 cm dia tubewell pipe; 100 m deep

Sl. No.	Item	Quantity	Rate Rs	Cost Rs
1.	Selection of site for well drilling		LS	1,000
2.	Drilling 45 cm dia bore up to 104 m	104 m	45/m	4,680
3.	20 cm MS-ERW B-class pipe	100 m	180/m	18,000
4.	Workshop labour charges for cutting slots on the pipe	20 m	50/m	1,000
5.	Bail Plug	1 no.	LS	300
6.	Centre guides	3 no.	30 each	90
7.	Well cap	1 no.	LS	50
8.	Iron clamp	1 no.	LS	200
9.	Pea gravel of suitable size including transportation	15 m^3	100/m^3	1,500
10.	Development by air compressor for 10 hours		60/hr	600
11.	Development by vertical turbine pump for 30 hours		30/hr	900
12.	Submersible pump of suitable size and hp with switch, starter and other accessories		LS	16,000
13.	Construction of pump house		LS	4,000
			Total	48,320
	Add 10% of the total cost towards transportation costs, contingencies and increase in prices			4,832
				53,152
			Grand total, say, Rs	54,000

Appendix

Table D.5 Detailed estimate for a borewell drilled with DTH (Ingersoll Rand) rig in hard rock areas of Karnataka: 15 cm dia bore 60 m deep

Sl. No.	Item	Quantity	Rate Rs	Cost Rs
1.	Selection of site for well drilling		LS	1,000
2.	Drilling 15 cm bore up to 60 m deep	60 m	140/m	8,400
3.	Casing pipe	20 m	250/m	5,000
4.	Transportation		LS	600
5.	Yield test		LS	1,000
6.	Submersible pump with switch, starter, etc.	1 No.	LS	10,000
7.	Construction of pump house		LS	3,000
			Total Rs	29,000

Average yield $\approx$ 14 m³/hr. Can irrigate $\approx$ 2.5 ha of crop like maize, ragi, sunflower, soyabeans, chillies, etc.

$$\therefore \text{Capital cost} = \frac{29,000}{2.5}$$
$$= \text{Rs 11,600 per ha}$$

While for tubewell irrigation in alluvial areas the capital cost is around Rs 2,000 per ha.

For irrigation with sprinklers the capital cost will go up by another Rs 5,000 per ha.

The cost of sprinkler irrigation in hard rock areas works out to $\approx$ Rs 35 per ha-cm., while the cost of tubewell irrigation in alluvial areas is $\approx$ Rs 10 per ha-cm and with sprinklers $\approx$ Rs 30 per ha-cm.

Appendix E
Geological Time Scale

(Reveals the history of rock formations and unfolds the development of life with culmination at last with the modern man)

Group/Era	System/period (European)	Series/Epoch	Age in years from present Million, (M)	Main life	Rock group	Indian formations	
						Name of formation	Distribution
Cenozoic	Quaternary	Holocene (Recent)	25,000 years	Modern man			Kashmir sediments, Laterites in west coast, river alluviums
		Pleistocene	1 M	Primitive man		Sandstones, limestones, shales, laterites, gravel and marl	
	Tertiary	Pliocene	15 M	Mammals, toothless birds and flowering plants			Siwaliks, sandstones of Kathiawar, Kutch and Tamil Nadu
		Miocene	35 M				
		Oligocene	50 M				
		Eocene	70 M			Deccan traps	Vesicular basalts; found in Gujarat, Rajasthan, Maharashtra and parts of Deccan.
Mesozoic	Cretaceous		120 M	Reptiles and birds	Aryan Group	Upper Gondwana system	Sandstones of Kathiawar and Kutch, Cuttack, Rajahmundry Vijayawada, Sriperambadur, Uttatur (Trichi district)
	Jurassic		150 M			Middle	
	Triassic		190 M			Lower	

Appendix

Era	Period	Age	Life	Group	System	Description
Palaeozoic	Permian	220 M	Amphibians primitive plants and fishes	Dravidian Group		Upper Carboniferous period Spiti and Kashmir
	Upper Carboniferous	280 M				
	Devonian	320 M				
	Silurian	350 M	Invertebrates			Not represented in peninsular India
	Lower Ordovician	400 M				
	Cambrian	500 M				
Proterozoic	Precambrian	1500 M	Primitive simple form of life	Purana Group	Vindhyan system	Sandstones and limestones in Rajasthan, Bihar, M.P., Cuddapah, Kurnool and Godavari valley
					Cuddapah system	Sandstones, limestones, shales, slates and quartzites in Cuddapah, Kurnool and Rajasthan
						Eparchaean unconformity
Azoic	Archaean	2,000 M	No fossil evidence of life	Archaean	Dharwar system	Dharwar and lesser Himalayas
						Archaean crystalline complex—mostly gneisses, granites and schists; Bundelkhand gneiss, Bengal gneiss and schistose gneisses of the Peninsula; pegmatites, applites, basic dykes and charnockites.

Appendix F

Unit Conversion Factors

Conversion Factors—FPS to MKS Units

LENGTH

1 in = 2.54 cm
1 ft = 0.305 m
1 mile = 1.609 km
1 m = 3.281 ft
1 yd = 0.9144 m
1 km = 0.6214 mile
 = 0.54 nautical mile
1 naut. mile = 1.852 m

AREA

1 in^2 = 6.452 cm^2
1 ft^2 = 0.0929 m^2
1 cm^2 = 0.155 in^2
1 m^3 = 10.76 ft^2
 = 1.094 syd
1 acre = 0.4047 ha
 = 4047 m^2
 = 43560 sft
1 ha = 10^4 m^2
 = 2.471 acres
1 mile^2 = 2.59 km^2
 = 640 acres
1 km^2 = 100 ha
 = 0.3861 mile^2

VOLUME

1 cft = 28.32 litres
 = 0.02832 m^2
 = 6.24 imp. gal
 = 7.48 US gal
1 litre = 1,000 cc
 = 0.22 imp. gal
1 barrel = 42 US gal
1 imp. gal = 1.2 US gal
 = 4.546 litres
1 US gal = 3.79 litres
 = 0.833 imp. gal
1 m^3 = 35.31 cft
 = 220 imp. gal
 = 264 US gal
 = 1,000 lit
1 cc = 0.061 in^3
1 acre-ft = 43,560 cft
 = 1,233.5 m^3
 = 2.71 × 10^5 imp. gal
1 in^3 = 16.387 cc (ml)
1 cyd = 0.7646 m^3
1 ha-cm = 100 m^3
1 acre-ft = 0.1234 ha-m
 = 1230 m^3
 = 2.71 × 10^5 imp. gal
1 ha-m = 10^4 m^3
 = 8.14 acre-ft

VELOCITY

1 ft/sec = 30.48 cm/sec

Appendices

1 cm/sec = 0.0328 ft/sec
1 m/sec = 3.281 ft/sec
1 mph = 1.467 ft/sec (fps)
= 1.609 km/hr (kmph)
= 0.8684 knot
1 knot = 1.69 fps
= 0.515 m/sec
1 km/hr = 0.2778 m/sec
= 0.9113 fps
= 0.6214 mph
1 m/day = 22.9 gpd/ft^2

ACCELERATION DUE TO GRAVITY (g)
1 ft/s^2 = 0.305 m/s^2
g = 32.2 ft/sec^2
= 9.81 m/sec^2

FLOW RATE (DISCHARGE)
1 cfs (cusec) = 0.0283 m^3/sec (cumec)
= 28.3 lps
= 374.03 imp. gpm
= 449 US gpm
= 1.983 acre-ft/day
= 724 acre-ft/year
1 imp. gpm = 0.0757 lps
= 0.0757 × 10^{-3} m^3/sec
= 1.2 US gpm
1 US gpm = 0.063 lps
= 0.063 × 10^{-3} m^3/sec
1 m^3/sec = 35.31 cfs
= 19.01 × 10^6 gpd (imp)
= 13200 imp. gpm
= 15800 US gpm
= 70 acre-ft/day
= 3.05 × 10^6 ft^3/day
1 lps = 0.03531 cfs
1 m^3/day = 2190 imp. gpd
1 imp. mgd = 695 imp. gpm
= 3160 lpm
= 0.0527 m^3/sec
1 acre-ft/day = 188.57 imp. gpm
= 271542 imp. gpd
= 1233.5 m^3/day

PRESSURE OF SHEAR INTENSITY
1 kg/cm^2 = 14.23 psi
1 atm = 14.7 psia.
= 34 ft of water
= 30 in of mercury
= 76 cm of mercury
= 101.32 kN/m^2
= 1013.2 mb (millibars)
1 bar = 14.5 psi
1 millibar = 0.0143 psi
1 psi = 6.895 kN/m^2
= 0.7031 m of water
1 psf = 47.88 N/m^2
1 atm = 1.033 m-atm
1 m-atm = 14.223 psia
= 0.9678 atm
= 10 m of water
1 m-atm = 0.967 atm
= 73.5 cm of mercury
= 32.8 ft of water
= 982 mb

MASS
1 slug = 32.2 lb
= 14.6 kg
1 lb = 453.6 gm
1 kg = 2.205 lb
= 0.06852 slug
1 ton = 1000 kg

FORCE (WEIGHT)
1 lb = 4.448 N
= 16 oz
= 7,000 grains
1 gm = 1543 grains
1 kN = 224.8 lb
1 qn = 981 N

MASS DENSITY
1 gm/cc = 1.94 slugs/cft
= 1,000 kg/m^3
1 slug/ft^3 = 515.4 kg/m^3

SPECIFIC WEIGHT
1 gm/cc = 62.4 pcf
1 pcf = 157.1 N/m^3

WORK OR ENERGY
1 ft-lb = 1.356 N.m (J)
= 3.77 × 10^{-7} kW hr
1 BTU = 778 ft-lb
= 1055 N.m (J)
1 kWhr = 2.66 × 10^6 ft-lb

DYNAMIC VISCOSITY
1 lb-sec/ft^2 = 47.88 N.s/m^2
= 478.8 poise
1 centipoise = 0.01 poise
1 N.s/m^2 = 10 P

KINEMATIC VISCOSITY
1 ft^2/sec = 0.093 m^2/sec
= 929 stokes
1 m^2/s = 10^4 St
1 c St = 10^{-6} m^2/s

MOMENT OF INERTIA
1 in^4 = 41.67 cm^4
1 cm^4 = 0.024 in^4

PERMEABILITY (also see Table 4.4)
1 darcy = 0.966 × 10^{-3} cm/sec
= 0.987 × 10^{-8} cm^2
1 lpd/m^2 = 1.16 × 10^{-6} cm/sec
1 m/day = 1.16 × 10^{-3} cm/sec
= 1,000 lpd/m^2
= 20.44 imp. gpd/ft^2
= 24.54 US gpd/ft^2
= 0.017 US gpm/ft^2

POWER
1 hp = 550 ft-lb/sec
= 1.014 metric hp
= 746 watts

1 metric hp = 75 m-kg/sec
= 736 watts
= 542.8 ft-lb
= 0.986 hp

MODULUS OF SECTION
1 in^3 = 16.4 cm^3
1 cm^3 = 0.061 in^3

TEMPERATURE
(°F − 32)$\frac{5}{9}$ = °C
460 ÷ °F = °R (British system)
273 + °C = °K (Metric system)

WATER QUALITY
1 ppm = 1 mg/l
1 grain/US gal = 17.1 ppm
1 taf = 735 ppm
1 me/l = 1 epm
1 ppm = 1.56 μ mho/cm
TDS in ppm = 0.64 EC in μ mhos/cm

TRANSMISSIBILITY
1 m^2/day = 67.05 imp. gpd/ft
= 80.52 US gpd/ft
= 0.056 US gpm/ft
π = 3.1416...
e = 2.7183...
$\log_{10} e$ = 0.4343
$\log_e 10$ = 2.303
$\log_e x$ = 2.303 $\log_{10} x$
milli = 10^{-3}
micro = 10^{-6}
hecto = 10^2
kilo = 10^3
mega = 10^6
1 micron = 10^{-3} mm
= 10^{-6} m
1 million = 10^6
1 lakh = 10^5

Appendices

Conversion Factors—MKS to SI

[SI—'Systeme' International d' Unites' (International System of Units)]

Quantity	Symbol	Dimensions	MKS units	Conversion factor (multiply MKS to convert to SI)	SI units
Length	l	L	m	—	m
Mass	m	M	$kg_m = \frac{kg}{g}$	9.807	kg
Time	t	T	sec	—	s
Force	F	MLT^{-2}	kg (kg_f)*	9.807	N (newton)
Work or Energy	W, E	FL	m-kg	9.807	J (=N-m) (Joule)
Power	P	FLT^{-1}	m-kg, sec / metric hp	9.807 / 736	W (=J/S =N-m/s) (watt)
Moment or Couple, BM, Torque	M, T	FL	kg-m	9.807	N-m
Mass density	ρ	ML^{-3}	kg_m/m^3 / t/m^3	9.807 / 1000	kg/m^3
Weight density (specific weight)	$\gamma (=\rho q)$	FL^{-3}	kg/m^3	9.807	N/m^3
Pressure, Shear intensity, Bulk modulus	p, τ, K	FL^{-2}	kg/cm^2	9.807×10^4	N/m^2
Dynamic Viscosity	μ	FTL^{-2}	$kg/sec/m^2$	9.807	
		$ML^{-1}T^{-1}$	poise (dyne-sec/cm^2) / centipoise	0.1 / 10^{-3}	$N\text{-}sm^2$ (kg m-s)
Kinematic viscosity	ν	L^2T^{-1}	stoke (cm^2/sec)	10^{-4}	m^2/s
Surface tension	σ	FL^{-1}	kg/m / dyne/cm	9.807 / 10^{-3}	N/m

*1 (kg_f) in engineering is the force required to accelerate a mass of 1 kg_m at 9.807 m/sec^2; 1 kgf = 9.807 N.

1 N = 10^5 dynes. 1 kg = 980,665 dynes (exactly).

1 Standard atmosphere ≈ 100 kN/m² ;

1 Pa = 1 N/m²

1 bar = 10^5 N/m²

    = 100 kN/m²

    = 1 atm

    = 75 cm of Hg

1 bar = 10^5 Pa

1 Joule = 9.48×10^{-4} BTU (Energy—FPS)

1 kilocal = 3.97 BTU (Heat energy—FPS)

(1,000 calories)

1 Watt = 3.413 BTU/hr (power—FPS)

1 BTU = 1,054 joule

1 kilocal = 4,187 J

1 cal = 4.18 J

Universal gas constant, R = 8312 J/mol °K

Gas constant for air ;

R (for air) = 287 N-m/kg °K (or J/kg °K)

$\rho_{freshwater}$ = 1000 kg/m³, $\gamma_{f.\ water}$ = 9.81 kN/m³

$\rho_{seawater}$ = 1025 kg/m³

Atm. pr. grad. = 1 mb per 9 m

Geothermal grad. = 0.7 to 1°C per 100 m

Selected References

Chow, V.T., 'On the determination of transmissivity and storage coefficients from pumping test data', *Trans. Am. Geophy. Un.* 33: 397—404, 1952.

Ahmed, N., 'Dyanamics of ground water with special references to tube wells', *Proc. Ankara Symposium on Arid Zone Hydrology*, UNESCO (Paris) 77-98, 1953.

Ahmed, N. and Sunada, D.K., 'Nonlinear flow', *Jour. Hyd. Div.*, Proc. ASCE, Vol. 95, pp. 1847-57, Nov. 1969.

Allen, M.J. and E.E. Geldreich, Bacteriological criteria for ground water quality, *Ground Water* 13 (1): 45–51, 1975.

Anderson, T.W., 'Electrical-analog analysis of the hydrologic system, Tuscon Basin, Southeastern Arizona', *USGS Wat. Sup. paper*, 1939-C.

Anon, 'The corrosion and incrustation of well screens', *Minn., Bull.* Edward E. Johnson Inc., St. Paul, 834, 1955.

———, 'Ground water and wells'. UOP Johnson Div., 315 North Pierce st., St. Paul, Minn., 1966.

ASCE, *The Ground Water Management,* ASCE (New York), 1970.

Banmann P., 'Experiments with fresh-water barrier to prevent sea water intrusion', *Jour. Am. Wat. Works Asso.*, Vol. 45, 521-34, 1953.

Bear, J. and Dagan, G., 'Intercepting fresh water above the interface in a coastal aquifer', *IASH Publ. No. 64*, 1963.

———, 'Moving interface in coastal aquifers', *Jour. ASCE, Hyd. Divn.*, Hy-4, 1964.

———, 'Solving the problem of local interface upconing in coastal aquifer by the method of small perturbations', *J. Hydraul. Res.* 6 (1): 16–44, 1968.

Bear, J., 'Hydrodynamic dispersion', in *Flow through porous media*, R.J.M. De Weist (Ed.), Academic Press (New York) 109-199, 1969.

Boreli, M., 'Free surface flow toward partially penetrating wells'. *Trans. Am. Geophys. Un.*, Vol. 36, 664–72, 1955.

Boulton, N.S., 'The drawdown of water table under non-steady conditions near a pumped well in an unconfined formation', *Proc. Inst. of C.E.* (London) 3. Part III: 564–79, 1954.

———, 'Unsteady radial flow to a pumped well allowing for delayed yield from storage', *ASSOC. Int. Hydrol.* (Rome) 37: 472-77, 1955.

———, 'Analysis of data from non-equilibrium pumping tests Allowing for Delayed yield flow storage, a discussion', *Proc. Inst. C.E.* (London), Vol. 28, 1964.

Boulton, N.S. and T.D. Streltsova, 'New equations for determining the formation constants of an aquifer from pumping test data', *Wat. Resour. Res.* 11: 148-153, 1975.

Cooper, H.H. et al., 'Response of a finite diameter well to an instantaneous charge of water.' *Water Resources Research*, 3 (1) pp. 263-269, 1967.

Das, R.N., 'Finite element analysis of unconfined flow towards wells' Ph. D. thesis, Indian Institute of Technology, Kharagpur, 1975.

Dey, A.K., *Geology of India*, National Book Trust of India, New Delhi, 1968.

Dobrin, Milton B., *Introduction to geophysical prospecting*, McGraw-Hill Kogakusha, Ltd., 3rd Edn. 1981.

Doneen, L.D., 'The influence of crop and soil on percolating waters', *Proc. 1961 Biennial Conference on Ground Water Recharge*, 1962.

Ferris, J.G. et al., 'Theory of aquifer tests', *USGS water supply paper 1536-E*, 69-174, 1962.

Ferris, J.G., 'Ground Water,' in *Hydrology*, Wisler and Brater (Ed.), Wiley, (New York), 198-272, 1949.

Ferris, J.G., 'A quantitative method for determining ground water characteristics for drainage basin', *Agri. Eng.* V. 31 (6), 1950.

Foley, F.C., W. C. Walton, and W.J. Drescher, 'Ground-water conditions in the Milwaukee-Waukesha area, Wisconsin', *USGS Wat. Sup. paper 1229*, 1953.

Forchheimer, P., *Groundwasserbewegung, in Hydraulik*, 3rd ed., B.G. Teubner (Leipzig), 51-110, 1930.

de Glee, G.J., 'Over ground waterstroomingen bij wateronttrekking door middle van putten', J. Waltman Jr. (Delft), 175, 1930.

———, Berekening methoden voor de Winning van ground water. In Drink water voorziening 3e Vacanite Cursus, 38–80. Moorman's periodicke pers, The Hague, Netherlands, 1951.

Glover, R.E. and G.G. Balmer, 'River depletion resulting from pumping a well near a river, *Trans. Am. Geophys. Un.*, Vol. 35, 468-70, 1954.

Glover R.E., 'The pattern of fresh water flow in a coastal aquifer, *Jour. Geophys. Res.*, Vol. 64, No. 4, 1959.

Guymon, G.L., 'Digital computers and drainage problem analysis: Part III-Finite element method', in *Drainage for Agriculture*, J. Van. Schilfgaarde (ed.), Agronomy Monograph, No. 17, Am. Soc. Agron., Madison, Wisconsin, 587-607, 1974.

Hantush, M.S. and C.E. Jacob, 'Non-steady radial flow in an infinite leaky aquifer', *Am. Geophys. Un. Trans.* 36: 95-100, 1955.

Hantush, M.S., 'Analysis of data from Tests in Leaky Aquifer', *Trans. Am. Geophys. Un.*, Vol. 37, 1956.

———, 'Hodraulics of wells', in *Advances in Hydroscience*, Vol. 1, V.T. Chow (ed) Academic Press, New York and London, 281-432, 1964.

Houtkamp, H. and Jack, G., 'Geohydrologic well logging,' *Nordic Hydrology*, 3 (3), 162-182, 1972.

Hubbert, M.K. 'The theory of ground-water motion', *Jour. Geol.* Vol. 48, 785-944, 1940.

Irmay, S., 'On the hydraulic conductivity of unsaturated soil, *Trans. Am. Geophy. Un.*, Vol. 35, 463-67, 1954.

Jacob, C.E., 'Flow of ground water', in *Engineering Hydraulics*, Hunter Rouse (ed.), Wiley, New York, 1960.

Kozeny, J., *Theorie and Berechnung der Burnnen*, Wasserkraft und Wasserwirtschaft, Vol. 28, 88-92, 1933.

———, Das Wasser in Boden, Groundwasserbewegung, in Hydraulik, Springer-Verlag, Vienna. 380-445, 1953.

Selected References

Krishnan, M.S., *Geology of India and Burma*, Higginbothams (P) India Ltd., Madras, 1968.

Kruseman, G.P. and N.A. De. Rider, 'Analysis and Evaluation of Pumping Test Data', *Inst. for Land Recla. aud Improvement*, Wageningen, The Netherlands, 1970.

Lohman, S.W., 'Ground water hydraulics', *USGS prof. paper 708*, 1972.

Maucha, Rezso, 'The graphic symbolisation of the chemical composition of natural waters, *Hidrol Kozlony*, Vol. 29, 1949.

Muskat, M., *The flow of homogeneous fluids through porous media*, McGraw-Hill, New York, 1937.

Piper, A.M., 'A graphic procedure in the geochemical interpretation of water analysis', *Trans. Am. Geophys. Un.* Vol. 25, 1944.

———, 'A graphic procedure in the geochemical interpretation of water analysis. *USGS. Grnd. Wat. Note 12*, 1953.

Prickett, T.A., 'Modelling Techniques for Ground water evaluation', in *Advances in Hydroscience*, Vol. 10, V.T. Chow (ed.), Academic Press, New York, 1-143, 1975.

Rorabaugh, M.I., 'Graphical and theoretical analysis of step drawdown test of artesian well', *Proc. ASCE*, Vol. 79, Separate no. 362, 1953,

———, 'Ground Water Resources of the North-eastern part of the Louisville Area, Kentucky', *U.S. Geol. Sury. Wat. Sup. paper* 1360-B, 1956.

Rumer and Harleman, 'Intruded salt water wedge in porous media', *Jour. ASCE Hyd. Divn.*, Hy-6. Vol. 89, 193-220, 1963.

de Rider, N.A., 'The use of computers in water resources development and water supply planning, *Geologic En. Mijnbouw*, Vol. 51 (i), 1972.

Satish Chandra and P.K. Pande, 'Recharge studies for a basin using mathematical model, Water For Human Needs', *Proc. of the 2nd World Cong. on Wat. Res.*, Vol. III, IWRA—CBIP, New Delhi, 1975.

Skibitzke, H.E., An equation for potential distribution about a well being bailed. *U.S. Geol. Sury.* open file report, 1958.

———, 'Electronic computers as an aid to the analysis of hydrologic problems', *Int. Assoc. Sci. Hydrology, Publ.* 52, 1961.

Stallman, R.W., 'Numerical analysis of regional water levels to define aquifer hydrology', *Trans. Am. Geophys. Un.*, Vol. 37, No. 4, 1956.

———, 'From geologic data to aquifer analog models', *Geol. Times J*, 5(7), 1960.

Stiff, H.A., (Jr.), 'The Interpretation of chemical water analysis by means of patterns', *J. Petrol. Tech.* Pt. 15, 1951.

Sunada, D.K., 'Laminar and turbulent flow of water through porous media', Final report, CER-68-69 DK 533, Colorado State University, Fort Collins, 1969.

Suter, M., 'The peoria recharge pit: its development and results, *Proc. ASCE*, Vol. 82, no. IR 3, 1956.

Theis, C.V., 'The relation between the lowering of the piezometric surface and the rate and duration of discharge of a well using ground water storage, *Trans. Am. Geophys. Un.* 16: 519-24, 1935.

Thiem, G., *Hydrologische Methoden*. Gebhardt (Leipzig), 1906.

Tyson, H.N. and E.N. Weber, 'Ground water management for the nation's future-computer simulation of ground water basins', *Jour. of Hyd. Divn., Proc. ASCE*, Vol. 90, No. HY4, July 1964.

UNESCO, *The development of ground water resources with special reference to deltaic areas*, United Nations, New York, 1963.

Verruijit A., *Theory of ground water flow*, The Macmillan Press Ltd., London, 2nd Edn., 1982.

Volker R.E., 'Solutions for unconfined non-Darcy seepage', *J. Irrig. Drain Div. ASCE*, 101, IR1 : 53-65, 1975.

Wadia, D.N., *Geology of India*, Macmillan, London, 1969.

Walton, W.C., 'Selected analytical methods for well and aquifer evaluation', *Illinois State Water Survey Bulletin*, No. 49, 1962.

——, *Ground Water Resource Evaluation*, McGraw-Hill, New York, 1970.

Wen Hsiung Li *et al.*, 'A new formula for flow into partially penetrating wells in aquifers', *Trans. American Geophysical Union*, Vol 35, No. 5, pp. 805-12, 1954.

Wenzel, L.K., 'The Thiem method for determining permeability of water-bearing materials and its application to the determination of specific yield'. *U.S. Geol. Sury. Wat. Sup.* paper 679-A, 1936.

——, 'Methods of determining permeability of water-bearing materials with special reference to discharging well methods', *U.S. Geol. Sury. Wat. Sup.* paper 887, 1942.

Wenzel, L.K. and A.L. Greenlee, 'A method for determining transmissibility and storage coefficients by tests of multiple well systems', *Trans. Am. Ceophys. Un.*, Vol. 24, 1943.

Wilcox, L.V., 'The quality of water for irrigation use', *U.S. Dept. Agri. Tech. Bull.* 1962, Washington D.C., 1948.

Williams, D.E., 'Digital computer models and ground water basin management', *Int. Sym. on Dev. of Grnd. Wat. Reso.*, Proc. Vol. 3, Madras, 1973.

Advances in Hydroscience (V.T. Chow, ed), Vol. 1, Academic Press, pp. 281-432, New York, 1964.

Baig, M.Y.A., 'Direct Slope Technique of Determining Absolute Resistivity', *Journ. of the Inst. of Engrs. (India)*, Civil, Vol. 61, Sept. 1980, UDC 624.131.3, Calcutta-700 020.

Bermes, B.J., 'An Electric Analog Method for Use in Quantitative Hydrologic Studies', *U.S. Geol. Surv. Phonix, Arizona*, 1960.

Bulusu, K.R., 'Defluoridation of Waters using combination of Aluminium Chloride and Aluminium Sulphate', *Journ. Inst. of Engrs. (India), Environmental Engg. Divn.*, Vol. 65, Part EN 1, October 1984, Calcutta 700 020.

Bureau of Reclamation, 'Ground Water Manual', *U.S Dept. Interior*, pp. 480-1977. Revised reprint, John Wiley & Sons, New York 1981. A Water Resources Technical Publication.

Cooper, H.H., Jr., et al., *Sea Water in Coastal Aquifers*, *U.S. Geol. Surv. Wat. Sup.* paper 1613-C, pp. 84, 1964.

Glover, R.E., Ground-water Movement, *USBR Engg. Monograph*, No. 31, pp. 67, Denver, 1964.

Glover, R.E., 'The pattern of Fresh water flow in a coastal aquifer', In Sea Water in Coastal Aquifers, *U.S. Geol. Surv. Wat, Sup. paper* 1613-C, pp. C32-C35, 1964.

Horton, R.E., 'An approach toward a physical interpretation of infiltration capacity,' *Proc. Soil Sci. Soc. Am.* 5: 399-417, 1940.

Hantush, M.S., 'Hydraulics of Wells', *Advances in Hydroscience* (V.T. Chow, ed), Vol. 1, Academic Press, pp. 281-432, New York, 1964.

Karplus, W.J., 'Analog Simulation', *McGraw Hill Book Company*, New York, 1958.

Kruseman, G.P. and de Ridder, N.A., 'Analysis and Evaluation of pumping Test

Selected References

Data', *Intl. Inst. for Land Reclamation and Improvement, Bull.* 11, Wageningen, pp. 200, The Netherlands, 1970.

Mooney, H.M. and Wetzel, W.W., 'The potential about a point Electrode and Apparent Resistivity curves for a Two-, Three- and Four-layer Earth', *Univ. Minnesota Press*, Minneapolis, 1956.

Orellana, E. and Mooney, H.M., 'Master Tables and Curves for Vertical Electrical Sounding over Layered Structures', *Interciensia, Madrid, Spain*, 1966.

Prickett, T.A., 'Modelling Techniques for Groundwater Evaluation', *Advances in Hydroscience*, Vol. 10, Academic Press Inc., New York, 1975.

Prickett, T.A., Type-curve solution to aquifer tests under water table conditions, *Ground Water*, Vol. 3, No. 3, pp. 5-14, 1965.

Roragaugh, M.I., 'Graphical and Theoretical Analysis of Step-drawdown Test of Artesian Well', *Proc. ASCE*, Vol. 79, Sept. no. 362, pp. 23, 1953.

Stallman, R.W., 'Numerical Analysis of Regional Water Levels to Difine Aquifer Hydrology', *Trans. Am. Geophys. Union*, Vol. 37, No. 4, 1956.

Southwell, R.V., Relaxation Methods in Theoretical Physics', *Oxford University Press*, London, 1946.

Walton, W.C. and Prickett, T.A., 'Hydrogeologic Electric Analog Computers', *Journ. Hyd. ASCE*, Vol. 89, No. Hy-6, 1963.

Tagg, G.F., 'Interpretation of Resistivity Measurements', *Trans Am. Inst. Min. Met. Eng.*, Vol. 110, Geophysical Prospecting, pp. 135-147, 1934.

Compagnie Générale de Géophysique, 'Abaques de sondage électrique', *Geophys. Prospect.* Vol. 3, suppl. 3, pp. 1-7 plus charts, 1955.

Compagnie Générale de Géophysique, 'Master Curves for Electrical Sounding', *European Association of Exploration Geophysicists*, The Hague, 1963.

Sankar Narayan, P.V. and Ramanujachary, K.R., 'An Inverse Slope Method of determining Absolute Resistivity', *Geophysics*, Vol. 32, No. 6, 1967.

Bibliography

1. *Analysis and evaluation of pumping test data*, G.P. Kruseman and N.P. de Rider, International Institute for Land Reclamation and Improvement, Wageningen, The Netherlands, 1970.
2. *A guide for estimating irrigation water requirements*, Technical Series No. 2, July 1971, Water Management Division, Ministry of Agriculture (Dept. of Agriculture), New Delhi-110001.
3. *Concepts and models in ground water hydrology*, Patric A. Domenico, McGraw-Hill Book Company Inc., New York, 1972.
4. *Elements of Photogrammetry*, B. Paul R. Wolf, McGraw-Hill Book Company Inc., New York, 1974.
5. *Hydrogeology*, S.N. Davis and R.J.M. de Wiest, John Wiley & Sons, New York, 1966.
6. *Ground water—a selected bibliography*, compiled and edited by Frits van der Leeden, Water Information Centre, Water Research Building, Manhasset Isle, port Washington, New York, 1971.
7. *Ground water and water well journals*, Jay H. Lehr (Ed.) 500 W. Wilson Beidge Road, Suite 135, Worthington, OH 43085.
8. *Ground water hydrology*, D.K. Todd, John Wiley & Sons, New York, 2nd Edn., 1980.
9. *Ground water resource evaluation*, William C. Walton, McGraw-Hill, New York, 1970.
10. *Ground water and wells*, Edward E. Johnson Inc., St. Paul, Minnesota, 1966.
11. 'Geophysics and ground water,' *Water Well Journal*, Columbus, Ohio, July and August 1971.
12. *News from Israel*, Vol XIX, No. 24, Consulate of Israel, Bombay, 1972.
13. *Planned utilisation of ground water basins*, Coastal plain of Los Angeles Country, Bull. No. 104, Califonia Dept. of Water Resources, Sacramento. Sept 1968.
14. *Scientific allocation of water resources*, Nathan Buras, American Elsevier, New York, 1972.
15. *Selected analytical methods for well and aquifer evaluation*, William C. Walton Bull. 49, Illinois State Water Survey Division, Urbana, 1967.
16. *Symposium on efficiency of water distribution and use on the land*, CBIP publication no. 84, New Delhi.
17. The phenomenon of dowsing, R.D. Char, *Deccan Herald Weekly Magazine*, March 7, 1976.
18. 'Training course on ground water techniques' (First Course: Sept-Oct 1970)

—Lecture notes (unpublished), UNDP Ground Water Project, Thanjavur, Tamil Nadu (South India).

19. *Water Well Technology*, M.D. Campbell, Jay H. Lehr, NWWA, Worthington, USA, 1973.
20. *A guide to application of irrigation water*, Donald J. Minehart, USAID/FARM Irrigation Advisor, UPAU, Pant Nagar, District Nainital, U.P., 1967.
21. *A manual of conservatian of soil and water*, United States Department of Agriculture, Oxford Book Company, Calcutta, 1964.
22. *Hand book of subsurface geology*, Carl A. Moore, Harper & Row, New York, and John Weatherhill Inc., Tokyo, 1963.
23. *Handy data for the sprinkling expert*, Perrot-Regnerbau GMBH & Co. Calw, West Germany, Dec. 1963.
24. *How to make bench terraces*, Informatton Leaflet No. 47, Farm Information Unit, Directorate of Extension, Ministry of Food and Agriculture, New Delhi, Jan. 1963.
25. *Outlines of geophysical prospecting—A manual for geologists*, M.B. Ramachandra Rao, Prasaranga, University of Mysore, 1975.
26. 'Slug Tests for Preliminary Evaluation of Bore Well Yields, 'K.R. Karanth etc., *Seminar on water well drilling in hard rock areas of India*, Institution of Engineers (India), Mysore Centre, Bangalore, 1974.
27. *Small wells manual*, U.P. Gibson and R.D. Singer, Department of State-Agency for International Development, Washington, D.C., Sept. 1969.
28. 'Some remarks on ground water conditions' in *Crystalline rock areas*, AKE. I. Moller, Hydrogeologist, UNICEF, Water supply section, New Delhi, April 1973.
29. *The Madras agricultural journal*, Vol. 59; No. 2, Special Issue on 'Ground water and water wells', Agricultural College and Research Institute, Coimbatore-641003, south India, Feb 1972.
30. *Study of oil and gas series well logs*, MIR Publishers, Moscow, 1971.
31. *Water Resources Department*, AFPRO, Publ. No. 9, C–52, New Delhi South Extension II, New Delhi–110049.
32. *Water For Human Needs, Proc. of the 2nd World Cong. on Wat. Res. Vol. III, Development and Meteorology*, IWRA—CBIP, New Delhi, Dec 1975.
33. '*Water well drilling in hard rock areas of India*' Proc. of Seminar, Bangalore, Institution of Engineers (India), Oct 1973.
34. *Ground water resources, Vol. 1 to 4, Int. Symp. on Dev. of Grnd. Water Res.*, Madras, India, Nov 1973.
35. *Hydrogeological Map of India*, Scale 1: 5000000 B.K. Baweja, Chief Hydrogeologist and Member CGWB, Jamnagar House, Mansingh Road, New Delhi–110001, 1976.
36. *Polluted Ground Water*, A review of significant literature by D.K. Todd and D.E. Orren McNulty, Water Information Centre, Inc., 7 High St., Huntington, New York, N.Y. 11743: 1976, with Bibliographies and References.
37. '*Ground Water Drilling*', O.P. Handa, Oxford & IBH Publishing Co., 66 Janpath, New Delhi–110001, 1985.
38. '*Ground Water Manual*', Bureau of Reclamation, U.S. Dept. Interior, 1977, Revised reprint, John Wiley & Sons, New York, 1981. A Water Resources Technical Publication.

39. *'Ground Water Recovery'*, L. Huisman, The Macmillan Press Ltd., London, 1972.
40. *'Theory of Ground Water Flow'*, A. verruijt, The Macmillan Press Ltd., London, 2nd Edn., 1982.
41. *'Ground Water Hydrology'*, Herman Bouwer, McGraw-Hill, Kogakusha, Ltd. Int. Student Edn., 1978.
42. *'Ground Water and Tube Wells'*, Satya Prakash Carg., Oxford & IBH Publishing Co., 2nd Edn., New Delhi, 1984.

Index

Acid treatment, 435
Aerial photography, 63
  authority in India, 67
  C-factor, 74
  crab, 74
  mosaic, 67
  photo interpretation, 64, 68
  relief displacement, 71
  specifications, 75
  scales, 65
  stereoscopic, 66
Analog models, 255
  dipole pumping test, 272
  direct simulation, 271
  mechanical, 255
  viscous flow (Hale Shaw), 256
Aquiclude, 89
Aquifuge, 89
Aquifer, 89
  leaky artesian, 164
  mechanical analysis of aquifer material, 78
  permeability, 89
  performance tests, 178
    of multiple aquifer, 185
  properties, 78
  specific yield, 81
  storage coefficient, 85
  transmissibility, 89
  types of, 40
  water table, 136
Aquitard, 89
Automatic water level recorder, 52

Barometric efficiency, 118
Bench terracing, 520
Blasting techniques, 417

Cavity wells, 406

Collector wells, 390, 395
Consumptive use, 34
  efficiency, 48
Conjunctive use, 17, 481
Contour bunds, 518
Crops details, 523

Darcy's law, 90
Dowsing, 337
Dug-cum-bore well, 14

Effective root zone, 44
Electrical resistivity, 304
Evaporation, 29
  Penman's equation, 32
  soil, 29
Evaporimeter, Pan, 29
  piche, 30
Evapotranspiration, 31
  estimation of, 31
  equations, 32

Finite element method, 488
Fishing operations, 428
Flow net analysis, 99

Geochemical studies, 344
Geophysical techniques, borehole, 321
  cement bond logging, 337
  downhole photography, 337
  electrical logging, 321
  fluid logging, 332
  focussed resistivity log, 36
  induction logging, 332
  microcaliper log, 336
  radioactive logging, 330
  sonic logging, 332
  spontaneous potential—SP, 325
Geophysical techniques, surface, 304

electrical resistivity, 304
   use of, 307
gravity, 320
magnetometer, 320
NGRI resistivity meter, 306
Seismic refraction, 315
Soil temperature, 320
Terrameter, 306
Ground water, 1, 15
   dating of, 335
   development, 2
   exploitation, 13
   flow direction, 100
   investigation, 15
   legislation, 17, 457
   movement, 99
   pollution, 17, 457
   potential in India, 4
   theory, 115
   unsteady confined flow, 120
Ground water basin
   management of, 474
   mathematical modelling, 482
   water balance studies, 475, 478
Ground water pollution, 457
   grouting, 459
   sealing of abandoned wells, 457
   well distance from contamination source, 458
Ground water recharge, 463
   ditch and flooding system, 464
   injection wells, 464
   spreading, mechanics of, 469
   in stream channels, 468
Ground water table
   contour maps, 99
   fluctuations of, 52, 186
Ghyben-Herzberg principle, 285, 286

Hydrogeology, 58
Hydrometeorology, 22

Image wells, 127
Infiltration, 49
   measurement of, 51
Infiltration gallery, 394
Irrigation, depth of, 44
   economics of, 386
   efficiency, 44
   interval, 44
   requirement, 47
Irrigation system, 493

   basin, 500
   border, 493
   corrugation, 499
   drip, 513
   furrow, 497
   sprinkler, 501
   economics of, 503

Land subsidence, 88
Leaching requirement, 47, 362
Logging and sampling, 413

Open wells, 239
   recuperation test, 241
Orifice meter, 440

Permeability, 89
   conversion factors 95
   index, 356
   laboratory, 94
   unsaturated, 41
Pumps, airlift, 448
   centrifugal, 450
   jet, 445
   plunger (hand or motor), 445
   submersible, 447, 449
   deep well vertical turbine, 446
Pumpsets, selection of, 444

Radioactive tracers, 334
Rainfall, mean areal depth, 25
   a.a.r. of India, 1
Raingauge, 22
   density, 24
Remote sensing, 321
R.C. network analog, 257

Salt tolerance of crops, 361
Schlumberger arrangement, 305
Sea water intrusion, 286
   control of, 294
   dispersion, 291
   interface, slope, shape of, 289
   salt water wedge, 288, 290
Seepage velocity, 90
Seismic refraction, 315
Sodium absorption ratio (SAR), 353
Soil and water conservation, 517
Soil moisture, 38
   equilibrium points, 42
   extraction, 43, 46
   movement of, 40

Index

PF, 44
Soils, saline and alkaline, 361
   gypsum, 363
Solar radiation, 27
Step-drawdown test, 231, 234
Storage coefficient, 85
Sunshine recorder, 27

Tensiometer, 45
Tidal efficiency, 118
Tile drainage, 516
Transmissibility coefficient, 89
Tube (bore) wells, 377
   economics of irrigation from, 386
   irrigation from, 384
   yield test, 438

Unsaturated flow, 40

V-notch 90°, 441

Water balance studies, 478, 485
Water quality, 346
   chloride concentration, 344
   drinking water standards (WHO), 346
   for industrial uses, 348
   incrustation and corrosion, 364
   irrigation, 350
   plots, 353
   Sea water contamination, 364
Water requirement for
   industries, 350
   rural water supply, 347
   live stock, 347
Water resources, world's, 1
Water sampling, 345
   chemical analysis, 365
   major constituents, 352
Water table contour maps, 99
Water well
   cavity wells, 406
   design of, 371, 379
   development of, 429
   efficiency, 138
   maintenance, 434
   selection of site, 60
   spacing of, 62
Water well drilling, 403
   air-rotary, 410
   boring, 404

   cable-tool (percussion), 412
   core drilling, 407
   driving, 404
   jetting, 406
   reverse circulation, 410
   rotary, 408
   rotary-cum-hammer (DHD, DTH), 413
Well development, 429
Well disinfection, 434
Well hydraulics, 135
   application of formulae, 185
   bailer method, 184
   barrier and recharge boundaries, 127, 154, 160, 163, 189
   Chow's method, 153
   Dupuit's (Thiem) equation, 136
   Island aquifers, 296, 297
   Jacob's method, 161
   Jacob's correction, 124
   laws of times, 158
   open wells, 239
   multiple well systems, 214
   partial penetration, 216
   slug method, 184, 442
   steady radial flow, 136
      in leaky artesian aquifer, 164
   Theis recovery, 126
   testing for yield, 438
   unsteady radial flow, 140
      in leaky artesian aquifer, 164
   upconing, 300
Well, gravel packed, 373
Well losses, 224
Well points, 405
   driving of, 425
Well screens
   corrosion resistant, 375
   design of, 372
   installation of, 423
   recovering of, 426
   selection of, 373
   slot size, 372
Wenner arrangement, 305

Yield test, 438
   bailer method, 443
   orifice meter, 440
   slug method, 184, 442